DEVELOPMENTS IN ADHESIVES—1

A. Stereoscan photograph of a single dot, 40×.

B. Stereoscan photograph of cross section of a single dot, 56×.

FRONTISPIECE

(Photographs: Courtesy H. Rose and Staflex International Ltd.)

DEVELOPMENTS IN ADHESIVES—1

Edited by

WILLIAM C. WAKE

Ph.D., D.Sc., C.Chem., F.R.I.C., F.P.R.I.
Formerly Assistant Director (Research) Rubber and Plastics Research Association of Great Britain
Visiting Professor, The City University, London

APPLIED SCIENCE PUBLISHERS LTD
LONDON

APPLIED SCIENCE PUBLISHERS LTD
RIPPLE ROAD, BARKING, ESSEX, ENGLAND

ISBN: 0 85334 749 2

WITH 49 TABLES AND 65 ILLUSTRATIONS

Printed in Great Britain by Galliard (Printers) Ltd, Great Yarmouth

INTRODUCTION

The scientist is accustomed to keeping abreast of new work by reading the journals of his speciality. In the field of *adhesion* science this is done but in the field of *adhesives* it is frequently the patent abstracts and trade leaflets that provide guidelines and, as relevant information from these sources appears at irregular intervals, it is more difficult to obtain a balanced picture of the state of the art than of the science.

What I have done, therefore, is to select topics which are at the present time of particular interest in the application of adhesives technology and persuade someone personally active with that topic to review it. This book is concerned with practice rather than theory but, I hope, with informed practice. At a recent conference on Adhesion and Adhesives a discussion developed on test procedures and the indications derived therefrom for surface preparation to give optimum test results. An apposite retort from the speaker (Ir. P. F. A. Bijlmer of Fokker–VFW) was to the effect that he was responsible for making aeroplanes—not test pieces. I make a similar point here by saying that these chapters are written by the doers. As is pointed out in Chapter 1, with proper surface preparation, the use of corrosion-resistant primers, the absence of metal fasteners and the right adhesive, there have been no bond failures or corrosion problems in the Fokker F27 and F28 aircraft although the former has now been in service for over 15 years.

REQUIREMENTS OF AN ADHESIVE

The properties that govern adhesive behaviour and utility, apart from the obvious one of molecular weight, are the glass transition temperature, the surface free energy, the solubility parameter and the morphology of the solidified material. They are conveniently regarded as separate quantities or qualities because their relative importance relates to different aspects of adhesive behaviour although they are, in fact, inter-related. The T_g and morphology depend directly on molecular weight whilst surface energy and solubility parameter can be derived from each other. It is the province of the organic chemist and the polymer chemist to prepare materials with selected values of these properties. Given that the organic chemist provides the monomers, polymerisation by condensation or addition methods is available to turn these into adhesives. Although most fusible textile adhesives today are based on vinyl chloride copolymers they are not the materials currently being evaluated by the manufacturers. The developments are of the copolyesters and copolyamides discussed in Chapter 8 with one or other of these types of materials providing the next generation, though which will be favoured is still uncertain. The copolyesters, like the polyurethanes and the thermoplastic rubbers of Chapter 9, have an interesting morphology based on domains of harder, polar units in a matrix of softer, less-polar material. Such material can show more than one transition temperature as each phase acquires or loses molecular mobility. The phases may be regarded as having different solubility parameters and can interact with surfaces in different ways. We are able to pass, by means of this two-phase concept of adhesive structure, from a consideration of fusible textile adhesives to adhesives for joining dissimilar materials, in particular when one of these is a polyolefin. Quite apart from altering the nature of the surface of the polyolefin, itself a subject of active development as Chapter 10 recounts, manufacturers are endeavouring to satisfy a requirement for an adhesive. The formulations which are being tried have two-phase structures though on a coarser scale than the domain structures of the polyurethanes, copolyesters or thermoplastic rubbers. They seem to favour a matrix of polyolefin with a dispersion of a chlorinated polyolefin and frequently an adhesion promoter specific to the non-olefinic surface to which adhesion is required.

Pressure-sensitive adhesion is exhibited if the T_g of the formulated adhesive is suitably below room temperature. This by itself is not a sufficient condition as reading either Chapter 4 or 5 will show, but it is an essential condition. To contain T_g within a desirable range may require raising it or lowering it. Random copolymerisation enables either of these actions to be realised and the newer materials seem to be terpolymers. In the field of acrylics the number of monomers currently commercially available is large, as can be gauged from Table 1 of Chapter 6, so that most of the desirable properties can be built into the molecule and the use of external plasticisers or tackifiers rendered unnecessary. This is an important feature in making the transition from solvent-based to emulsion-based systems because a mixture of dispersions of polymer and tackifier does not behave when dry in the same way as a solution of the mixed materials (Chapter 5).

If attention is switched to structural adhesives, it need not be switched from acrylics for within their compass lies metal–metal joint strengths of $15\,MN\,m^{-2}$ and even associated with quite good peel resistance (*see* Table 5, Chapter 6) and cyanoacrylates seem to vie with epoxies for underwater fastenings of engineering significance.

However, most structural applications of adhesives have been made with epoxy resins and hence the importance of Chapters 2 and 3. It seems to be a fact that architects and engineers have mainly employed new technologies and new materials in connection with ecclesiastical buildings and bridges. In the former, the advantages of grand aesthetic effect tempt the artist in the architect to do new and imaginative things and in the latter the strength and service demands have always stretched the designs and materials available to their extremes. Recent design developments in bridging have not always had happy results. Motorway bridges have been in the news over the last two or three years and problems of underdesign and hence strengthening or of cracking can be remedied by adhesives as is outlined in detail in Chapter 2.

If one were to consider the total range of adhesives used only in conjunction with textiles, one would still be concerned with the chemistry of almost all the adhesives mentioned in the non-textile chapters of this book. Acrylics are used for joining and flocking textiles, polyurethanes for bonding textiles. The latter frequently comprise a woven facing textile bonded to a knitted backing as are used for car upholstery. Epoxy resins, we learn from Chapter 7, form a component of the two-part *Shoaf* dip which, cured by an isocyanate will give a urethane group. Resorcinol resins, however, introduced widely in the fifties, still function as the primary adhesive in highly-stressed flexible applications of the tyre cord, Vee-belt reinforcement and Hovercraft skirt. In 1945 the need for improved tyre cord adhesives could have been reasonably foreseen as also their application to Vee-belts and conveyors but the emergence of the Hovercraft application as a high-stress requirement was unexpected. Dracones were another post-war invention dependent upon adhesives and textiles. There may be other applications unexpectedly round the corner of time future.

INDUSTRIAL CONTROLS

A development brought about largely by external pressure is the projected transfer of our adhesive technology away from a solvent-based one. To some extent the major developments of recent years such as the use of hot-melt adhesives or liquid acrylics have already provided stepping stones. But there are difficulties of technology and the alternatives are, as Chapter 11 points out in several places, not necessarily free from hazard. Very many adhesives require special priming of the substrate and, apart from concrete and siliceous materials, primers even more than the adhesives themselves have been solvent based. Moreover, substrate wetting without the aid of surfactant molecules is what is specifically required of the primer. Encapsulation of hazardous materials in microcapsules is a possible and enticing development that has figured in articles and in patents rather more than in practice. The requirements outlined in Chapter 11 may have a greater impact on Developments in Adhesives than we yet realise.

W.C.W.

CONTENTS

LIST OF CONTRIBUTORS

D. W. Aubrey, B.Sc., Ph.D., C.Chem., F.R.I.C.

National College of Rubber Technology, The Polytechnic of North London, Holloway, London N7 8DB, UK.

P. Bosworth, B.Sc., Ph.D., C.Chem., F.R.I.C.

Lime Cottage, Lime Walk, Dibden Purlieu, Southampton SO4 5RB, UK.

A. R. Carter, L.R.I.C., A.B.S.I.

Adhesion & Solings Research Department, SATRA, Satra House, Rockingham Road, Kettering NN16 9JH, Northants, UK.

J. L. Cotter, B.Sc., Ph.D.

Royal Aircraft Establishment, Materials Department, Farnborough, GU14 6TD, Hants, UK.

P. J. C. Counsell, B.Sc., Ph.D.

Evode Holdings Ltd, Common Road, Stafford ST16 3EH, UK.

W. A. Dukes, B.Sc.

Adhesives Group, Ministry of Defence, Propellants, Explosives and Rocket Motor Establishment, Powdermill Lane, Waltham Abbey EN9 1BP, Essex, UK.

E. W. Garnish, B.Sc., A.R.C.S.

Epoxy Products Centre, Plastics Division, Ciba-Geigy Plastics and Additives Company, Duxford, Cambridge CB2 4QA, UK.

P. C. HEWLETT, B.Sc., Ph.D., C.Chem., F.R.I.C.

Cementation Research, Denham Way, Maple Cross, Rickmansworth WD3 2SW, Herts, UK.

A. J. KINLOCH, B.Sc., Ph.D.

Adhesives Group, Ministry of Defence, Propellants, Explosives and Rocket Motor Establishment, Powdermill Lane, Waltham Abbey EN9 1BP, Essex, UK.

F. R. MARTIN, B.Ch.E., M.Sc., Ph.D.

Loctite (Ireland) Ltd, Whitestown Industrial Estate, Tallaght, Co. Dublin, Eire.

D. PETTIT, B.Sc., C.Chem., A.R.I.C.

Adhesion and Solings Research Department, SATRA, Satra House, Rockingham Road, Kettering NN16 9JH, Northants, UK.

J. D. N. SHAW, M.A.

Cementation Chemicals, Denham Way, Maple Cross, Rickmansworth WD3 2SW, Herts, UK.

L. SMITH, B. A., D. Phil., C.Chem., M.R.I.C.

Staflex International Ltd, 14 Centre Way, Clavering Industrial Estate, London N9 0AG, UK.

R. S. WHITEHOUSE, B.Sc., Ph.D.

Evode Holdings Ltd, Common Road, Stafford ST16 3EH, UK.

D. B. WOOTTON, M.A.

John Bright & Brothers Ltd, Fieldhouse Mills, P.O. Box No. 26, Rochdale OL12 0AB, Lancs, UK.

Chapter 1

THE DURABILITY OF STRUCTURAL ADHESIVES

J. L. COTTER
Royal Aircraft Establishment, Farnborough, UK

SUMMARY

Factors affecting the long term durability of adhesively bonded metal assemblies exposed to natural environments including the nature of the adhesive, the stress applied to the adhesive bond, the operational environment and quality of the metal prebond surface preparation are discussed with particular reference to aluminium and titanium alloys. Details are given of an alkaline hydrogen peroxide etch process for titanium alloys which leads to improved environmental performance, particularly in warm, moist conditions, of adhesive to titanium bonds.

INTRODUCTION

By far the largest single problem facing adhesives scientists and technologists today is that of the long term durability of adhesive bonds exposed to natural environments. Structural adhesives are used extensively in airframe construction principally with aluminium alloys, and aircraft are required to function in many diverse natural environments ranging from low-temperature arctic conditions to tropical, high-humidity climates. Whereas a few years ago it was customary to consider the effective life of a civil aircraft as being 15 years, the new generation of wide-bodied jets will have anticipated lifetimes of up to 30 years (90 000 flying hours). The current emphasis is on longer life and safer aircraft and this of necessity demands that structural adhesives be able to maintain structural integrity of bonded components for at least several decades.

The problem of maintaining structural integrity of adhesive bonded assemblies is, of course, not a recent development, and it is significant that early natural weathering trials of adhesive bonds were conducted in areas of high humidity.[1-4] Today, however, the retention of strength of adhesive bonds exposed to hostile environments has assumed even greater importance with the introduction of fibre-reinforced materials, as a reinforcement or replacement for metal structural components, and the requirement to bond these materials to metals with adhesives. The structural integrity of adhesive bonded metal-to-metal and metal-to-composite assemblies must now be viewed against a background of high initial and running costs of modern transport aircraft.

It is worth mentioning that although the expression 'durability of structural adhesives' is in general use, we are concerned not only with the structural integrity of the cured adhesive but also with the environmental stability of the bond between the adhesive and its substrate. Although structural adhesives are well able to accommodate current aircraft design loads, which are generally limited to about 15–20% of ultimate static strength of the bond for a fatigue life requirement of 10^7 cycles, degradation in strength of bonded joints that can occur on exposure to high humidity and which is particularly pronounced when the bond is under stress does give cause for concern and has spurred the development of new surface prebond preparations for metal bonding.

SURFACE PREPARATION OF ALUMINIUM FOR ADHESIVE BONDING

Although a wealth of information on the weathering behaviour of adhesive bonded metal-to-metal assemblies has been accumulated over the past 25 years,[1-9] a detailed understanding of the mechanism of degradation is still lacking; perhaps this should not be too surprising in view of the lack of clear understanding of the true nature of forces involved in, and mechanisms underlying, the processes of adhesion. It is well known, however, that the immediate pre-bond condition of the metal surface greatly determines the durability of the adhesive bond. Figure 1 illustrates the effect of different surface pre-treatments on the resistance to high humidity of aluminium lap shear test pieces bonded with an epoxy-polyamide adhesive. Best results were obtained with surfaces etched in chromic–sulphuric acids and poorest results with surfaces prepared by a simple degreasing process, and it was concluded that any chemical etching of the metal surface is likely to be

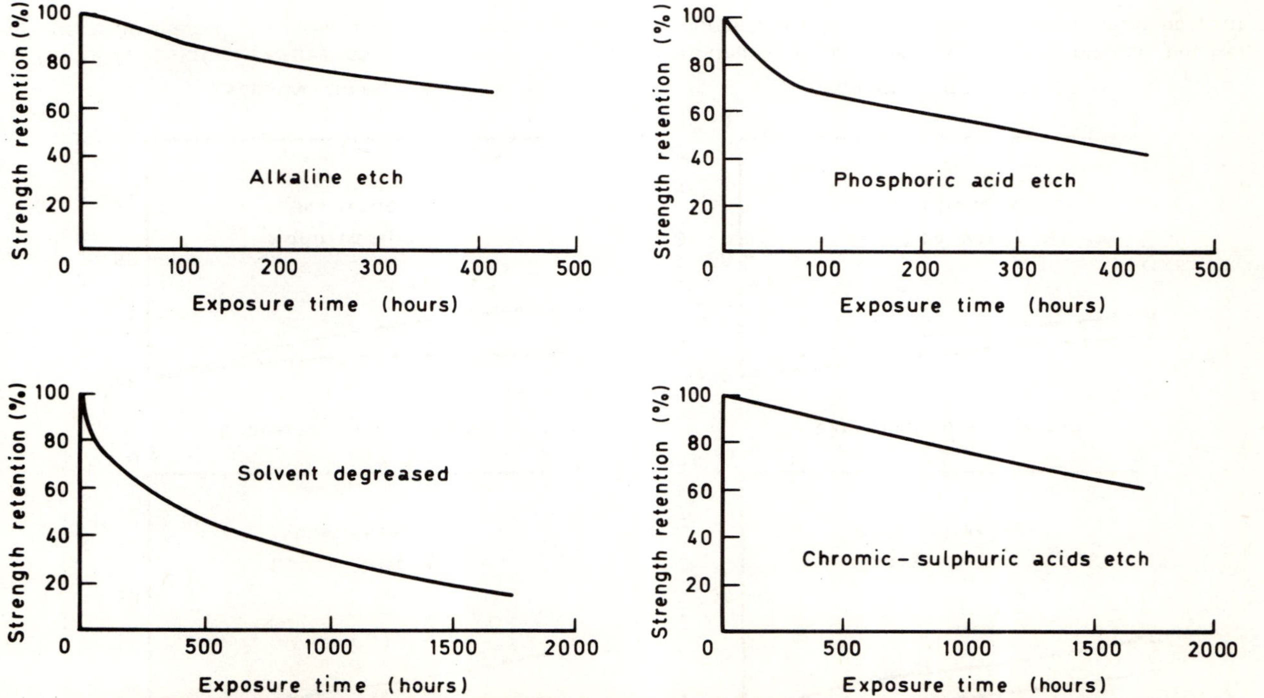

FIG. 1. Effect of surface pre-treatment on the strength of aluminium single overlap joints, bonded with an epoxy-polyamide adhesive, exposed to high humidity (43 °C; 97 % relative humidity).

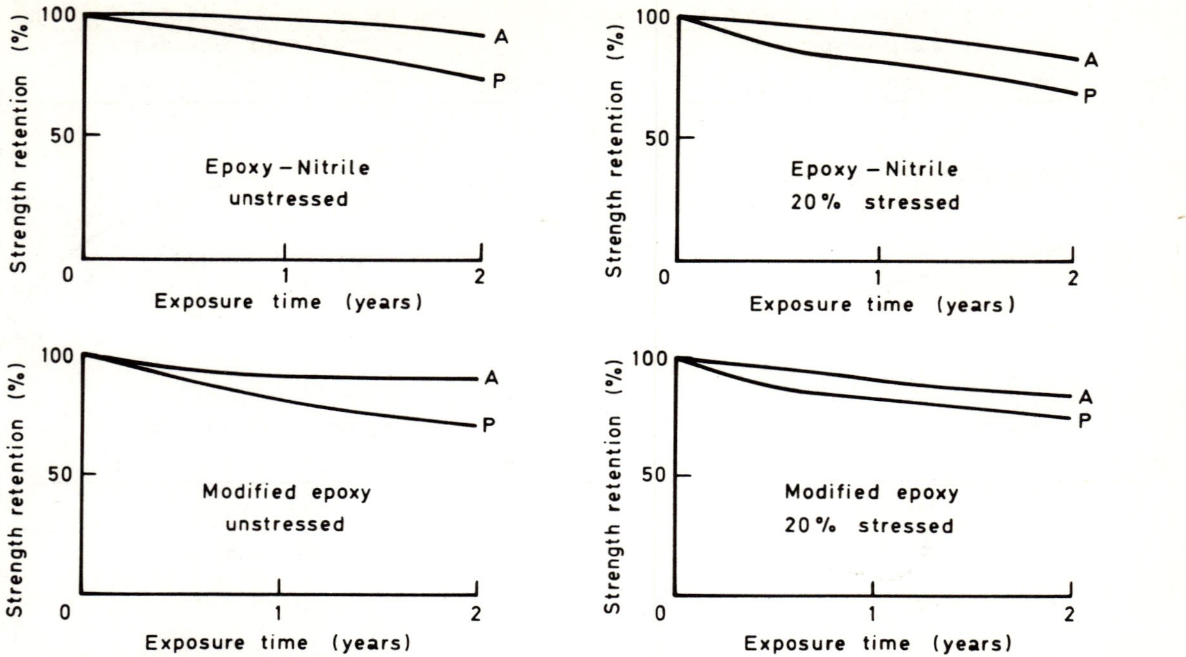

FIG. 2. Effect of surface pre-treatment on the strength of aluminium double overlap joints exposed to a tropical, hot–wet, environment at Innisfail, Australia; P: pickled specimens (chromic–sulphuric acids); A: anodised specimens (chromic acid).

beneficial. It was further noted that although different surface pre-treatments imparted greatly different strength-retention characteristics to the adhesive bonds, initial strengths of degreased specimens were only about 8 % below those of specimens etched in chromic acid.[10] In a natural weathering situation, bonds formed with anodised aluminium have proved more durable than those formed with surfaces pickled in chromic acid (Fig. 2).

The results of accelerated laboratory ageing clad aluminium test pieces bonded with a modified epoxy, 120 °C curing, adhesive for three surface pre-treatments are summarised in Fig. 3 and confirm the efficiency of phosphoric anodising pre-treatment[11] in promoting superior durability in high humidity (Boeing Spec., 5555).

The effect of chemically etching aluminium surfaces is to remove old and cohesively-weak oxide layers and to deposit, in a controlled manner, a uniform oxide surface, typically 1 × 10-nm to 4 × 10-nm thick in the case of chromic acid–sulphuric acid etch and up to 200-nm and 3000-nm thick when anodised in phosphoric and chromic–sulphuric acids, respectively.

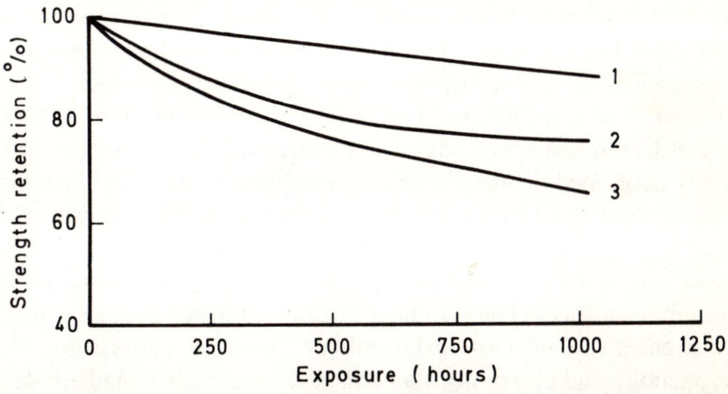

FIG. 3. Effect of surface preparation on the strength of aluminium single overlap joints, bonded with a modified epoxy adhesive, immersed in deionised water at 50 °C; 1: phosphoric acid anodised; 2: chromic–sulphuric acids anodised; 3: chromic–sulphuric acids etched.

The nature of the oxide film on freshly-etched aluminium is the subject of some controversy, Bowen[12] considers it to be boehmite ($Al_2O_3 \cdot H_2O$) whereas other workers[13–15] consider the fresh oxide to be γ Al_2O_3. However, it is generally agreed that ageing chromic acid etched aluminium in high humidity results in growth and transformation of a thin oxide film to a thicker hydrated oxide layer which is not a durable bonding surface.

Contrary to the view that the oxide layer on freshly-etched aluminium develops in the chromic acid, evidence is available[15] which suggests that during the latter part of the etch cycle the metal surface is essentially free of oxide and that the oxide layer is developed in the subsequent rinsing operation. Thus, the quality of the rinse water together with temperature and duration of rinsing become vitally important in finalising the composition and thickness of the oxide layer. Better adhesion was obtained on chromic acid etched aluminium surfaces rinsed in tap water than with distilled or deionised water, and this was attributed to stabilisation by multivalent ions of the surface oxide against conversion to a thicker hydrated oxide layer unfavourable for adhesive bonding.[16,17] More recent work[14] considers that in distilled water at 60 °C bayerite ($Al_2O_3 \cdot 3H_2O$) is rapidly formed as a thick cohesively-weak layer, and at 60 °C in tap water hydration of the aluminium surface is probably prevented by divalent ions diffusing into the oxide layer.

Aluminium etched in chromic acid and rinsed in cold tap water presents a high-energy surface covered with microscopic etch pits[18] with sharp metal ridges surrounding each cavity. Such surfaces are particularly favourable for adhesive bonding as they are readily wetted by adhesives to produce high strength bonds, but in humid conditions bonds can degrade by oxidation of the metal ridges to form a low energy surface.[19]

In response to a specific need to improve the durability of assemblies bonded with modified epoxy adhesives cured at 120 °C, the Boeing Aircraft Company developed a phosphoric acid anodising process for the adhesive bonding of aluminium alloys. In-service failures of bonds with new, rubber modified, epoxy adhesives occurred at low stress levels, were interfacial in character and once again focused attention on the nature and preparation of aluminium surfaces. The standard chromic acid etch process which had proved so effective in promoting durability of bonds with hot curing, 175 °C, nitrile-phenolic and epoxy-novolac adhesives was clearly inadequate with the new modified epoxy formulations curing at lower temperatures.[11] Development of the phosphoric acid pre-treatment involved use of lap shear and fracture mechanics type (wedge test) specimens bonded with a range of commercially-available adhesives and aged in natural and high-humidity environments. Extensive testing revealed that phosphoric acid anodising produced surfaces capable of forming bonds more durable than those obtained using chromic acid etch and anodising processes. This improvement in durability was attributed to the oxide film produced in the phosphoric acid pre-treatment being thinner, more open and resistant to hydration than anodic films produced in chromic acid.[20] It is now considered that environmental problems associated with these newer adhesives can be effectively controlled by use of an optimised phosphoric acid anodising pre-treatment for aluminium alloys together with application of a corrosion-inhibiting primer and adhesives capable of sustaining design loads when plasticised under conditions of high humidity.[21]

The evidence from in-service experience, field and laboratory evaluations of surface pre-treatments and surface studies of aluminium alloys is in accord with a general mechanism of failure of adhesive bonds involving diffusion of water through the adhesive to the metal oxide–adhesive interface to produce cohesively-weak hydrated oxides which readily fail under stress. Corrosion initiated at crevices in the exposed metal surface can then accelerate the de-bonding process.

It is pertinent at this point to consider the use of clad and unclad alloys in adhesive bonding applications. In aircraft construction clad alloys are used to provide corrosion protection whereas unclad materials are utilised to save weight and are protected in other ways. The core material is cathodic with respect to commercially-pure aluminium cladding, and consequently when metal fasteners are used in the presence of an electrolyte the anodic cladding corrodes along the interface between cladding and adhesive destroying the bond. The problem of corrosive delamination is considered so serious in the USA that there is almost total commitment to unclad alloys in bonding operations, and corrosion protection is afforded by using thin unsealed oxide films, produced by anodising procedures, and corrosion-inhibiting primers. In Europe, clad alloys are extensively used in aircraft construction and with adequate protection, including anodising in chromic acid, use of corrosion-resistant primers, sealed bond lines, the absence of metal fasteners through bonded areas and correct choice of structural adhesive, very durable bonds can result as exemplified by the absence of bond failures and corrosion problems in the Fokker F27 and F28 aircraft with over fifteen and six years' service life respectively.[22]

WEATHERING BEHAVIOUR OF ADHESIVELY-BONDED ALUMINIUM

In order to obtain detailed information on the long term durability of adhesive bonds exposed to natural environments weathering trials[8] involving bonded aluminium double overlap joints, stressed and unstressed, peel and honeycomb specimens were undertaken at the RAE, Farnborough, and at the Joint Tropical Research Unit (JTRU) sites at Innisfail (hot–wet) and Cloncurry (hot–dry), Australia. Mean annual temperatures at the hot–wet, hot–dry and temperate sites are 24 °C, 26 °C and 9 °C, respectively, with corresponding annual rainfalls of 323 cm, 28 cm and 48 cm. At the tropical, hot–wet site the mean relative humidity is 83 % while at the desert site it is only 34 %. Double-overlap test specimens were fabricated with clad aluminium alloy (BS 2L73) that had been etched in chromic–sulphuric acids according to DEF STAN 03-2/1;† additionally, some specimens were anodised in chromic acid according to DEF STAN 151.† Figures 4–6 summarise the weathering behaviour of unstressed overlap specimens bonded with a range of commercially-available adhesives, and it is apparent that the durability of metal-to-metal bonded joints is greatly influenced by the nature of the adhesive. At the hot–wet site the marked susceptibility of an epoxy-polyamide adhesive to a humid environment is in sharp contrast to the excellent strength retention of test pieces bonded with nitrile-phenolic, epoxy-novolac and vinyl-phenolic adhesives. The superior durability of the last three mentioned adhesives is maintained at the temperate and desert locations, but the fall in strength of epoxy-polyamide bonds at the latter site is only about 10 % at the end of six years' exposure and must be related to the low relative humidity of the environment.

The combination of environment and stress (Figs. 7–9) is particularly deleterious to bonded joints but the effect is more pronounced with certain adhesives.

At the jungle location, stressed epoxy-polyamide bonds suffered a rapid decrease in strength and no specimens survived two years' exposure; at the same site vinyl-phenolic bonds exhibited little overall loss in strength up to two years' exposure but thereafter, in contrast to behaviour at the other sites, loss of strength was rapid and no test pieces survived six years' exposure. At the temperate site stressed epoxy-polyamide bonds proved more durable than at the jungle site but overall performance of the adhesive was unsatisfactory. Stressed vinyl-phenolic bonds weathered better at the temperate site than at the jungle location, whereas overall performance of a modified epoxy adhesive was better at Innisfail than at Farnborough. All adhesives performed best at the desert site although epoxy-polyamide bonds were the least durable. The poor performance of this adhesive at all

† UK Ministry of Defence, Defence Standards, 03-2/1 (Cleaning and Preparation of Metal Surfaces) and 151 (Anodising of Aluminium and Aluminium Alloys).

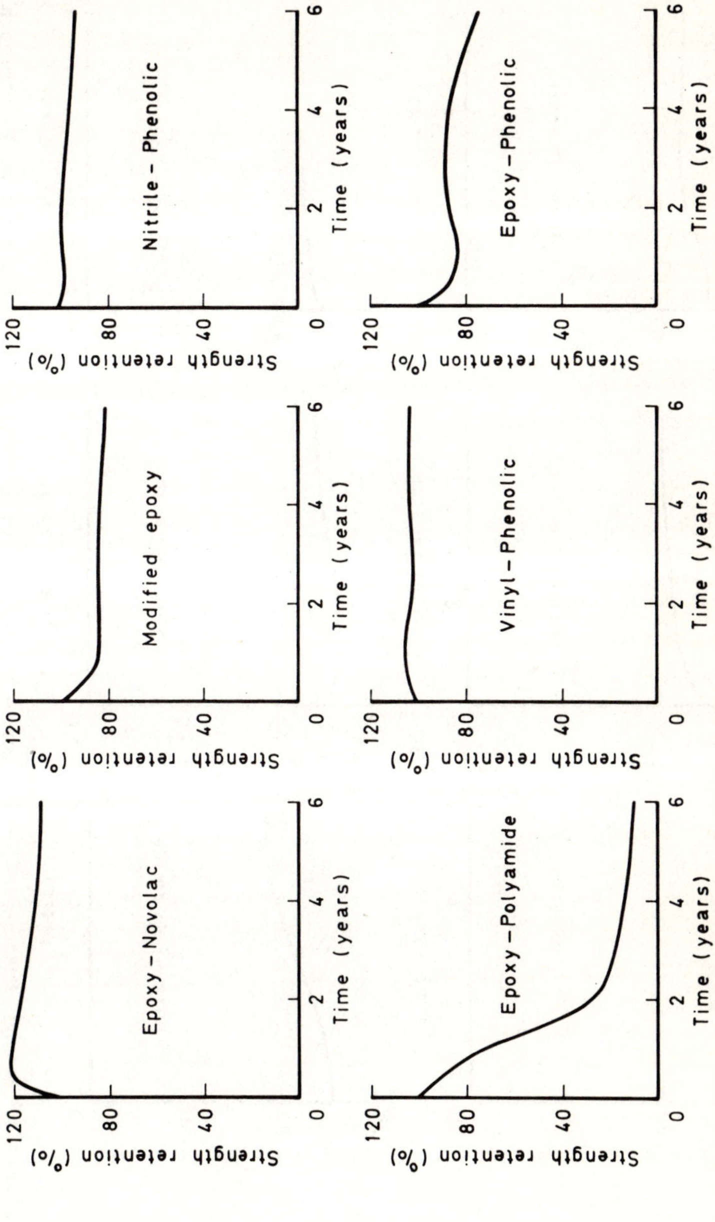

FIG. 4. Strength retention profiles of unstressed aluminium double overlap joints exposed at hot–wet site, Innisfail, Australia.

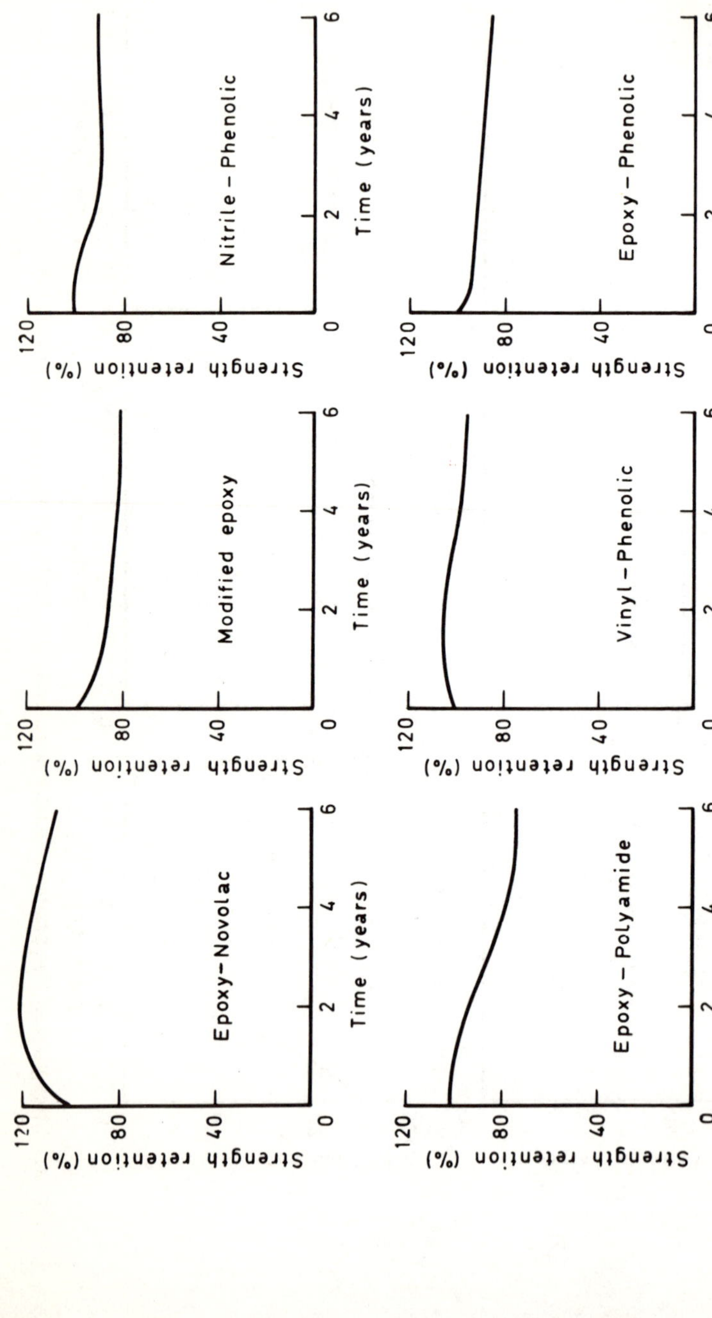

FIG. 5. Strength retention profiles of unstressed aluminium double overlap joints exposed at RAE, Farnborough.

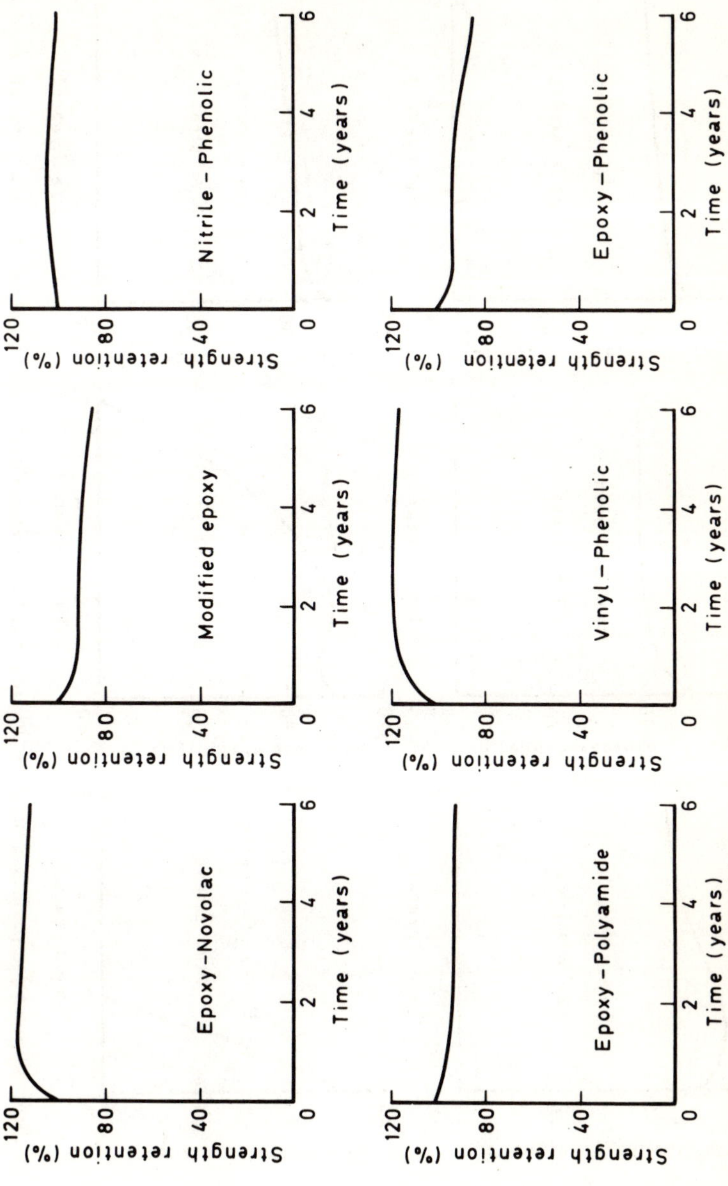

FIG. 6. Strength retention profiles of unstressed aluminium double overlap joints exposed at hot–dry site, Cloncurry, Australia.

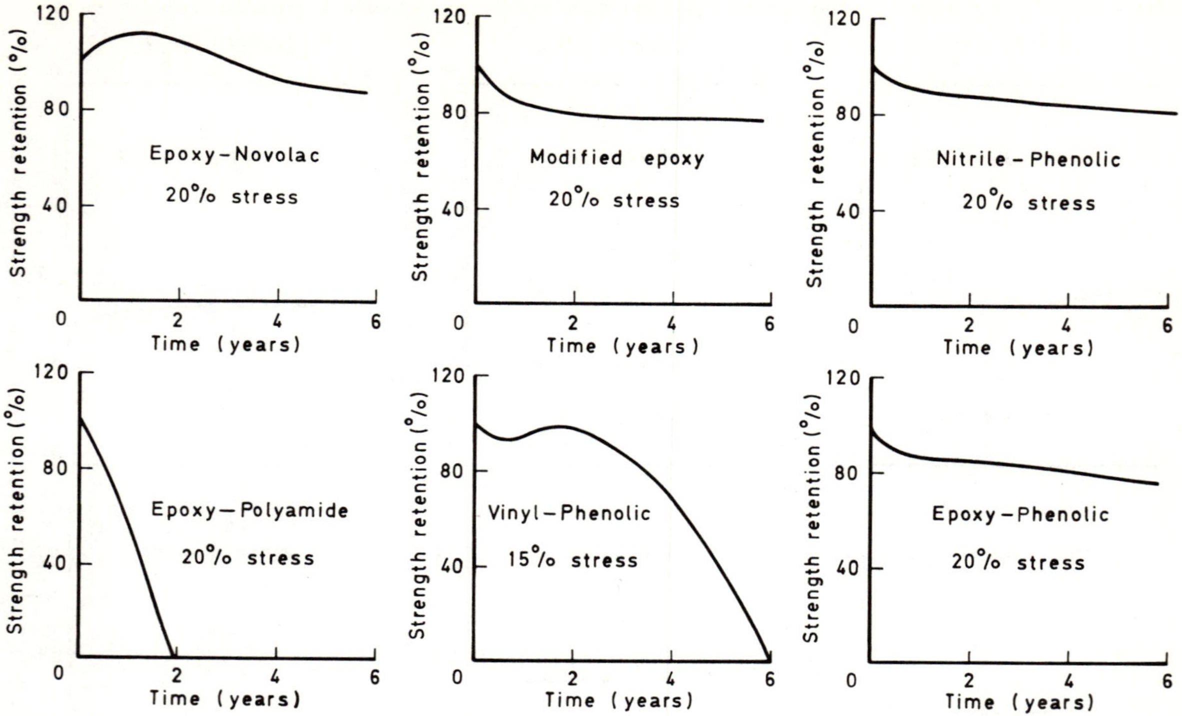

FIG. 7. Strength retention profiles of stressed aluminium double overlap joints exposed at hot–wet site, Innisfail, Australia.

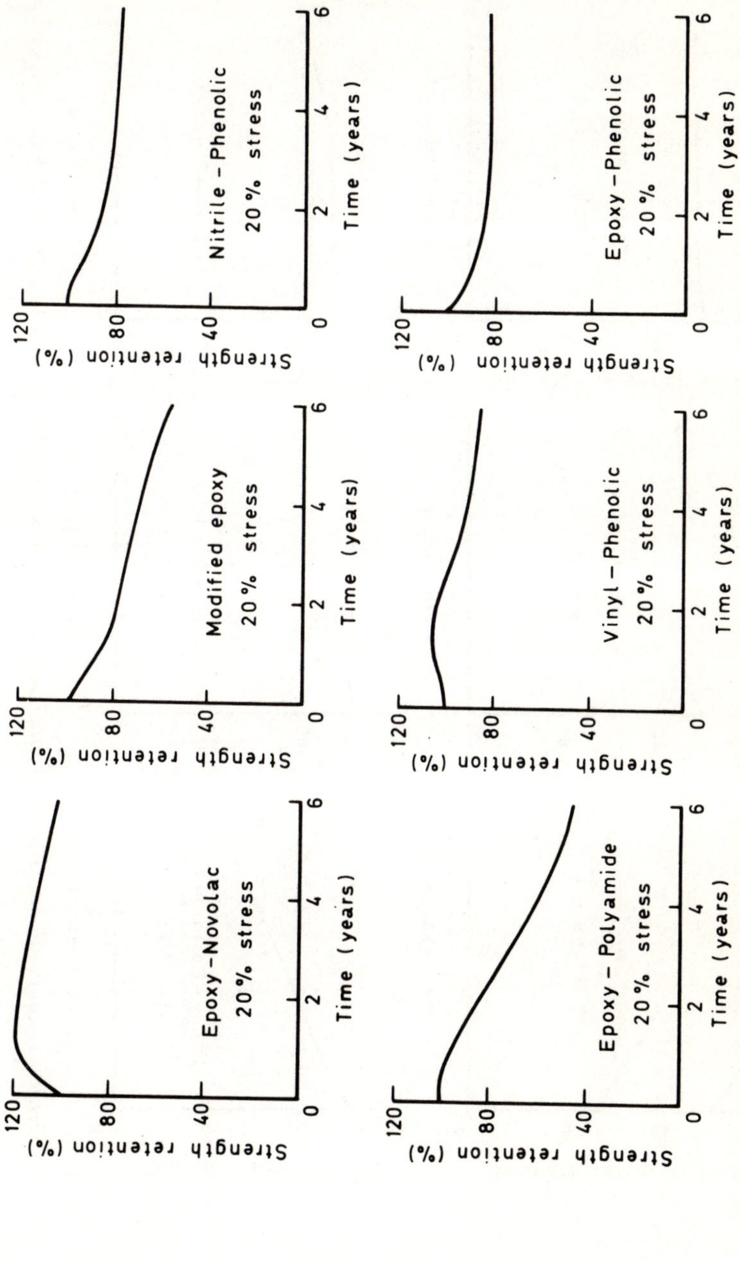

FIG. 8. Strength retention profiles of stressed aluminium double overlap joints exposed at RAE, Farnborough.

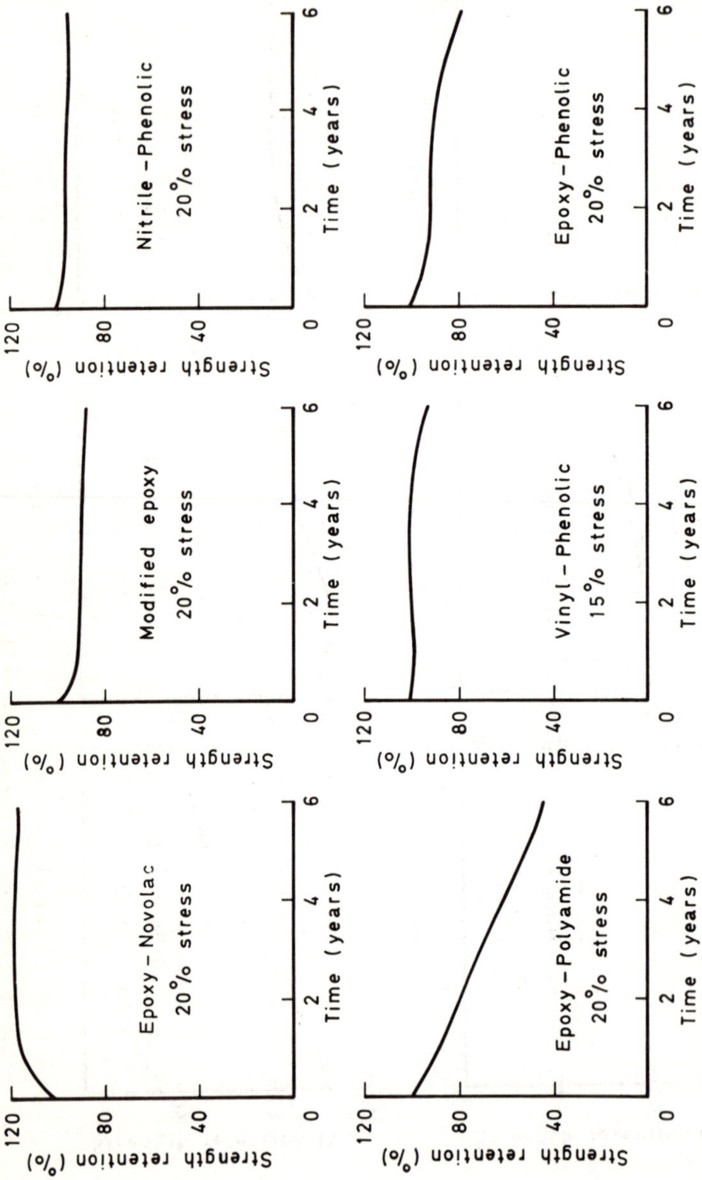

FIG. 9. Strength retention profiles of stressed aluminium double overlap joints exposed at hot–dry site, Cloncurry, Australia.

sites is almost certainly related to the ability of the cured adhesive to absorb large amounts of water when exposed to humid environments.[10]

SURFACE PREPARATION OF TITANIUM FOR ADHESIVE BONDING

The high-specific strength and stiffness of titanium coupled with its excellent resistance to corrosion renders the metal attractive for use in aircraft construction. Principally because of weight-saving factors, carbon fibre composite is being used to reinforce conventional metal structures in aircraft, and in such applications involving composite-to-metal adhesive bonding thermally-induced stresses, arising in the bonded component as a result of differences in the thermal coefficients of expansion of metals and composite, must be accommodated. Such stresses are much less evident in titanium reinforced with composite than with stiffened aluminium structures and make titanium a more attractive proposition for adhesive bonding with fibre-reinforced materials than aluminium. These considerations have led to renewed interest in the surface preparation of titanium for adhesive bonding an interest rendered all the more necessary in the light of reported in-service failures of adhesively bonded titanium sandwich assemblies.[23,24] Work in the USA at Picatinny Arsenal and the Gillette Research Institute showed that titanium surfaces etched with alkali are composed largely of rutile TiO_2, and the less-stable anatase surface is produced using a phosphate–fluoride etch process; in the presence of moisture, however, adhesive bonds to rutile were found to be not as stable as those to anatase. In addition, anatase slowly reverts to the rutile form of the oxide and is accompanied by a decrease in surface volume of about 8 % which in an adhesive bond situation must result in stressing at the interface and weakening of the bond. The problem of producing durable adhesive bonds to titanium was seen as being that of stabilising surface films of anatase against reversion to unfavourable rutile TiO_2. This reversion, it was claimed, could be controlled by incorporating sodium sulphate in the phosphate–fluoride etch solution, and a ten-fold improvement in durability was reported.[25]

It has long been known that hydrogen peroxide and sodium hydroxide mixtures produce a black coating of oxide on titanium[26] and that depending on the concentration of reactants, length of time of etch and temperature surface layers with colours varying from yellow, blue, grey to black can be produced on titanium alloys.[27] Strongest adhesive bonds were

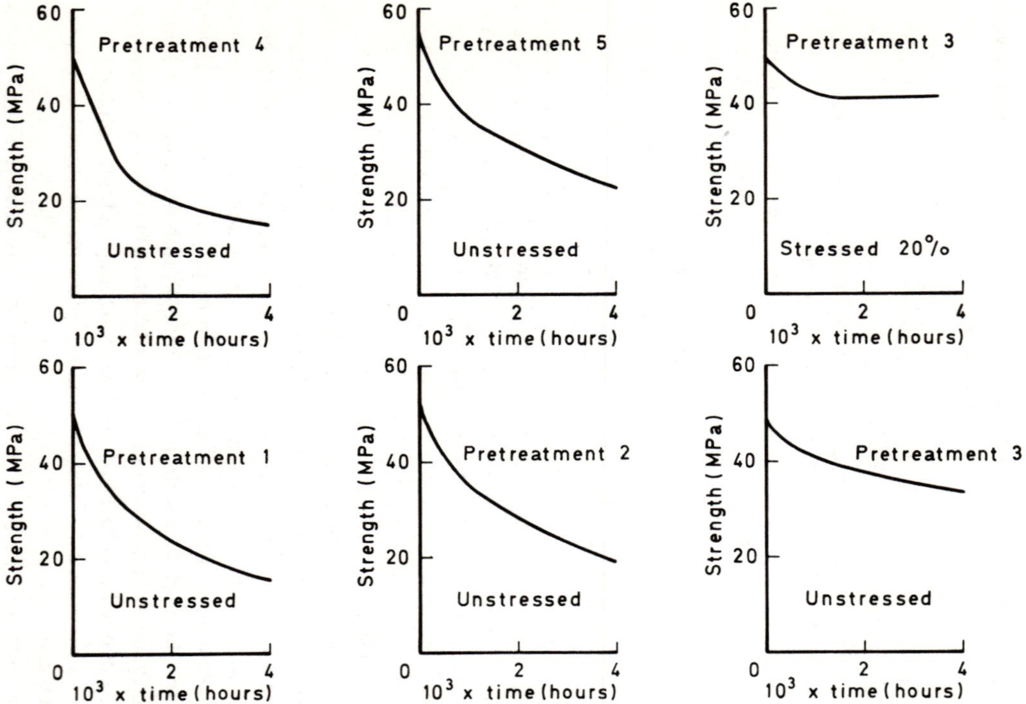

FIG. 10. Effect of surface preparation on the strengths of double overlap titanium test specimens bonded with a modified epoxy adhesive after exposure to 95–100% relative humidity at 45°C.

formed with the dark grey surface produced by chemical oxidation of the metal,[28] and this finding was confirmed by other workers.[29] The optimum composition of the etch solution and processing conditions necessary to obtain high-strength durable bonds to titanium were determined at the RAE, and this was followed by a detailed commercial assessment of the alkaline hydrogen peroxide process in relation to other prebond surface preparations for titanium.

Results of accelerated weathering of double overlap and peel specimens of titanium for a range of surface pre-treatments, details of which are given in the Appendix, are summarised in Fig. 10 and Table 1 respectively.

TABLE 1
STRENGTH RETENTION (%) OF TITANIUM 90° PEEL SPECIMENS BONDED WITH A MODIFIED EPOXY ADHESIVE AFTER EXPOSURE TO HIGH HUMIDITY (95–100% RELATIVE HUMIDITY; 45 °C; INITIAL PROOF FORCE = 100)

Exposure time (h)	*Pre-treatment*[a]			
	1	2	3	5
500	13	47	99	89
1 000	13	9	99	89
4 000	0	0	72	52

[a] See Appendix for details.

Out of twelve double overlap specimens, stressed at 20% of mean control strength, for each pre-treatment (1–4) the numbers that survived the 1000-h exposure to high humidity were four, one, ten and zero, respectively, and with pre-treatment 4 the strengths of all peel specimens had fallen to zero by 500 h. The superiority of the alkaline hydrogen peroxide etch process in imparting improved durability to adhesive bonds with titanium led directly to field trials to further assess the efficiency of the pre-treatment.

WEATHERING BEHAVIOUR OF ADHESIVELY BONDED TITANIUM

The alkaline hydrogen peroxide etch process was used in fabricating titanium (6Al-4V) double overlap and peel test pieces which were then exposed to natural weathering at JTRU and RAE sites. Results at the end of

TABLE 2

STRENGTH RETENTION (%) OF TITANIUM DOUBLE OVERLAP SPECIMENS BONDED WITH A MODIFIED EPOXY ADHESIVE, EXPOSED TO NATURAL WEATHERING (INITIAL PROOF FORCE = 100)

Stress level	*Site and exposure (years)*					
	RAE (temperate)		*Innisfail (hot–wet)*		*Cloncurry (hot–dry)*	
	0·5	1	0·5	1	0·5	1
0	93	94	97	94	101	99
5	97	98	95	94	102	99
10	98	99	98	95	99	100
20	96	98	98	97	100	100

one year of exposure are available for overlap specimens and are summarised in Table 2.

For purpose of comparison, aluminium double overlap specimens bonded with the same modified epoxy adhesive used in fabricating titanium specimens and cured at 120 °C, were simultaneously exposed at the three sites. Aluminium test pieces were etched in chromic–sulphuric acids according to DEF STAN 03-2/1, and exposure results are summarised in Table 3.

Titanium test pieces have performed extremely well at the three exposure sites in contrast to the behaviour of adhesive bonds to aluminium at the jungle location where lap joints experienced a fall in strength of 43 % by the end of the first year of exposure. Although the results of natural weathering of titanium test pieces pre-treated with alkaline hydrogen peroxide look very promising, more data on the environmental performance of the etch process is required before a full assessment of its efficiency in promoting

TABLE 3

STRENGTH RETENTION (%) OF ALUMINIUM DOUBLE OVERLAP SPECIMENS, BONDED WITH A MODIFIED EPOXY ADHESIVE, EXPOSED TO NATURAL WEATHERING (INITIAL PROOF FORCE = 100)

Stress level	*Site and exposure (years)*					
	RAE (temperate)		*Innisfail (hot–wet)*		*Cloncurry (hot–dry)*	
	0·5	1	0·5	1	0·5	1
0	90	92	88	79	95	97
10	86	93	86	57	96	96

durability of adhesive bonds to titanium can be made. The exposure trial with bonded titanium is scheduled to continue for a further five years, and work to characterise titanium-bonding surfaces and to render the alkaline hydrogen peroxide etch process suitable for large-scale industrial application is continuing.

With particular reference to aluminium and titanium alloys, an outline has been given of the important factors affecting the durability of adhesive to metal bonds including the quality of the metal pre-bond surface preparation, the nature of the adhesive, the operational environment and level of stress applied to the bond. Most bond failures in service are interfacial in character and are often initiated at fasteners and exposed sites, but durability can often be markedly improved by application of corrosion-inhibiting primers. It is noteworthy that the strengths of chromic acid pickled aluminium joints bonded with an epoxy based, 145 °C cured, adhesive and protected with a paint system survived ten years' weathering in a central European climate without any loss in strength.[30] There is little doubt that if the extent of chemical bonding between adhesive and substrate could be increased, then improvements in the weathering performance of bonded structures would result. Silane coupling agents which are considered to provide a bridge between adhesive and substrate and to interact chemically with them, have been particularly successful in promoting durability of bonded structures.[31]

Much information on the natural weathering behaviour of adhesive bonds is now available, and in view of ever-increasing costs and the time-consuming nature of field trials it is pertinent to consider if the necessary environmental information could be obtained from accelerated laboratory ageing tests. At present, there is no accelerated ageing method for providing a reliable assessment of the long-term durability of adhesive bonds exposed to natural environments, and indeed the validity of attempting to deduce the durability of an adhesive bond from results of accelerated ageing at temperatures above those encountered in service has been questioned.[32]

In the absence of reliable accelerated ageing tests, natural weathering exposures of adhesive bonds are essential if detailed information on environmental performance is required; additionally, information derived from field trials provides a base against which the results of accelerated tests can be assessed. Accelerated ageing methods, particularly when based on crack propagation techniques reported by the Boeing Aircraft Company can, however, provide much useful information, in a short time, on the effectiveness of surface preparation and performance of bonded joints under stress in humid environments.

CONCLUSIONS

The durability of bonded structures is greatly influenced by the nature of the adhesive, the best performers in all climates are epoxy-novolac and nitrile–phenolic formulations. Epoxy-polyamide adhesives do not weather well and show a marked susceptibility to moist environments.

A tropical, hot–wet, climate is the most damaging to bonded structures and the combination of high humidity and applied stress is particularly deleterious.

During exposure to natural environments, the failure mode of aluminium-bonded joints gradually changes from wholly cohesive, within the adhesive, to include increasing amounts of interfacial failure.

Corrosion does not appear to play a major role in the loss of strength of metal-to-metal bonds, and more durable bonds are formed with anodised than acid-etched aluminium surfaces.

Although a 'bare is best' and 'clad is bad' philosophy is prevalent in the USA regarding the relative merits of using clad or unclad aluminium alloys in bonding applications, European experience with clad alloys has been that careful attention to surface preparation and choice of adhesive, the use of corrosion-inhibiting primers and sealed gluelines, and avoidance of mechanical fasteners through bonded areas can result in very durable structures.

A pre-bond surface preparation for titanium (6Al-4V) involving etching in alkaline hydrogen peroxide imparts improved environmental resistance to bonded titanium structures exposed to laboratory and natural high-humidity environments.

REFERENCES

1. Leigh, J. H. and Savage, J., *RAE Technical Report* 1168, 1952.
2. Eickner, H. W., *WADC Technical Reports*, 54–447, Part 1, 1955; Part 2, 1956; Part 3, 1959.
3. Sharpe, L. H., *Symposium on Structural Adhesives Bonding*, Stevens Inst. of Technology, Hoboken, N.J., 1965.
4. Hause, C. I., Pagel, W. C. and McKown, A. G., *ASTM STP No.* 401, 1966, p. 94.
5. Scott, J. A., *ASTM STP No.* 401, 1966, p. 16.
6. Wegman, R. F., Bodnar, W. M. and Bodnar, M. J., *Picatinny Arsenal Technical Report* 3689, 1968.
7. Swanson, F. D. and Gregornik, N. W., *Adhesives Age*, **14**(10), 1971, p. 18.

8. HOCKNEY, M. D. G., *Technical Reports* (*a*) 70081 (1970); (*b*) 72100 (1972); (*c*) 73013 (1973). Royal Aircraft Establishment, Farnborough.
9. MINFORD, J. D., *Treatise on Adhesion and Adhesives* Vol. **3,** ed. Patrick, R. L., 1973, Marcel Dekker Inc., New York, p. 79.
10. BUTT, R. I. and COTTER, J. L., *J. Adhesion*, **8,** 1976, p. 11.
11. BETHUNE, A. W., *SAMPE Quarterly*, **11,** 1975, pp. 3–4.
12. BOWEN, B. B., *Symposium on Adhesion and Adhesives*, Univ. of Pittsburgh, 1974.
13. WEBER, K. E. and JOHNSTON, G. R., *SAMPE Quarterly*, Oct. 1974, p. 16.
14. MCCARVILL, W. T. and BELL, J. P., *J. Adhesion*, **6,** 1974, p. 185.
15. MCCARVILL, W. T. and BELL, J. P., *J. Applied Polymer Sci.*, (*a*) **18,** 1974, p. 335; (*b*) **18,** 1974, p. 343; (*c*) **18,** 1974, p. 2243.
16. WEGMAN, R. F., *Adhesives Age*, **11,** 1967, p. 20.
17. WEGMAN, R. F., BODNAR, W. M., BODNAR, M. J. and BARBARISI, M. J. *SAMPE Journal*, 1967, p. 35.
18. BIJLMER, P. F. A. *J. Adhesion*, **5,** 1973, p. 319.
19. SMITH, T., *Symposium on Durability of Adhesive Bonded Structures*, Dover New Jersey, 1976, p. 21.
20. MCMILLAN, J. C., QUINLIVAN, J. T. and DAVIS, R. A., *SAMPE Quarterly*, **6,** 1976, pp. 3 and 13.
21. NOLAND, J. S., *Adhesion Science and Technology*, Part A, 413, ed. Lee Lieng-Huang, 1975, Plenum Press, New York.
22. KOETSIER, J., *Proceedings of the National SAMPE Technical Conference*, **7,** 1975, p. 126.
23. HAMILTON, W. C., LYERLY, G. A. and WEGMAN, R. F., *Picatinny Arsenal Technical Report*, 4185, 1971.
24. WEGMAN, R. F., ROSS, M. C., SLOTA, S. A. and DUDA, E. S., *Picatinny Arsenal Technical Report*, 4186, 1971.
25. WEGMAN, R. F., HAMILTON, W. C. and BODNAR, M. J., *Proceedings of the National SAMPE Technical Conference*, **4,** 1972, p. 425.
26. MAZZA, F., *Chimica and Industria*, **43,** 1961, pp. 1293 and 1424.
27. BIANCHI, G., MAZZA, F. and TRASATTI, S. *Proceedings 2nd International Conference on Metallic Corrosion*, New York, 1963.
28. CAPRIOLO, I. and TURRIS, F., paper presented at the *Titanium Symposium*, Milan, September 1969. Translated by Preece, R. L. (IMI).
29. ALLEN, K. W., ALSALIM, H. S. and WAKE, W. C., *J. Adhesion*, **6,** 1974, p. 153.
30. PETERKA, J., *Adhäsion*, **7/8,** 1975, p. 200.
31. CASSIDY, P. E. and YAGER, W. J., *Tracor Inc. Technical Report T75*, 7537, 1969.
32. GLEDHILL, R. A. and KINLOCH, A. J., *J. Adhesion*, **6,** 1974, p. 315.

APPENDIX

Titanium Pre-treatments for Adhesive Bonding

1. (i) Vapour degrease in trichloroethylene.
   (ii) Clean in 'Stripalene 532' at 70 °C.

(iii) Rinse in running tap water (cold).
(iv) Etch in a solution of 12% by volume HNO_3 (70%) and 1% by volume 40% HF solution at 20 °C for 30 min.
(v) Rinse in running tap water (cold).
(vi) Phosphate–fluoride film forming in a solution of 50 g/l of trisodium phosphate, 20 g/l potassium fluoride and 33 ml/l HF acid (40%) operated at 20 °C for 2 min.
(vii) Cold rinse in tap water.
(viii) Hot water soak at 60 °C for 15 min.
(ix) Scrub with a stiff bristle brush in cold running water to remove any fluoride smut not removed in the hot water soak.
(x) Rinse in distilled water.
(xi) Dry in warm air.

2. Identical with process 1 except that in item (iv) 7·5 g/l of anhydrous sodium sulphate is added to the etch solution.

3. (i) Trichloroethane or trichloroethylene degrease.
(ii) Wipe with ethyl methyl ketone or iso-butyl methyl ketone, to remove paint and identification marks.
(iii) Vacu-blast, preferably, with 120 grit alumina (aloxite) at 70–80 lb in^{-2} in pressure (a suitable laboratory alternative is to scrub with industrial scotch brite charged with scouring powder).
(iv) Hot-water wash.
(v) Immerse in an alkali-based cleaner (an aqueous solution of 'Stripalene 532' at 3% for 20 min at 70 °C; or 3% 'ICI ADA' salts for 10–20 min at 60 °C or 'Diversey 808' for 1–2 min at 60 °C, or any other similar commercial preparation).
(vi) Hot-water wash for 20 min.
(vii) Dry.
(viii) Immerse for 20 min at 65–70 °C in:
sodium hydroxide, 20 g (0·5 mole);
hydrogen peroxide 30% (100 vol), 22·5 ml (0·2 mole);
water to make 1000 ml.

NOTE: Take bath to operating temperature before adding hydrogen peroxide, with stirring and immediately prior to inserting metal.

(ix) When metal blackish in colour remove.
(x) Hot-water rinse.
(xi) Hot-water wash for 15–20 min.
(xii) Dry in warm air.
(xiii) Bond.

4. (i) Specimens cleaned as detailed in items (i), (ii) and (iii) of pre-treatment 1.
   (ii) Etch for 2 min at 20 °C in a solution of 350 ml/l of 70 % w/w HNO_3, 35·7 ml/l of 40 % w/w HF acid and 22·5 g/l anhydrous sodium sulphate.
   (iii) Etch process completed as detailed in terms (v)–(xi) of pre-treatment 1.
5. (i) Vapour degrease in trichloroethylene.
   (ii) Vacu-blast with 180–220 alumina grit.
   (iii) Clean in 'Stripalene 532' at 70 °C.
   (iv) Rinse in running tap water.
   (v) Remove grit by scrubbing with bristle brush in cold running water.
   (vi) Dry in warm air.

Chapter 2

STRUCTURAL ADHESIVES USED IN CIVIL ENGINEERING

P. C. HEWLETT† and J. D. N. SHAW‡
†*Cementation Research Ltd, Rickmansworth, UK*
‡*Cementation Chemicals Ltd, Rickmansworth, UK*

SUMMARY

Adhesives are now an integral part of modern construction practice. The types of materials available and their performance specification/suitability often make selection difficult. This chapter is concerned with adhesives used for structural purposes which is perhaps the most demanding area of application. Their uses for concrete bonding, segmental construction and repair—both internal and external—are reviewed. The properties limiting their application are mentioned and shortcomings highlighted. This may stimulate development work where needed. The chapter has been written for the practising engineer, specifier and construction materials technologist who wishes to obtain a perspective of applied adhesives usage in the construction industry.

INTRODUCTION

Synthetic polymers having good adhesive capability form the basis of many proprietary construction and building materials. Adhesives, sealants, admixtures, surface finishes, resin concretes, bonding agents, fixings and cladding are a few of the routine applications using adhesive polymers that we take for granted. Equally varied are the polymer compositions that are used, for instance, epoxides, polyesters, polyurethanes, acrylics, styrene butadienes, polyvinyl acetates, furanes, polyvinyl chlorides and various formaldehyde resins such as melamines and phenolics.

Several descriptive and detailed accounts have been published covering the general application and usage of epoxide resins in construction.[1–6] Similarly, practical advisory leaflets covering general adhesives in building are available.[7]

However, many of the uses are non-structural in the sense that the adhesive composition does not transmit or accommodate significant stresses. In recent years there has been both interest in and application of synthetic polymer adhesive compositions for truly structural situations. The engineer is faced with real difficulty in selecting a material on a chemical basis and it is essential to specify the end application as far as possible in terms of performance and use in the conditions found on a construction site. It is necessary to understand the performance of the synthetic resin adhesive under continuous loading and dynamic stress, factors which are essential to the safe usage of the adhesive. Of further importance to the engineer is the life expectancy of the synthetic resin adhesive in terms of the durability of more traditional structural materials such as concrete and steel.

To the engineer then, the adhesive should exhibit, firstly with regard to workability and cure, the following:

(i) Material consistency that allows ease of use under the prevailing temperature and general weather conditions found on site.
(ii) Adequate useable time.
(iii) Cure that is independent of temperature and humidity.

Secondly, regarding performance:

(i) Little or no reduction in adhesion to concrete as a function of time or on exposure to moisture/humidity.
(ii) Negligible creep deformation.
(iii) Good temperature resistance.

Thirdly, with regard to speed of construction:

(i) A cure rate that does not hinder a construction programme, which means systems exhibiting high early strength.

It is not commonly known to the structural and civil engineer, that the range of materials types is so broad and often the method of function less known. It is the purpose of this chapter to select some constructional applications and particularise the test data that relates to the material used, especially the adhesive structural requirements.

It has been necessary to omit several topics in a chapter of this length and end uses such as glued, laminated structures, patching and grouting compositions, as well as resin anchors and rock bolts, are not dealt with.

RESIN ADHESIVES FOR CONCRETE BONDING

Wet, fresh or plastic concrete has little tendency to adhere to hardened concrete. This practical fact was recognised in the crafts of screeding, floor laying and rendering. Good sub-surface preparation usually aimed at increasing the mechanical bond is a pre-requisite. However, such preparation is not always possible and is invariably expensive. It is desirable to interpose at the bonding interface a layer of material that adheres well to the hardened concrete and also to the fresh concrete being placed on it. Concrete-to-concrete bonds using epoxy adhesives have mechanical properties greater than those of concrete itself.

Typical average properties of such adhesives are:

(i) Compressive strength = 77 MN m^{-2} (concrete maximum value = 84 MN m^{-2}, but typically 40–60 MN m^{-2}).
(ii) Shear strength = 35 MN m^{-2}.
(iii) Tensile strength = 20–28 MN m^{-2}.

When cured, epoxide resin bonding agents resist very well the alkali from concrete (by comparison polyvinyl acetate compositions are suspect), as well as moisture. The adhesive transmits compressive, tensile and shear forces excellently, but situations causing peel stresses should be avoided. Where these forces cannot be avoided, special flexible compositions may be needed. Consider the two situations outlined below.

Concrete/Concrete Bonding

New to old concrete bonding is required in the repair or enlargement of existing concrete structures and where the extended structure is expected to act as a monolithic unit. The old concrete is cleaned so that loose contaminants are removed, the adhesive applied, and whilst still tacky, new concrete or mortar is applied and worked at least in part into the adhesive layer, so giving both adhesive and mechanical bond.

Typical situations that may benefit are, application of thin toppings to an existing concrete base, impact resistance is also improved by incorporating a bonding adhesive. Conduit walls, dam buttresses, pillar supports and placed extruded curves. The jointing of pre-cast structural elements. These units may be small or up to 100-ton sections in the case of bridges. Usually the joints are thin compression joints with the adhesive acting as the load distributor and gap filler.

Compared to cement mortar joints the use of a resin adhesive leads to time saving through faster cure (strength build-up with time) and gives higher tensile and shear strengths. Aesthetically, the joints also appear as thin lines rather than thick bands of mortar.

Metal-Concrete Bonding

Epoxide adhesives have the distinct advantage of being able to bond dissimilar materials. On construction sites fastening steel into concrete is a commonplace requirement. This may be achieved either by coating steel and casting concrete against it whilst still tacky or by coating steel and applying a layer of sand to the resin before it has hardened. The rough sand interface now integrates with the cast concrete.

Typical uses are:

(i) Securing steel bolts and fastenings in concrete.
(ii) Bonding steel tiles.
(iii) Bonding metal road studs.
(iv) Bonding steel reinforcement wires, and post tensioning tendons to concrete.
(v) Formation of composite 'T' beams in which the adhesive acts as a shear connector between metal and concrete, so improving stress distribution.
(vi) Bonding of steel sheets or plates to beams, so improving flexural strength of composite.
(vii) Rock bolt fastening for stabilising fractured rock formations local to excavations such as tunnels and shafts.

Strengthening a metal-reinforced concrete structure has always been difficult but with the advent of epoxide resin mortars used to fasten extra reinforcement to the structure, it is now more commonplace. For instance, if there is sufficient room, it is possible to bond an extra layer of reinforced concrete to that already existing. Alternatively, steel plates may be bonded externally to the concrete in areas of high tensile or shear stress.[8,9]

In order to be able to effect additional reinforcement repairs, be they concrete-to-concrete or steel-to-concrete, it is necessary to use resins that have exceptional properties at elevated temperatures. Many of the properties traditionally used to characterise epoxy mortar, such as reactivity, short-term strength and adhesion, are now regarded as only partially adequate for structural repair. The high-temperature response is considered the most critical parameter for all cold curing binders, and not only for epoxy resins. Creep deformation as a function of steadily increased temperature is a criterion of suitability. Work published by Heugenschmidt[10] using rectangular epoxy prisms, indicated that highly cross-linked systems start to lose their strength at temperatures above *ca.* 45–55 °C. However, thin layers of adhesives in bonded joints give better response. Compression tests on diagonally bonded concrete prisms have shown that 25 % of short term ultimate load can be sustained up to 95–100 °C. Having performed static failure tests, as well as creep and dynamic fatigue tests, in no case was failure related to breakdown of adhesion. The control beam had almost identical deformation and load bearing capacity to the bonded beams. The author[10] recommends the highly cross-linked epoxide resins used as structural load bearing adhesives for pre-cast concrete elements for bonding steel to concrete and steel to steel. Creep data for these materials are shown in Figs. 1–3.[11]

It is becoming evident that epoxy-based adhesives are suitable for various uses within civil engineering. Materials such as concrete and steel can be bonded structurally. The possibility of reinforcing concrete structures simply and effectively using epoxy adhesives, could be of specific interest to civil engineers.

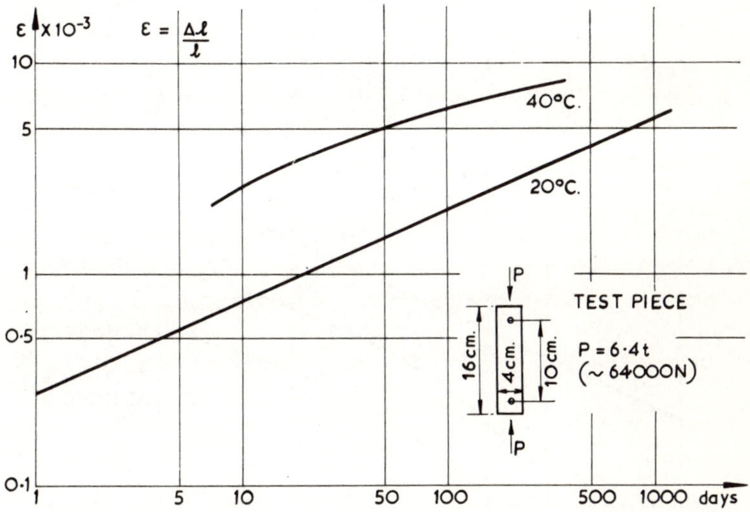

FIG. 1. Creep deformation curves for an epoxy mortar at a compressive stress of 40 MN m^{-2} at 20°C and 40°C. (Courtesy Concrete Society, A. Letts and Shell Chemical Co. Ltd.)

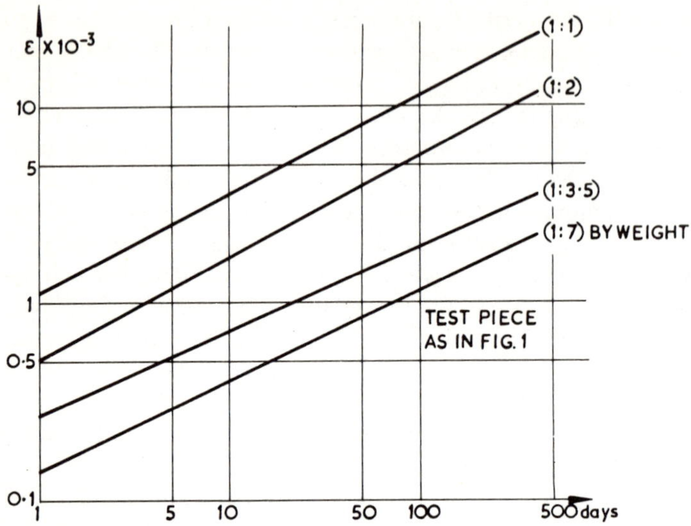

FIG. 2. Creep deformation as a function of filler ratio (binder to filler ratio given in parts by weight) at 20 °C compressive stress 40 MN m^{-2}. (Courtesy Concrete Society, A. Letts and Shell Chemical Co. Ltd.)

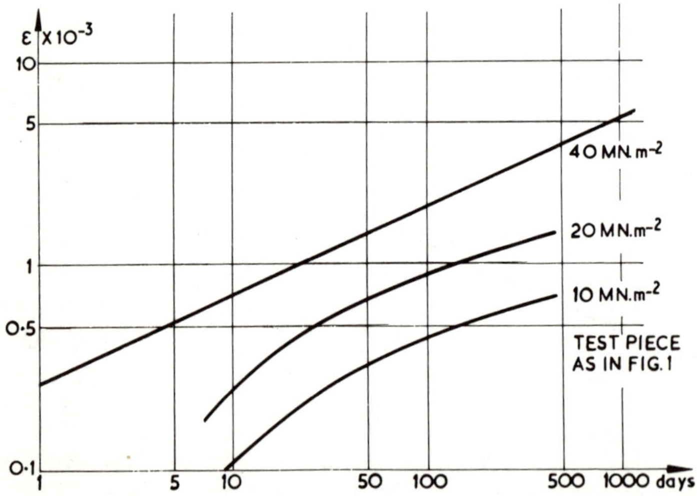

FIG. 3. Creep deformation as a function of compressive stress measured at 20 °C. (Courtesy Concrete Society, A. Letts and Shell Chemical Co. Ltd.)

An interesting and aesthetically-satisfying application of epoxide resins acting as semi-structural bonding adhesives was in the slab glass of the lantern of the Cathedral of Christ the King, Liverpool.[11] The pieces of glass were bonded together by means of an extrudable epoxide resin grout which fastened them into the concrete frames. These windows had to withstand wind forces up to 150 miles/h, and remain weatherproof. Conventional cement mortar grout was unsuitable in part due to inadequate strength, and the comparison is given in Table 1.

TABLE 1

STRENGTH COMPARISON OF EPOXY AND CEMENTITIOUS MORTARS

Materials	*Tensile strength* (MNm^{-2})	*Flexural strength* (MNm^{-2})	*Impact resistance* (*Nm*)
Epoxide resin filled composition	16	49	30·5
Cement mortar	3·5	7·0	16·0

The superior adhesion of epoxide resins to glass and concrete was an additional advantage. As a consequence of high mechanical properties of the resin mortar thinner grout sections could be used which was aesthetically desirable. The finished effect is shown in Fig. 4. The estimated durability of the epoxy mortar has been given as 1000 years![11]

Other bonding agents to epoxides are widely used, particularly for concrete/concrete bonding. The polyvinyl acetate (PVA) dispersions are the most widely used, although others such as styrene butadienes (SB) and natural rubbers, acrylates and water-dispersible epoxides, may appear as alternatives. Selected properties of PVA modified mortars in particular, their adhesion, is shown in Table 2.[12]

However, the bonding capacity of these resin dispersions is limited and not regarded as structural in the load bearing or transverse sense. They are typically used for bonding thin screeds and renders to a base concrete, improving the adhesion of plasters and concretes to difficult surfaces, such as dense brick and smooth concrete.

PVA bonding agents are often criticised as having poor adhesion in the wet compared to acrylics and SBRs since the latter are not readily hydrolysed by the water. Nevertheless data in Table 3 shows poor adhesion under wet conditions for the more stable dispersions.[12]

SBR latex dispersions, whilst stable to hydrolysis, are liable to oxidative

FIG. 4. Lantern of Christ the King Cathedral, Liverpool. Resin mortar bonded glass/concrete panels. 16 Sections in all; total section height 66 feet; 10 panels per section. Each panel measures 12 × 6 feet or 12 × 8 feet. Total area = 12 000 square feet of glass. Top: Exterior view; Bottom: Interior view. (Courtesy Shell Chemical Co. Ltd.)

TABLE 2
PROPERTIES OF PVA MODIFIED MORTARS

Property	*Relative change*[a]			
	Cure at 50% relative humidity		*Cure at 100% relative humidity*	
	$P/C = 0$	$P/C = 0{\cdot}2$	$P/C = 0$	$P/C = 0{\cdot}2$
Tensile strength	0·4	3·0	1·0	0·8
Tensile modulus of elasticity	0·2	0·3	1·0	0·04
Tensile elongation at break	1·0	30	1·0	45
Compressive strength	0·2	0·75	1·0	0·3
Compressive modulus of elasticity	0·35	0·2	1·0	0·1
Compressive deformation at break	2	5	1·0	4·5
Abrasion resistance			1·0	10
Impact resistance			1·0	1·2
Corrosion resistance			1·0	1
Adhesion	0·5	5	1·0	1
Thermal expansion	1	1	1	1

[a] Change in properties relative to non-polymer modified mortar cured and tested at 100% relative humidity. All samples cured 28 days. Tested at conditions similar to those used for curing.

TABLE 3
RELATIVE ADHESION CAPABILITY OF BONDING AGENTS

Polymer type	*Water resistance (wet rub)*	*Bond strength* (MNm^{-2})	
		Dry	*Wet*
Polyvinyl alcohol	soluble	0	0
PVA 1	re-emulsifies	0·38	0·37
PVA 3	500	0·36	0·03
VA/VeoVa[a]	>500	0·39	0
Acrylic	>500	0·29	0
Styrene butadiene	>500	0·38	0
None		0	0

[a] Vinyl versatate available from Shell Chemicals (UK) Ltd.

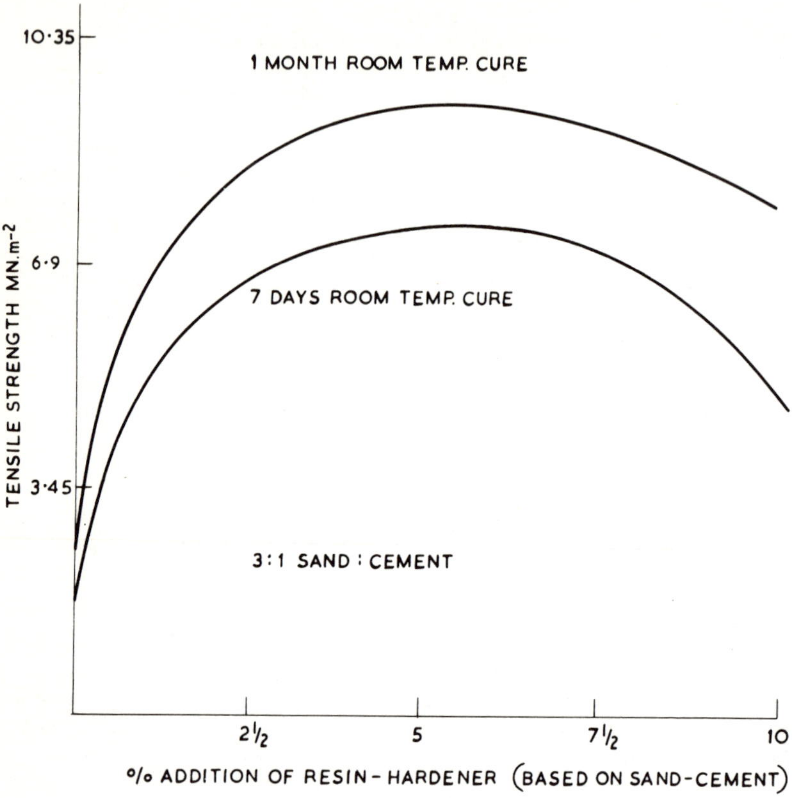

FIG. 5. Tensile strength of mortar containing water-dispersible epoxide. (Courtesy *The Consulting Engineer*.)

attack which may limit life expectancy. Modified SBR latex has been reported on indicating the superiority of SBR latex over vinyl acetate copolymers.[13]

Water-dispersible epoxide resins appeared some years ago but application has been limited, in part due to relatively high cost compared to other bonding agents, and admixture dispersions. These materials used neat as primary/bonding agents or integral with cement (admixtures) at approximately 20% w/w cement content, can markedly improve a sand:cement mortar's tensile and flexural strength.[14,15] Strength data relating to mortars containing water dispersible epoxides, are given in Figs. 5–7.

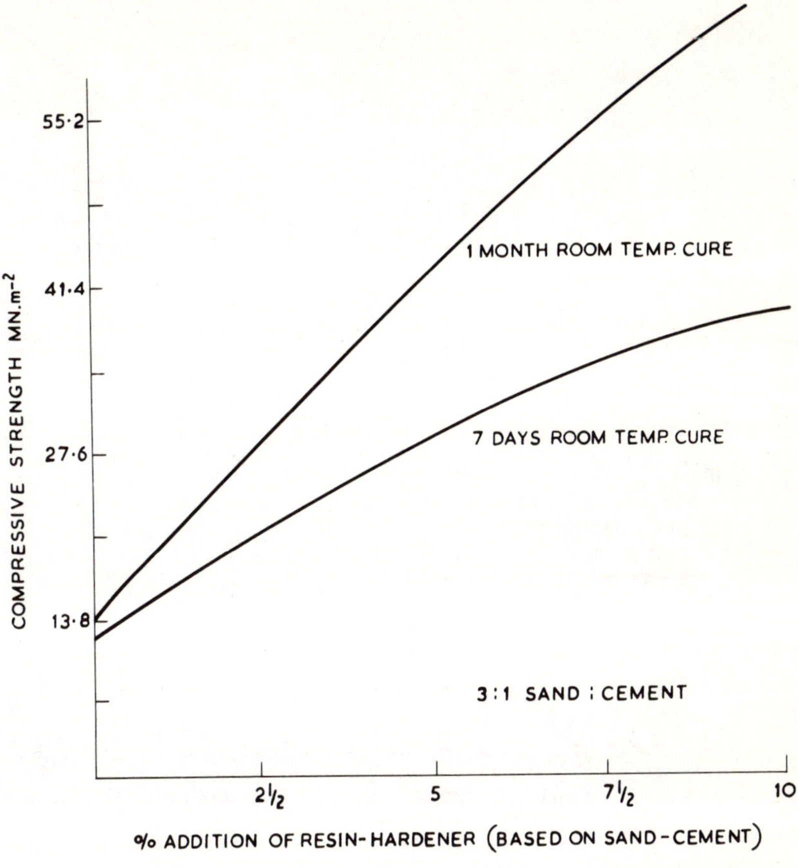

FIG. 6. Compressive strength of mortar-containing water-dispersible epoxide. (Courtesy *The Consulting Engineer*.)

Mortars containing water dispersible epoxides show enhanced workability or 'plasticising' effect, allowing low water contents (typically 0·35) to be used reducing drying shrinkage and improving density and porosity.

A comparative study of some 20 proprietary bonding agents has been carried out[16] and from the viewpoint of bond strength using a direct tension test, water dispersible epoxides, SBRs and a selected acrylic emulsion gave the best performance, i.e. tensile adhesion greater than 0·7 MN m^{-2} compared with approximately 0·2 MN m^{-2} for the control. The results, shown in Table 3, were obtained by measuring the adhesion of plaster to glass. To the engineer often faced with a confusing selection of

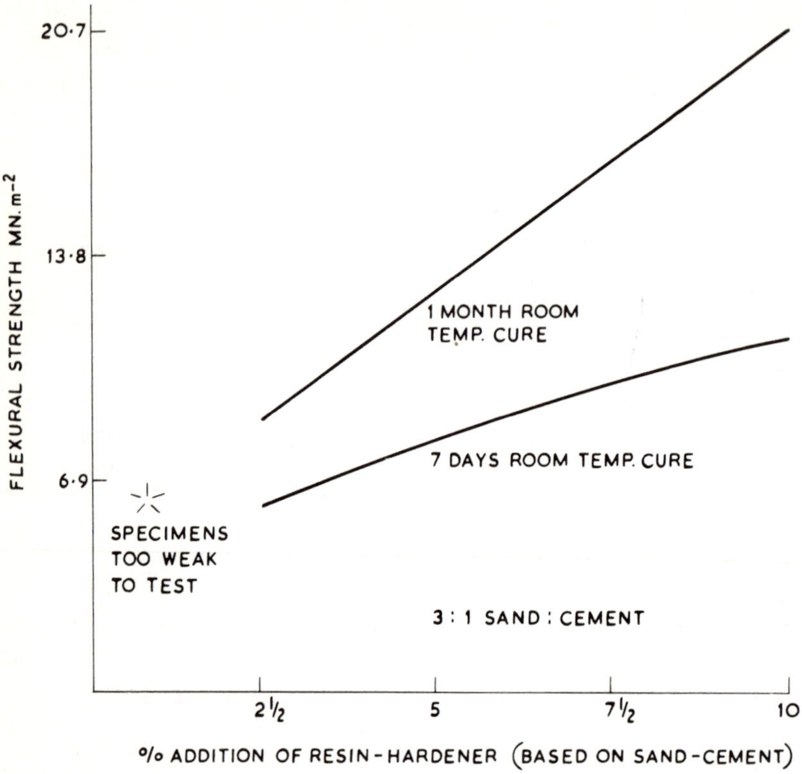

FIG. 7. Flexural strength of mortar containing water-dispersible epoxide. (Courtesy *The Consulting Engineer*.)

materials, it is worth highlighting the cost/performance advantage of acrylic dispersions. In addition they are less affected by humidity, have better resistance to fats, oils and hydrocarbon solvents, compared to SBRs, and give non-dusting, non-slip, drag free toppings when used as an admixture.[17]

USE OF RESIN IN SEGMENTAL CONSTRUCTION

The concept of constructing structures with long spans, in particular bridges, from pre-cast units, with post-tensioning, has become widely used over the last 20 years.

Originally pre-cast units were jointed with wide joints approximately

750- mm wide of cast-*in-situ* concrete and then designs progressed through narrower joints using high strength cementitious mortars, to very narrow joints designed as dry joints, which, however, required too high a degree of precision and the provision of shear keys until the structure was stressed, to be entirely practicable. The dry joint was then further modified by the use of epoxy resin adhesives to provide uniform shear connection over the whole mating area between the units and also completely to seal the joints between the units.[18] It is this concept of segmental concrete construction with post tensioning, employing epoxy resins, which is increasingly used in the erection of long span bridges although some engineers would prefer to use wider cementitious joints to enable slight changes in the direction of the construction without having to cast different units.

Where epoxy resins are used, the individual pre-cast units are usually cast so that the matching end of one unit suitably treated with a thin removable release coat is used as the form for the next unit, thus when they are assembled on site, the units match very well and the joint thickness can be kept within 1–2 mm and often well under 1 mm using epoxy resin joints instead of cementitious mortar joints. It is estimated that the time required to make a joint and leave it to cure before full pre-stressing with epoxy resins is about $\frac{1}{5}$ of what would be required with cement mortar joints.[19] The first reported structure where pre-cast units were bonded together with epoxy resins, and then post-stressed, was at Coventry Cathedral, constructed in 1960–62. Here slender cruciform columns that supported the inner ceiling panels were pre-cast in three sections of equal weight. These sections were stressed together by four stranded cables and a resin of capillary adhesive composition was used to provide uniform transfer of the compressive stress over the whole joint area. The thickness of the resin glue line was well under 1 mm and was therefore almost invisible. In this application the adhesive acted purely in compression under a working stress of not more than $4\,\mathrm{MN\,m^{-2}}$ (ref. 19).

The first major structure designed and constructed from pre-cast concrete segments with thin epoxy resin joints was in 1962 for a bridge over the River Seine at Choisy-le-Roi not far from Paris, designed and constructed by Entreprises Campenon Bernard.[20] This bridge was made up of four continuous box girders over three spans and the box girders were cast away from site as a series of long continuous units with curled joints located at $2\frac{1}{2}$-m centres. Debonding at joints during casting was achieved by spraying on a release coat which would be readily peeled off after the units were struck from the mould. The units were transported to site and lifted by cranes on to a retractable jig.

Before a decision was made to use epoxy resin adhesives on the Choisy-le-Roi bridge the designers needed to be reassured that the epoxy resin system would give good performance when applied on site, under the differing weather conditions that can be encountered during the construction period lasting several months.

In conjunction with the resin suppliers exhaustive testing in the main of a short term nature on the compressive strength, modulus of elasticity and shear strength of the adhesive system, and shear and bonding tests on the adhesive system in the actual concrete–concrete joint situations, were carried out. The results of these tests convinced the design engineers that the use of a specially formulated epoxy resin was practicable. On completion of construction of the bridge, load tests carried out indicated that the epoxy bonded segmental construction behaved exactly as a conventional *in situ* concrete structure thus fully justifying the engineers' faith in the new design concept for such a large and critical structure. Since 1962 numerous bridges in many parts of the world have been constructed using this construction technique employing resin adhesives. These include: Pont d'Oleron, Gironde and Pont Aval, near Paris, both constructed by Entreprises Campenon Bernard;[21] the Chillon Viaduct in Switzerland;[22] the River Ebro Bridge, Pamplona, Spain;[21] Kleinpolderplein Motorway Junction, Rotterdam, Holland 1968–70;[23] Thruway Bridge, Amsterdam, New York;[22] Rio Niteroi Bridge, Brazil, 1974–75;[24] Rawcliffe Bridge, Yorkshire;[18] and Marron Bridge, Cumbria.[24]

In addition to bridges, segmental concrete construction techniques using epoxy resin adhesives have been employed in the construction of a number of buildings of which the Sydney Opera House and the large sugar storage silos at Durban, South Africa, are probably the most notable. The whole shell roof structure of the Sydney Opera House constructed in 1963–67 is made up from curved hollow concrete ribs to form a series of continuous spherical surfaces. The ribs are made of pre-cast segments cast with matching faces and made longitudinally continuous by post-tensioning across transverse epoxy joints. The decision to use epoxy joints here was made because it enabled:

(i) the critical geometry of the curve of the rib to be accurately set out in the casting yard and duplicated during erection in the air;
(ii) the ribs to be erected and post-stressed much more quickly than if a cementitious mortar joint had been used.[19,25]

The sugar terminals constructed in Durban, South Africa, in 1965 employed epoxy resin jointed pre-cast units since it reduced the construction time and overall cost compared with construction methods used in previous storage silos, which had been cast *in situ* and involved the use of a great deal of expensive shuttering.[26]

At the time the Choisy-le-Roi Bridge was constructed the information available on the performance characteristics of epoxy resin systems suitable for use as adhesives for concrete in structural applications was limited and many engineers were asking a number of questions concerning the application of properties and the long term mechanical properties, many of which at that time could not be fully answered. Questions which need answering, include:

1. Can the formulated adhesive be easily applied in the thickness required without any run-off of the adhesive from the vertical surface?
2. How much time can elapse between mixing the adhesive, applying it to the prepared surface to be bonded, and bringing the units together? Three factors are important. The useable life at different site temperature conditions; the open time of the adhesive applied to the units; the influence of ambient temperature, and, where this is different from the ambient temperature, the surface temperature of the units to be bonded.
3. How long after the bonding at different ambient temperature does the adhesive develop sufficient mechanical properties to permit full post-stressing?

Adhesive systems are now available that satisfy these practical/construction questions.[24] Table 4 gives in detail the properties of a typical epoxy adhesive used for this application. Both the application properties and short term properties under certain weather conditions, of one single adhesive system will not always give acceptable application and curing characteristics under all temperature conditions.

Resin systems with similar mechanical properties but which cure at different rates at different temperatures are available, so that reasonably uniform construction times can be achieved by selecting the right adhesive system for temperatures prevailing.

In addition to the above properties engineers need reassurance on the mechanical properties of the adhesive, both in the short term when bonding under adverse conditions, such as to damp and cold substrates, and changes in properties long term. Of particular importance are the shear strength of the adhesive when bonding both to dry and very damp concrete and long term creep. Detailed shear testing in adverse conditions, both on site and in

TABLE 4

PROPERTIES OF EPOXY ADHESIVES USED IN SEGMENTAL CONSTRUCTION. SPECIFICATION SUMMARY

1. Usable life (5 kg pack)										40–50 min at 5 °C 15 min at 20 °C
2. Open time (2-mm layer applied immediately after mixing)										90 min at 5 °C 60 min at 20 °C
3. Compressive strength build up with time $MN\,m^{-2}$ by ISO/R604										
Cure temperature/time	4 h	6 h	8 h	10 h	16 h	24 h	3 days	7 days	14 days	
5 °C	—	—	—	—	20	20	30	40	60	
20 °C	15	35	65	70	—	—	75	75	80	
4. Modulus of elasticity after 7 days cure at 20 °C (ASTM DT 38-72)										6·7 $MN\,m^{-2}$
5. Shear strength of bonded concrete test specimens (test piece three 100 mm concrete cubes bonded in line)										
Cure										
24 h at 5 °C										5 $MN\,m^{-2}$ concrete failure
24 h at 20 °C										5 $MN\,m^{-2}$ concrete failure
6. Long-term water resistance (Test piece 4 × 4 × 8 cm cement mortar prisms bonded together and immersed in water after curing.) Test piece subjected to four point flexural loading after varying intervals of time										
28 days' immersion										100 % concrete failure
6 months' immersion										85 % concrete failure
2 years' immersion										85 % concrete failure
7. Shear strength versus temperature (Test piece—simple lap-shear in mild steel joint area 25 × 22 mm)										
Temperature	20 °C	30 °C	40 °C	50 °C	60 °C	70 °C				
Shear strength ($MN\,m^{-2}$)	6	7	13	10	7	4				

many laboratories, has demonstrated that when using correctly formulated epoxy adhesive resin systems for bonding concrete, failure always occurs in the concrete layer adjacent to the concrete/resin interface.[10,22,27] Creep of materials like epoxy resin systems, which are of lower elastic modulus than the conventional construction materials, concrete and steel, has always been of some concern and more so than when epoxy resin systems are being used as structural adhesives.

Work on the creep of epoxy resin systems by a number of workers[10,28] indicates that the creep of epoxy resin systems depends upon a number of factors, including precise chemical nature of the epoxy resin system, types of fillers used and filler loading, the dimensions of the epoxy resin system in the creep situation, temperatures of testing in relation to the temperature resistance of the resin system, and the maximum stresses in service in relation to the yield stress of the system. Having selected the best compromise with regard to ease of application, cure characteristics and ultimate mechanical properties, it is important to note that with a thin glue line at the relatively low stress levels normally encountered in segmental construction, creep should not be a problem and this has been confirmed by work carried out by a number of workers, such as Johnson, Miklofsky and Topaloff.[29–31]

RESIN BONDED EXTERNAL REINFORCEMENT

Over the past ten years there has been considerable interest in the use of epoxy resin adhesives to bond external steel plate reinforcements on to structures to increase their load bearing capabilities. This development was pioneered in South Africa and France and has been used for a variety of reasons:

1. To strengthen reinforced concrete beams which had been wrongly constructed. Steel reinforcements may have been wrongly placed in error, or there may have been insufficient or total absence of steel reinforcements.
2. Faulty design.
3. To increase the load bearing capacity of a structure above its original design strength.
4. To replace the existing steel reinforcement which is no longer adequate due to damage through external forces or corrosion.

Extensive testing on the bonding of external steel reinforcement has been carried out in many countries including South Africa, France, Switzerland, and the United Kingdom.[8,9,26,32,33] This work has demonstrated that significant increases in the strength of concrete beams can be readily achieved by simple practicable methods on site.

The main requirements in strengthening structures by external reinforcement is to achieve a high shear strength between the steel/resin adhesive interface. Work by several people has shown that when a thin layer of epoxy resin is used to bond concrete to steel, failure as expected, always occurs in the concrete adjacent to the concrete/resin interface and thus significant increases in the live load bearing capacities of structures can be achieved. One aspect of steel-plate bonding which was apparent in tests carried out in the UK Transport & Road Research Laboratory,[33] and by Bresson,[9] which gave rise to some concern, is that when failure of an externally plated beam occurs, it occurs much more suddenly than with most reinforced concrete structures. In typical under-reinforced concrete structures, as loads are gradually increased, hair-line tension cracks develop in the concrete which gradually widen and extend until the concrete fails. The wide cracks, however, are very evident well before the structure fails and repair work can normally be put in hand before catastrophic failure occurs. With similar structures, further reinforced by steel plate bonding the initial pattern of cracking observed is similar but as the loads increase the cracks do not widen significantly prior to failure. However, test results on beams reinforced with external steel plates are in close agreement with theoretical considerations, so that failure loads can be accurately predicted and avoided in practice.

Interesting examples of the use of this method of strengthening structures, include the repair of the approach road to Louis Botha Airport, Durban, South Africa.[34] The underside of the concrete beams were struck by the jib of a mobile crane passing underneath at speed. Considerable damage to several of the concrete beams was caused, and some of the steel reinforcement was ruptured. In order to repair the bridge as rapidly as possible, with little hindrance to traffic flow, all damaged concrete was removed and replaced by an epoxy resin mortar. A steel plate was then bonded on to the underside of the patched up beams to restore the bridge to its original design strength.

The same approach has been used in the maintenance of numerous railway bridges along South Africa's coastline. Many of the bridges constructed in 1920–30 have deteriorated due to severe concrete spalling and corrosion of the steel reinforcement in the corrosive salt laden atmosphere.[35]

The technique has also been used to strengthen the Pont d'Arceuil on the Autoroute du Sud, near Paris, France, in 1971.[36] Here the measured deflections after strengthening were less than those calculated which indicates that the structure is very capable of carrying the live loads envisaged. Another example of a resin-bonded externally-plated steel reinforcement was to strengthen the ground and the first-floor beams of the Telephone Exchange in Zurich, Switzerland.[37] The work was required so that additional heavy equipment could be installed on these floors and the use of this relatively new strengthening technique allowed strengthening with only a slight reduction in the head-room of the rooms. Extensive tests were carried out by the Swiss Federal Materials Testing Institute (EMPA), to prove the techniques prior to this work being carried out.[32]

In the United Kingdom, the use of this technique has, to date, been limited. One interesting example was the strengthening of an eight storey building constructed from prefabricated units.[35,38] Several years after the building had been completed, the British Building Research Establishment published papers[39] indicating that many tall buildings constructed from prefabricated reinforced concrete units, might not be entirely safe if subjected to high wind forces, which could occur perhaps once or twice in 50 years. The BRE suggested strength criteria to ensure complete safety of such buildings under high wind forces. By coincidence the publication of the BRE reports happened to occur almost simultaneously with the collapse of the Ronan Point building due to a gas explosion, so that the BRE strength criteria were rapidly incorporated in revised building regulations. As a result of both the Ronan Point disaster and the BRE findings, consultant structural engineers all over the UK were called in to examine existing tall buildings constructed from prefabricated concrete units and advise whether they required strengthening. It was found that many buildings did indeed require some further support. Nearly all were strengthened using conventional construction techniques but one eight-storey building was strengthened using steel plates bonded on to the structures using epoxy resin based adhesives. From examination of the building and reference to the original structural design data the Consultant Structural Engineer decided that if steel plates 25-cm wide × 25-mm thick for the lower storeys, tapering to 15-cm wide and 6-mm thick on the upper storeys were bonded continuously up the structure to the corners of the support columns around the two staircases and the two lift shafts, the building would be adequately strengthened.

Design considerations indicated that a minimum shear strength of $0{\cdot}2\,\mathrm{MN\,m^{-2}}$ over the whole area between steel plate and the existing concrete beams was required. Proving trials were carried out on site and shear strengths of approximately 3·5 $MN\,m^{-2}$ were achieved. On the basis of these trials, the strengthening method of employing epoxy resin-bonded steel plate was approved by the client—a new town local authority.

The steel plates were shot blasted and primed off site and then bonded on to the concrete columns, the bonding areas of which were prepared by bush hammering followed by wire brushing. The thickness of the epoxy adhesive layer varied from about 1–4 mm and during cure the steel plates 6-mm long, were held in place by chemical fixings. After cure the plates were jointed by welding on short overlapping steel plates so that the steel was continuous up the structure. Tests were carried out to investigate the effect of site welding on the steel/epoxy/concrete bond. This test showed that the bond was almost completely destroyed in the vicinity of the welding but that the damage to the bond had not spread very far away from the actual areas of welding. From structural considerations, the engineers were not worried provided welding did not damage more than 25 % of the total bonded area. In fact it is estimated that well under 10 % of the bonded area was damaged by welding. In addition to the bonding of the vertical steel plate strengthening, it was also considered necessary to ensure stronger mechanical connections between adjacent prefabricated units.

To achieve this, the 140-mm diameter cores 120-mm deep, were extracted across the joints between units. The holes were coated with a layer of epoxy resin adhesive composition of buttery consistency, thickness approximately 7 mm and pieces of prepared steel pipe 130-mm diameter × 11-cm long were inserted into these holes. After cure the pipes were filled with a cement mortar and the decorations made good. A similar technique was also used to anchor the bottom of the vertical steel plates in the basement of the building. Ten-centimetre diameter cores, 100-mm deep, were extracted and 140-mm long steel pipe sections bonded into the holes. Matching holes were drilled into the bottom steel plate which was inserted over the steel pipes and the two were welded together on site.

Recently, a major UK motorway junction consisting of four bridges at Quinton Interchange on the M5, seven miles west of Birmingham, has been strengthened using the steel plate bonding techniques. During a routine inspection of the Quinton Interchange, cracks were discovered on the soffit and sides of the four bridges. A subsequent check on the design calculations revealed that the bridges had been under-designed and this, moreover, had been aggravated by detailing errors.[40]

After consideration of how best to strengthen the bridges with as little hindrance to traffic flow as possible, it was decided to strengthen and one aspect which had concerned engineers was what effect would load cycling of the structure during the curing period of resin adhesive have on the long term adhesive strength of the resin system. Brief laboratory tests at the UK Transport and Road Research Laboratory indicated that strain cycling of single resin bonded lap metal to metal test specimens during the curing of the resin, reduced the shear strength of the adhesive by approximately 30 %. Other tests carried out show that the weakest link in the bonding of steel plates on to concrete structures is the shear strength of the concrete itself. Therefore, this fall off was considered acceptable.

The tensile strength of the concrete surface on the soffits of the bridge beams on to which the steel plates were to be bonded were measured using a simple pull-off test and consistent figures of about $3\,MN\,m^{-2}$ were obtained. On the basis of these site test results and allowing a safety factor of approximately two on the strength of the concrete, and also assuming only 75 % bonding between the steel and the concrete due to imperfections on the concrete surface, a maximum design stress of $1{\cdot}2\,MN\,m^{-2}$ was possible. However, it was calculated that if 250-mm wide plates were used the stress in practice should not exceed $0{\cdot}14\,MN\,m^{-2}$ thus giving a very significant safety factor.

As with epoxy resin adhesives used in segmental concrete construction, the engineers required as much information as possible on the short and long term properties of the adhesive systems proposed for the job. A wide variety of adhesive systems based on chemically quite different epoxy resin/hardener systems have been used for steel plate bonding applications but apart from their proven performance in service there is very little long term data on many of the adhesives. In fact, several systems which have been used have been shown in subsequent laboratory tests, to give inferior bond characteristics when applied under adverse conditions such as cold or damp. For this reason adhesive systems which are both proven in service and backed up with extensive laboratory testing of short and long term performance are used. In recent jobs in the UK, such as Quinton Bridges, the same adhesive system was used in segmental concrete bridge construction as the four span Marron Bridge in Cumbria in 1975–76.[24]

In general, the stress levels in the resin adhesive are extremely low and under these conditions creep of the resin system is not normally a problem.

RESIN CONCRETES HAVING STRUCTURAL REQUIREMENTS

Concretes and similar materials in which the aggregates are bound together with synthetic resins (polymer concretes, PC) or alternatively, bound with

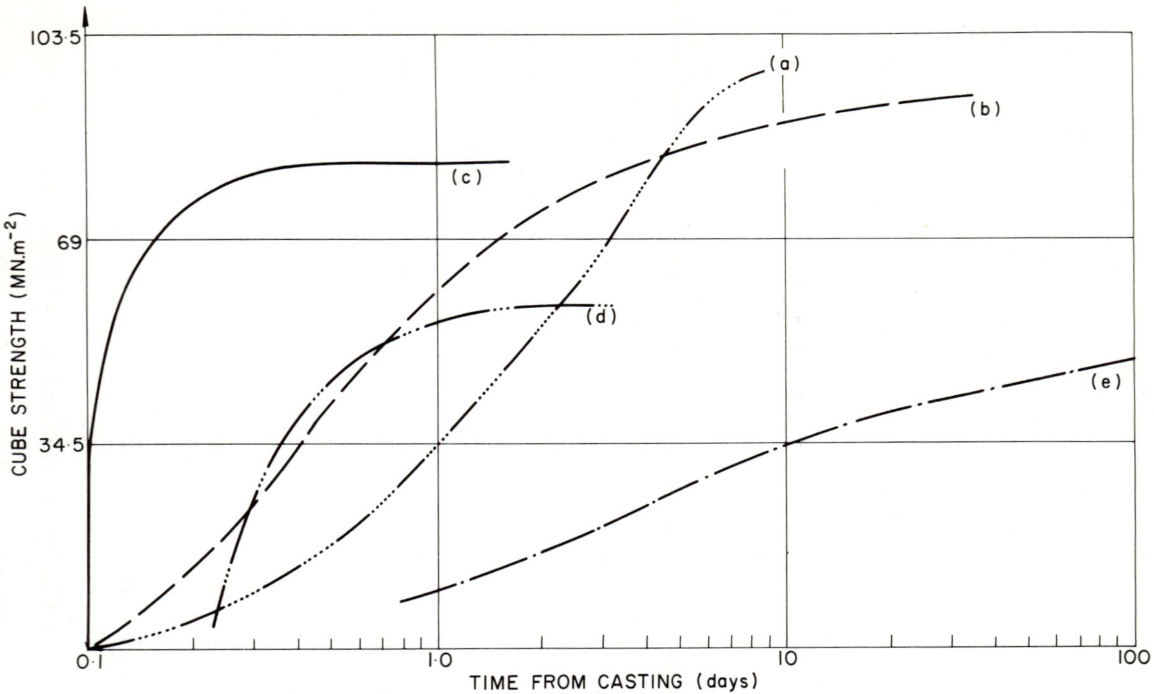

FIG. 8. Rate of strength gain for resin mortars (after L. H. McCurrich and W. M. Kay). (a) Epoxy mortar; (b) polyester mortar (slow setting); (c) polyester mortar (rapid setting); (d) high alumina cement concrete 0·5 w/c; (e) OPC concrete 0·5 w/c.

cement, but modified by polymer additions (polymer modified concrete, PMC), have found various uses in civil engineering. For instance, industrial flooring,[41] road joint nosings,[42] bridge and machine bearings, bearing pads for crane rails and machinery plinths as well as pre-fabricated wall panels[43] and many repair situations. This section of the chapter is aimed at highlighting some of the general engineering and structural properties of resin concretes of one kind or another, and putting them in relative context to structural concretes. However, the reader may wish to enlarge on the points made.[44,45]

Firstly, from the engineer's point of view, resin concretes can achieve high strengths quickly and the resulting cured material usually has good adhesion to most substrates. This strength/gain property is shown in Fig. 8[46] and comparative values are given in Table 5.

TABLE 5
COMPARATIVE STRENGTH PROPERTIES OF RESIN CONCRETE

Typical ultimate strength value for resin concrete ($MN m^{-2}$)	*Bulk property*
56–105	Compressive strength
7–21	Tensile
11–28	Flexural

Resin concretes are generally used for high performance applications and therefore hard, strong aggregates are often used with them. For instance, flint, silica sand and calcined bauxite. A maximum aggregate size is usually about 3 mm (cf. 9–37 mm for conventional concrete), since resin concretes are usually used in thin layers in part to minimise cost and also to reduce the effect of thermal shrinkage resulting from the exotherm during cure of the resin. To the engineer a strong but impact accommodating material has practical merit. Typical resin concrete has a Young's Modulus of 3500–21 000 $MN m^{-2}$ although tensile strain values of 1–10% typify the resins themselves and 0·3% may be all that is attainable for a resin concrete. Nevertheless this value is 20 times that of concrete (0·015% tensile strain). The impact strength too is better than ordinary concrete with values of 270 J (falling weight test[51])compared with 95 J for ordinary concrete.

In all structural applications strength retention at elevated temperatures is of concern. In this respect the engineer has to be cautious when specifying a resin concrete. At 20 °C resin concrete creeps more than conventional concrete, and as the temperature rises the creep effects become more

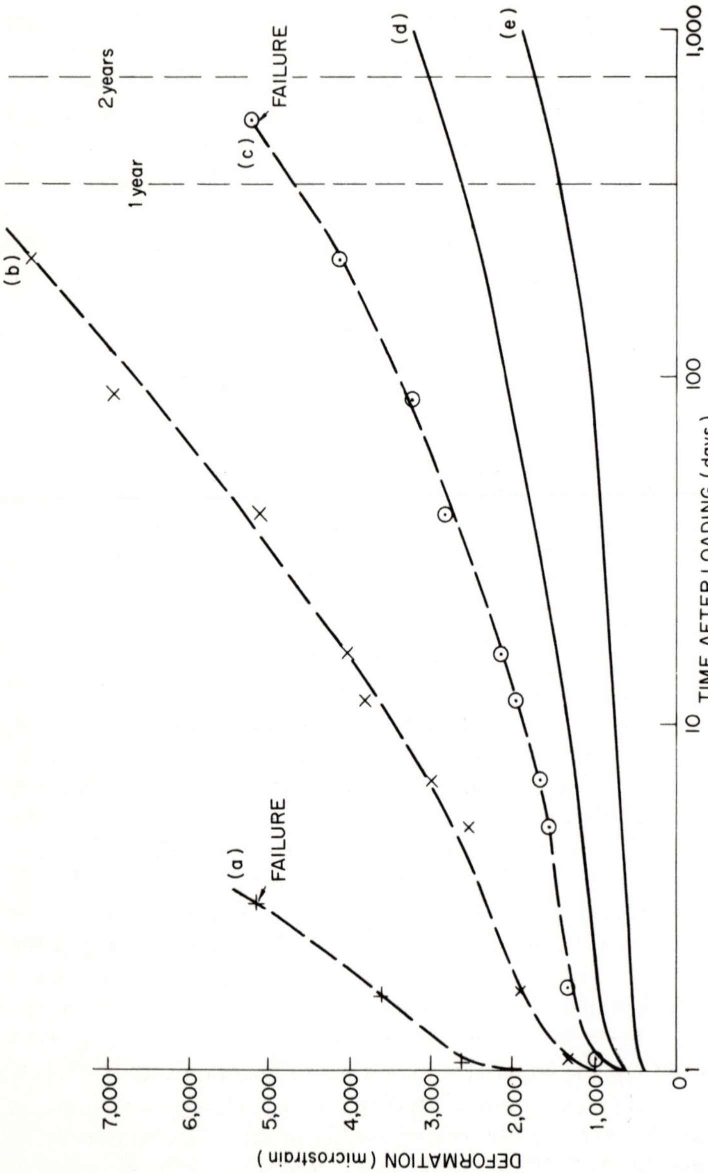

FIG. 9. Compressive creep of resin and cement concretes loaded at approximately 14 MN m^{-2} (after L. H. McCurrich and W. M. Kay). (a) = Polyester concrete mix 1 (65 °C); (b) = polyester concrete mix 2 (20 °C); (c) polyester concrete mix 1 (20 °C); (d) 6000/28 cement concrete (65 °C); (e) 6000/28 cement concrete (20 °C).

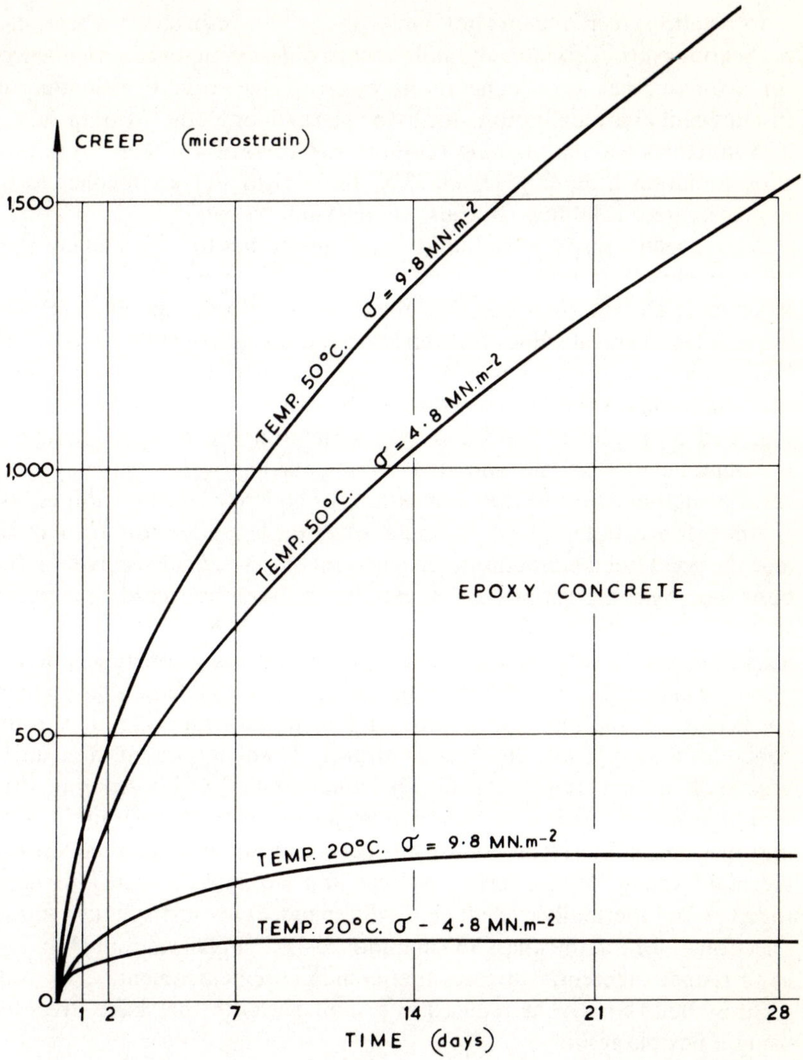

FIG. 10. Creep of resin concrete (after L. H. McCurrich and W. M. Kay).

pronounced (Figs. 9 and 10[46,47]). As a guide to the structural engineer, design structures at 20 °C for a mix made with a rigid low creep resin, the safe long term stress should not exceed 10% of the ultimate. However, recent data relating to a filled polyester chemical fixing composition indicates this recommendation might be unduly cautious.[48]

In practice, creep rates are not significant for thin resin concrete sections, i.e. bearing pads. Alternatively, and perhaps of less structural consequence, for a 6-mm thick resin concrete floor, creep due to vertical loading is insignificant as is indentation, due to long term floor loads.[46] For instance, a 25-mm thick pad maintaining a compressive stress of 7 MN m^{-2} results in a deformation after 2 years at 20 °C for a rigid polyester concrete of approximately 1500 microstrains, i.e. 0·04 mm.[46]

As well as the response of bulk strength properties to high temperature, the engineer is concerned about fire resistance of resin based compositions. Resin concrete will lose strength as the heat distortion temperature (HDT) is exceeded. Typical values for polyester and epoxy concrete is 120 °C or below. This means it is very likely that rapid strength loss will occur under fire conditions where temperatures may reach 1200 °C. This must be regarded as a real limitation on the structural use of resin concrete, although Levitt[49] has shown that external charring of the resin may yield some insulating effect to the resin within the concrete lessening this effect.

Another attribute of resin concrete is its excellent adhesion to concrete and the good bond strengths that result—much needed for speciality resin floor topping compositions. In general, bond strengths exceeding those of sound concrete are readily obtainable, i.e. 1·1–2·5 MN m^{-2}. However, the accumulation of shear stress in the bond at the resin concrete/concrete interface may result from differential thermal expansion. Resins have a high coefficient of thermal expansion 80–100 microstrains/°C, and resin concretes have typically 20–40 microstrains/°C which is about three times greater than concrete (typically 10 microstrains/°C[50]). Additionally, typical aggregates have a coefficient of expansion of 10 microstrains, which may also cause interfacial bond stress between resin and aggregate during thermal cycling. When resin concrete and normal concrete are used together and thermally cycled, i.e. road repairs, care must be taken in the selection of section thickness and maintained flexibility of resin and bond so as to reduce differential stresses that could cause detachment.

Stress build-up can be reduced in a similar way to shrinkage stress by using a flexible resin.

For non-structural applications a resin concrete may be bonded to conventional concrete using a flexible resin, but for semi-structural application where creep resistance is of high priority, it is better to use rigid resins. For either situation sufficient resins should be used to completely fill all the voids between the aggregate, i.e. 8–18 % resin by weight of total mix, or 17–34 % resin by volume of total mix. This corresponds to a voids factor of unity (Fig. 11).

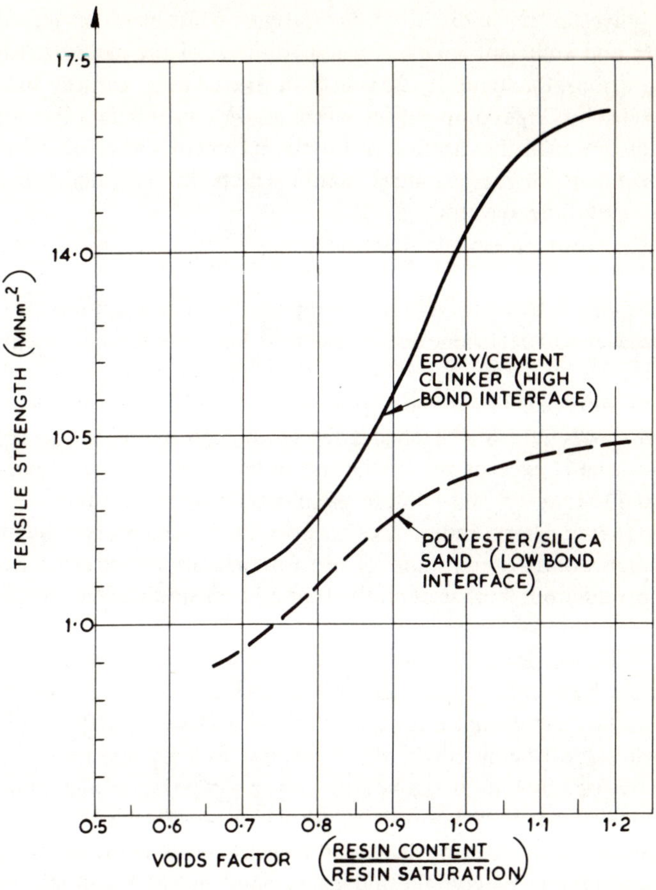

FIG. 11. Tensile strength of typical resin concretes related to resin content. Cure condition: polyester 14 days at 20 °C; epoxy 21 days at 20 °C. Test condition: strain rate approx. 3 % per min at 20 °C.

We would like to highlight two polymer concretes that have semi-structural application, and also exhibit related but technical novelty.

Estercrete†

This material[51] was first developed in 1962 and is disclosed in British Patent 1 065 053 Inorganic Polymer Structures. It comprises a mixture of treated Portland cement and insoluble free radical forming catalysts, further mixed

† Estercrete is a Registered Trade Name owned by Associated Portland Cement Manufacturers Ltd.

with a polyester resin dissolved in styrene. This blend is mixed with aggregate and sufficient water, causing solution of the catalysts together with liberation of calcium hydroxide from the cement, causing the whole mass to set solid. This composition combines the coincident polymerisation of polyester resin and hydration of Portland cement, the result of which is an amorphous polymeric matrix within which the hydrated cement gel forms a crystalline skeleton.

There are four broad mix designs:

1. *Flooring Mix*—used for thin toppings for concrete floors and roads as well as steel bridge decks. The base Estercrete resin/cement blend is mixed with crushed blue flint aggregate yielding a mortar density of 2003–2083 kg/m^3, compressive and tensile strengths at 23 and 4–5 MN m^{-2} are reached after seven days. Adhesion to concrete normally exceeds the tensile strength of concrete, i.e. greater than 0·69 MN m^{-2} and is often greater than 5·28 MN m^{-2}.
2. *Mortar Mix*—similar uses to normal cement mortar but having high bond strength and of use for pre-cast pre-stressed concrete production. Also used in the filling of tensioning ducts preventing corrosion of steel tendons.

   This is simply a sand mortar made with Estercrete as the binder (3:1 by weight), typical density is 2063–2083 kg m^{-3}. Compressive and tensile strengths are 34–35 and 4–5 MN m^{-2} respectively at 28 days. Adhesion to concrete is similar to flooring mix.
3. *Slurry Mix*—used as a bonding layer for the larger aggregate on to smooth surfaces that provides skid resistance, i.e. pedestrian areas, sealing and texturing of aircraft runways. This again is a mortar mix comprising Estercrete binder and sand but at 1:1 weight ratio.
4. *Concrete Mix*—usually used for non-load bearing situations as a result of doubtful creep properties. This is similar to conventional concrete with the cement and water fractions replaced by Estercrete binder, but mixed with coarse and fine natural aggregates (must be dry) in the ratio 370:109. However, it is usually used for non-load bearing situations as a result of doubtful creep properties. The density of this material is 2063–2195 kg m^{-3} and compressive strengths of 43–44 MN m^{-2} are reached at 28 days. As with previous mixes the adhesion to concrete is comparable to the flooring mix. The merit of this material is its capability to achieve significant strength quickly, coupled with good adhesion to cementitious substrates. The inherent versatility resulting from a

combination of Estercrete binder and various fillers affecting bulk properties and rheology presents novel prospects to the engineer.

U-crete†

This resin concrete is similar to Estercrete in that it too consists of a polymer based composition, in this instance a polyurethane and the ingredients of a normal concrete. The bulk properties of U-crete reflect this duality and are strain rate dependent. At low rates the polymer behaviour is said to dominate but at higher rates the material resembles concrete in being somewhat insensitive to change in strain rate. However, U-crete approximates to concrete-like behaviour in most practical situations and is independent of temperature within the range of 20–100 °C. Even so, the ultimate stress levels that can be used by the design engineer are those corresponding to the turnover point of the linear/elastic portion of the stress/strain curve. Typical ultimate values for grades of U-crete are given in Table 6.

TABLE 6

ULTIMATE STRENGTH OF POLYURETHANE CONCRETES[a]

Type of urethane concrete	*Elastic limit* ($MN\,m^{-2}$)
DC	15·20
FS	15·20
VA	17·25
WS	12·40

[a]Elastic limit in compression.[57]
T = 22 °C, (strain rate) = 8×10^{-3}/sec

Compressive strength/time are compared with normal concrete† in Fig. 12. The high early strength is apparent—70 % of ultimate within 24 h. However, the elastic modulus in compression is noticeably less than that for typical 27 $MN\,m^{-2}$ concrete. The values are given below in Table 7, together with tensile and flexural strengths.

The coefficients of thermal expansion of the U-cretes (12–43 micro-strain/°C) and in particular, that of the decking composition, are similar to those of concrete and steel. More so, it would seem than epoxy or polyester resin based concretes, of advantage to the engineer particularly in situations where thermal cycling might well occur.

† U-crete is a Registered Trade Name owned by Imperial Chemical Industries, Ltd.

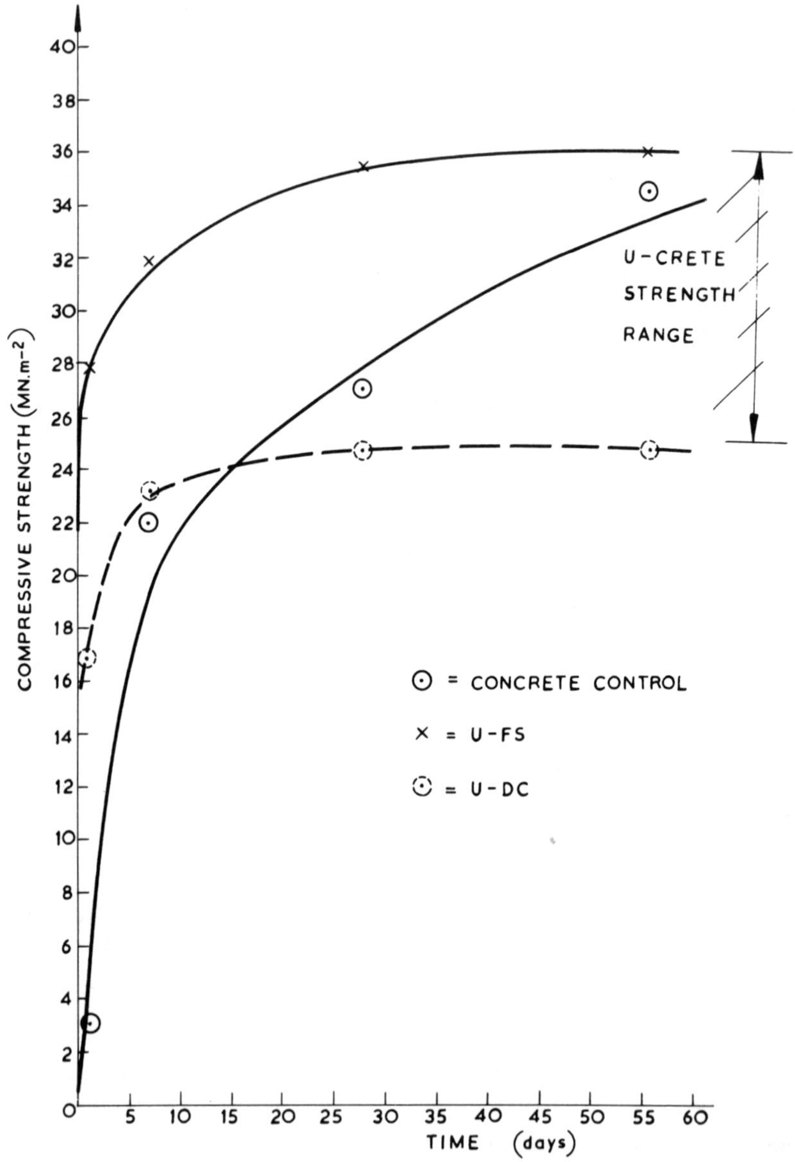

FIG. 12. Strength/time of U-crete.

TABLE 7
TENSILE PROPERTIES OF POLYURETHANE CONCRETES

Urethane concrete type	*Elastic moduli* $(MN\,mm^{-2}) \times 10^{-2}$	*Tensile strength* $(T = 22°C\ MN\,m^{-2})$	*Flexural strength* $(T = 22°C\ MN\,m^{-2})$
DC	18·6	—	3·35
FS	9·1	4·7	11·50
VA	9·9	5·24	10·70
WS	7·0	7·73	7·73
Concrete	207	2·07	3·45

An example is quoted[52] where a coating of U-crete on a steel vessel lid has withstood cyclic heating up to 130 °C for at least 2 years without distress. Finally, the adhesion of U-crete is excellent to both porous and impermeable substrates. For instance, 10 $MN\,m^{-2}$ bond values to mild steel have been reported.[53]

RESIN INJECTION FOR STRUCTURAL REPAIR

Concrete structures crack and that is a practical fact. If left in this condition further structural downgrading may occur which could be disastrous. This section of this chapter is concerned with what can be done to aid a cracked distressed structure, in particular using synthetic resins as a repair medium, be it as adhesive and/or sealant. We are interested in making fractured concrete monolithic again. Various aspects of this repair technique have been covered in previous publications.[54–59]

By means of this process resins are injected in a controllable manner so as to fill or treat crack or void, thereby reinstating the structure to its original design capability and/or preventing further downgrading of the structure. Resin injection will not impart a greater structural capability than was originally intended for the structure. In other words, if cracking occurred due to sustained over-load, then such overload has to be removed before resin repair can be effective. There are also situations where reinforcement may have been overstressed, causing permanent deformation. Such situations have to be assessed in detail before structural repair using this technique can be accepted with confidence.

Let us assume then, that the cause of failure has been determined and a strong but ductile adhesive can be inserted within the fractures, then there will be benefit. We are concerned here with the ways and means of carrying out the repair, the selection of adhesive and its properties, and an understanding of those factors that affect control and perhaps limit this repair technique.

Let us start by trying to describe these controlling factors. We are talking about placing liquid resins such as an epoxide or other in cracked concrete in order to impart structural strengthening. Of course, if the resin completely fills the gap then it will also act as a sealant but this is not its ultimate function. Because of the structural requirement the criteria governing effective materials performance are more stringent.

In very general terms these criteria can be listed:

1. Resins should be of low viscosity and cure in a controllable manner.
2. When set the resins should be dimensionally stable and adhere to wet or damp concrete, i.e. compatible with a highly-alkaline interface.
3. Resins should be capable of accommodating movement, particularly thermal, without resulting in fracture within themselves, detachment or inducing secondary cracking near the repaired zone.

It is worth digressing for a moment to consider how pressure can be built up in a crack, and the need to control it. Resin or liquid adhesive, may be represented as flowing into a simple uniform or horizontal gap or fissure as shown in Fig. 13. Penetration in this manner may be represented as:

$$Q \text{ (Injection rate cc sec}^{-1}\text{)} = \frac{2b^3 \pi \, \Delta p}{3\eta \ln (r_1/r_2)}$$

where $2b$ = crack width (cm)
Δp = pressure drop ($\mathrm{N\,m^{-2}}$)
η = resin viscosity ($\mathrm{Ns\,m^{-2}}$)
r_1 = radial penetration
r_2 = injection centre radius (cm)

The pressure falls off very quickly locally to the injection point whilst the resin is flowing freely. As r_1 increases, so does Δp in order to maintain Q. Unless pressure is monitored carefully and is capable of being quickly controlled, Δp can be destructively high. Another point to note is that for a given injection pressure, flow rate falls off markedly with radial penetration (Fig. 14, curve A), into the crack, resulting in very long injection times (curve B). This effect, of course, militates against batch mixing and makes it advantageous not to mix the two components until the moment of injection.

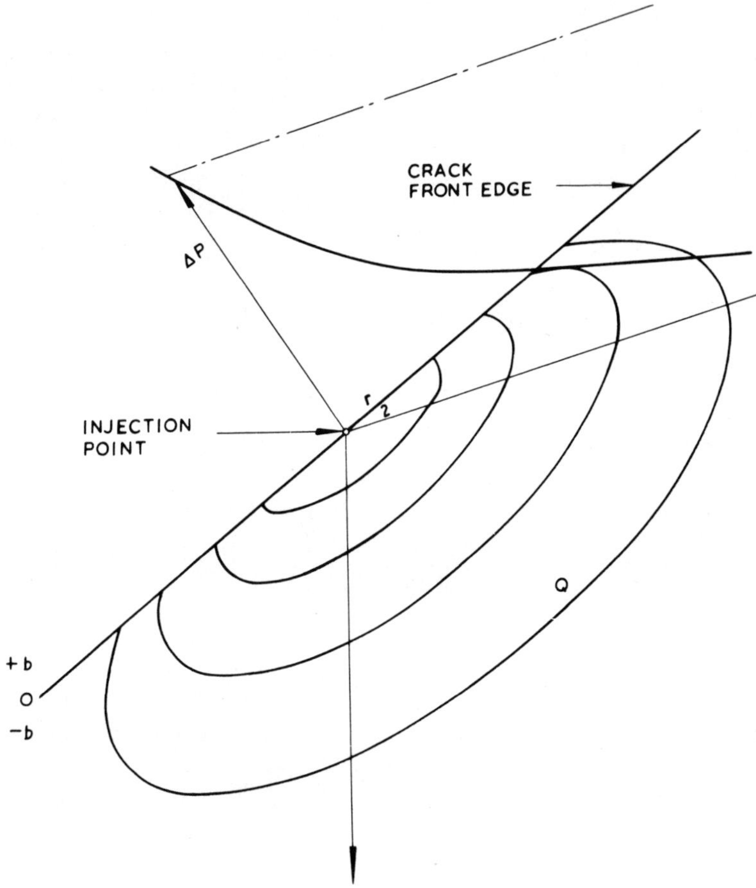

FIG. 13. Fluid flow in planar fissure (the injection stage).

For the resin, a non-solvented resin system of low viscosity, 0·40–1·50 Ns m^{-2}, depending upon ambient temperature, is required. This low viscosity allows adequate penetration even into the finest of cracks. There is little merit in attempting to treat cracks finer than 10–20 μm since most concrete will have cracks of this order due to normal hydration shrinkage. From the equation, the flow rate Q is linearly dependent upon the reciprocal of viscosity and halving the viscosity doubles the flow rate or perhaps more important, keeping all things constant, halves the injection pressure to maintain the same flow rate. Dependence of Q on $2b$ (crack

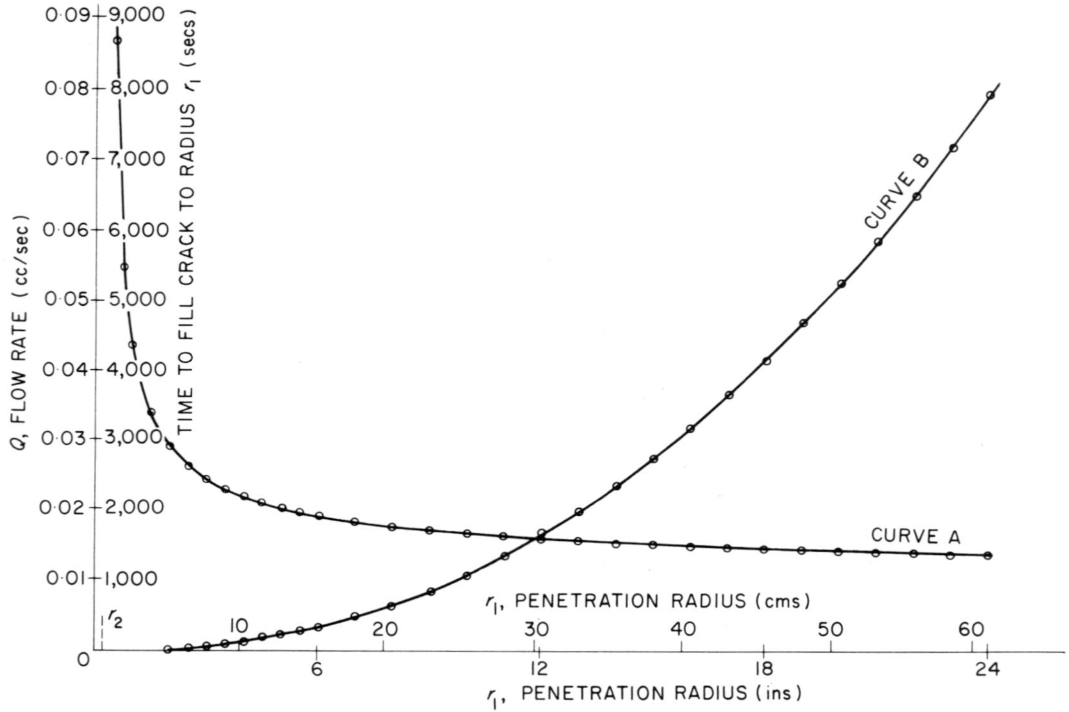

FIG. 14. Variation of flow rate or time to fill crack with extent of penetration. Curve A: instantaneous flow rate $Q(r_1)$ at penetration radius r_1, versus penetration radius r_1. Curve B: time required to fill a crack to radius r_1. Fixed data: total pressure drop $\Delta P = 0{\cdot}069\ \mathrm{MN\,m^{-2}}$; resin viscosity $= 0{\cdot}3\ \mathrm{Ns\,m^{-2}}$; crack width $= 100\,\mathrm{mm}$; open crack, air filled; $r_2 = 6{\cdot}3\,\mathrm{mm}$.

thickness is shown in Fig. 15). Increasing or decreasing the resin viscosity will displace this curve upwards or downwards in parallel to the one shown.

It is usually assumed that resin remains Newtonian during the injection stage. In all but the finest of cracks, less than 10 μm, there is some advantage in the resin having some shear or yield strength. Assuming a surface tension of at least 40 $MN\,m^{-1}$ and zero contact angle, then gravity drainage will occur from cracks greater than 10 μm. Let us then consider what happens for a non-Newtonian or Bingham type fluid—assuming such a resin could be formulated.

As before, it can be shown that a shear strength of approximately 5 $N\,m^{-2}$ is sufficient to support resin in cracks up to 1000 μm width and 60 cm deep without drainage. If we pursue this line of argument it can be calculated that maintaining a pressure in the region of 0·1–0·4 $MN\,m^{-2}$ there would be sufficient force to shear the resin and obtain penetration up to a radius of approximately 60 cm. Whilst these pressures are by our standards on the high side, they are often used and this opens up possibilities for formulating drainage-free repair resins. It also highlights the need to define working pressure limits—a point touched on earlier.

Shear strengths can be imparted to resins by incorporation of fine particle fillers as well as thickening agents. Usually, such modifications result in thixotropic behaviour rather than Bingham type.

Thixotropy would give rise to time dependent shear strength and it could well be that injection pressures required in practice would be lower than that calculated above and as a result be practicable.

Consider the various interfaces within a dry crack. Figure 16(a) shows the resin advancing under the applied pressure distorting the advancing contact angle θ'. The various interfacial tensions are also shown. To obtain good adhesion between the resin and the concrete surface, the following conditions should be satisfied:

1. $\gamma_{(resin/concrete)}$—should be low, or even zero.
2. $\gamma_{(resin/air)}$—should be low (less than critical surface tension for wetting).
3. $\gamma_{(concrete/air)}$—should be high (however, 2 and 3 are to some degree interdependent).

$$\gamma_{(resin/concrete)} + \gamma_{(resin/air)} \cdot \cos\theta = \gamma_{(concrete/air)}$$

where θ is the equilibrium contact angle.

For good wetting of the substrate, θ should be equal to zero or close to it. Some preliminary contact angle measurements using the sessile drop

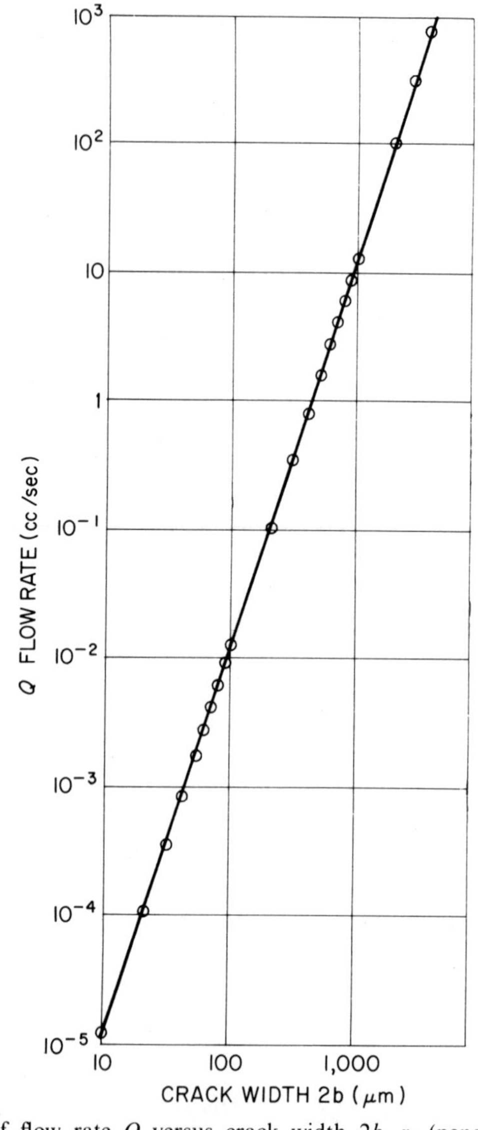

FIG. 15. Plot of flow rate Q versus crack width $2b$. r_1 (penetration radius) = 610 mm; viscosity of resin = 0·3 Ns m^{-2}; pressure drop over resin ΔP = 0·069 MN m^{-2}.

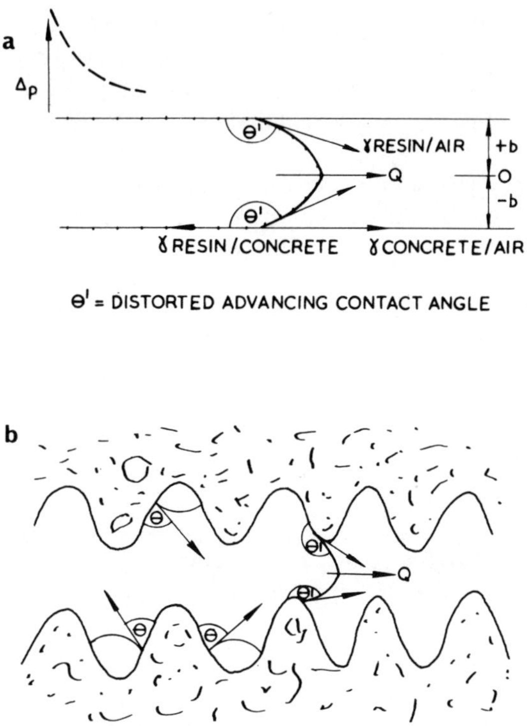

FIG. 16. Contact angles with cracks. (a): Wetting in idealised dry crack; (b) suggested wetting in non-idealised dry crack.

technique, indicated a value of θ between 18 and 27° on mortar and concrete surfaces.[60] Such measurements are very difficult since the high viscosity of the resins and the consequentially long period before equilibrium is reached, it could well be that the values quoted are high. Very little wetting phenomena work appears to have been carried out on cementitious materials—certainly more data are required. If θ is positive then perhaps a better representation taking into account surface roughness is given in Fig. 16. A possible consequence of θ being greater than zero is the tendency to occlude air at the resin/concrete interface resulting in reduced adhesion and reduced repair strengths.

Concrete is a mixed material comprising: coarse aggregate; sand; unhydrated cement particles; cement gel; capillary and gel pores; pore water; and entrapped air voids.

As concrete undergoes stress the micro-cracking already present at the coarse aggregate/mortar interface propagates in length, width and number, until 'bridging' between the propagating cracks occurs, giving rise to a continuous crack pattern and ultimate failure and development of a macro crack. Within this crack we would find exposed coarse and fine aggregate interspersed with a porous cement gel matrix. Concrete mixes vary depending on the structural application: however, it is true to say that the exposed aggregate surfaces would form the major part of the newly created surface of the crack and so dictate the chemical characteristics of the surface. The minor component consists of porous calcium silicate hydrate or cement gel.[61–63] In a general sense we may regard this inhomogeneous, rough and partially porous surface as being siliceous material and as such heavily hydrated. In common with other silica rich materials, such as the glasses, this adsorbed water is tenaciously held,[64] and could well be several hundred ångströms in thickness.[65] To all intents, therefore, the injected resin is spreading on a layer of adsorbed water molecules, and factors governing 'wetting' and adhesion will be dictated by this adsorbed water. It should be possible therefore to predict wetting performance from the critical surface tension (γ_c) of such a surface as that proposed above. According to Zisman[66] for a liquid to spread freely on a solid substrate, the surface tension of the liquid (γ) should be less than the critical surface tension of the solid. Work by Olsen and Osteraas indicates that γ_c for a water covered siliceous surface is about 19 $mN\,m^{-1}$ (ref. 66) and that for water itself, some 73 $mN\,m^{-1}$ (ref. 67). From our own measurements, the surface tension of the resins used for injection work are in the range 35–45 $mN\,m^{-1}$ (ref. 56) and without modification are likely to spread well. Resins containing surface tension modifiers, could be formulated and we ourselves do incorporate silane derivatives, but for the purpose of lowering the resin/solid interfacial tension.

In a process where the resin is being forced into contact with the substrate by the injection pressure, the need for spontaneous spreading is probably not paramount. On the other hand, good interfacial bonding between the resin and substrate is what will ultimately determine the strength properties of the injected mass. In order to maximise this bonding, the authors suggest that the resin should form true chemical links with the substrate. This linking may involve reaction at the interface resulting in chemisorption. This form of linkage is preferred[69] for good long-term bonding, particularly in wet or damp environments.

Let us consider the type of chemical modification that could be used. Organic silicone chlorides have been used to render glass hydrophobic by reaction with surface silanols at elevated temperatures.[70] Also, the adhesion of methacrylo-chromium chloride to glass is said to result from covalent bonds involving chromium, oxygen and silicon.[71,72] This subject has been reviewed by Rutzler.[73] More recently alkylalkoxy silanes[74] containing residual epoxide or amine reactivity, have been used as adhesion promoters in resin–concrete bonding systems. Our work is still active and it is too early to say whether incorporation of these additives results in improved adhesion although we have found that consistency of results is improved by their addition.

The authors consider that resins characterised by low resin-concrete interfacial tension and zero contact angle are a prerequisite to good permanent bonding. The resin–air tension (using a simple capillary model[66]) is of secondary importance although it has been implied that the greater this tension the more efficient is the pore filling capability. It is likely a compromise value for this tension would result in maximum adhesion.[75] What this value should be for concrete injection work is not known.

What of the curing and cured properties of the injected resins? A reasonable cure time in a thin film, which probably represents the situation in a crack is required. For instance, a resin giving short cure times in bulk would be a quick curing system (typically 16-min pot life for 100 g mass). For instance a 0·05 in film has a tack free gel time of about 3–3½ h at 25 °C reaching full cure in three days and six days at 4 °C. Once cured, the resin has excellent adhesive properties to both wet and dry concrete.

Direct tensile adhesion tests using cored specimens gave in the dry, consistent concrete (2·55 MN m^{-2}) and mortar (2·80 MN m^{-2}) failure. Resin properties once injected and bonding the structure together are often taken for granted or ignored. Set resin properties are of paramount importance. It is obviously desirable that shrinkage should be kept to a minimum otherwise shrinkage stresses could be set up, resulting in early detachment of the resins. It is for this reason that solvent free systems are preferred. Neat resin systems do shrink but values cover a wide range (0·1–10 % by volume). As a general rule, the shrinkage will occur towards the bulk of the material if the resin is unconfined. Shrinkage occurs in both the liquid and solid state, but for injection work it is the total value that is important. Addition of fillers may well reduce shrinkage, but this is not the practical approach for injection resins. Sufficient to say the shrinkage is undesirable. A study of shrinkage under confined conditions would be useful since the 'in-service' effect is not fully known. If shrinkage is unavoidable its effect could be reduced by using resins having lower moduli. The use of 'tough' rather than brittle materials might also be advantageous from other counts.

TABLE 8
MECHANICAL PROPERTIES OF RESINS USED FOR CONCRETE INJECTION

Bulk property	*Resin system*[a]	
	A	*B*
Compressive strength (ASTM D-695) ($MN m^{-2}$)	(67·5)	(113)
Compressive modulus (ASTM D-695) ($MN m^{-2}$)	(2×10^3)	($1{\cdot}84 \times 10^3$)
Tensile strength (ASTM D-638) ($MN m^{-2}$)	(47·6)	(62·4)
Elongation at break (%) (ASTM D-638)	—	2·5
Flexural strength (ASTM D-790) ($MN m^{-2}$)	(68·79)	(83·4)
Flexural modulus (ASTM D-790) ($MN m^{-2}$)	($2\text{–}2{\cdot}84 \times 10^3$)	($4{\cdot}6 \times 10^3$)
Heat distortion temperature (ASTM D-648) (°C)	~49	57–59

[a] *A* = Concresive 1050. *B* = Concresive 1380. Manufactured by Adhesive Engineering Co., San Carlos, California, and available as part of the Structural Bonding Process operated by Cementation Chemicals Ltd, Maple Cross, Rickmansworth, Herts., U.K.

The work on fractured reinforced beams[76,77] has shown that after resin injection repaired beams strengths were at least equal to those before fracture, the interposing of ductile resin aided stress distribution.

When cured, whilst extremely tough, the resins should remain non-brittle and accommodate some movement. This is necessary, since a repaired crack will move at the dictate of the structure without offering much restraint. A high modulus resin, when subject to strain, may cause stresses at the adhesive/adherend interface that could cause detachment or further cracking within the concrete adjacent to the first crack. The mechanical properties of typical injection repair resins used are shown in Table 8. Resins having a lower modulus than concrete are better able to accommodate movement strains.

There are few situations where a high modulus resin would be of benefit. Ideally, of course, the resin should match exactly the bulk properties of concrete so that injection results in a resin weld. As yet such resins do not exist. Ideally, the resin should move with the structure and so have its thermal coefficient of expansion matched to that of concrete. In general epoxide resins are about five times more expansive than concrete. It is probably lower temperatures that could cause detachment. Filling a resin usually lowers the thermal coefficient[78] but this is not practicable for injection. An alternative is to avoid high modulus materials (in compression less than or equal to 30×10^2 MN m^{-2}) since crack movement could result in further crack propagation by the resin acting as a wedge.

It is worth mentioning a little about crack preparation prior to injection, since this too may use both thermoplastic and thermoset adhesives. It has been common practice when preparing a crack for injection to remove loose material and locate injection nipples by drilling out first. In anything but really wide cracks this procedure is no longer recommended since it is expensive and drilling can cause dust and detritus to be lodged in the crack reducing adhesion as well as reducing accessibility into the crack. Depending on crack width, a thermoplastic seal (0·2–1 mm) or alternatively, a filled polyester (1–6 mm) a fast setting cement or coarse grained filled polyester may be used (Fig. 17(a) and (b)).

Each sealing method has its own injection point location technique and post injection drainage stop. The thermoplastic is simply applied hot to the crack, the crack having previously been taped at the injection centres. The thermoplastic cools almost immediately effecting a seal. The tapes are removed, exposing a short length of crack which becomes the injection centre. The mixing head of the injection gun has a self-sealing rubber gasket on the outlet and this forms its own injection cell. As the injection proceeds along the crack the holes are sealed using solid wax which is rubbed over the surface, or by rods or suturing of injection tubes. After an entire crack length has been treated the thermoplastic is simply removed with a spatula or on complicated shapes by sand blasting, leaving the surface almost unblemished.

An alternative to resin injection is resin impregnation using a reduced pressure/suction technique[79] known as Balvac. However, the structural resins used are similar to those described above and subject to the same restraints.

It is not always possible to be confident that all the cracks in a structure have been treated, but total treatment may not always be necessary. With the exception of cracks resulting from continuous tensile stress, or where

FIG. 17(a). Resin being injected into sealed crack via self metering pump/mixing head (note separate feed lines). (Courtesy Cementation Chemicals Ltd.)

cracking may have caused reinforcement to be over-stressed, partial resin injection may be all that is necessary. However, non-destructive methods of determining the extent of resin penetration are being considered but coring seems to be the most acceptable proof at the moment.

Looking ahead a little, there may be developments in the area of injected sealants. Being applied in thin gap situations it will be necessary to make very flexible materials since large strains may occur. New permanently flexible systems such as the polyurethane/epoxide blends warrant

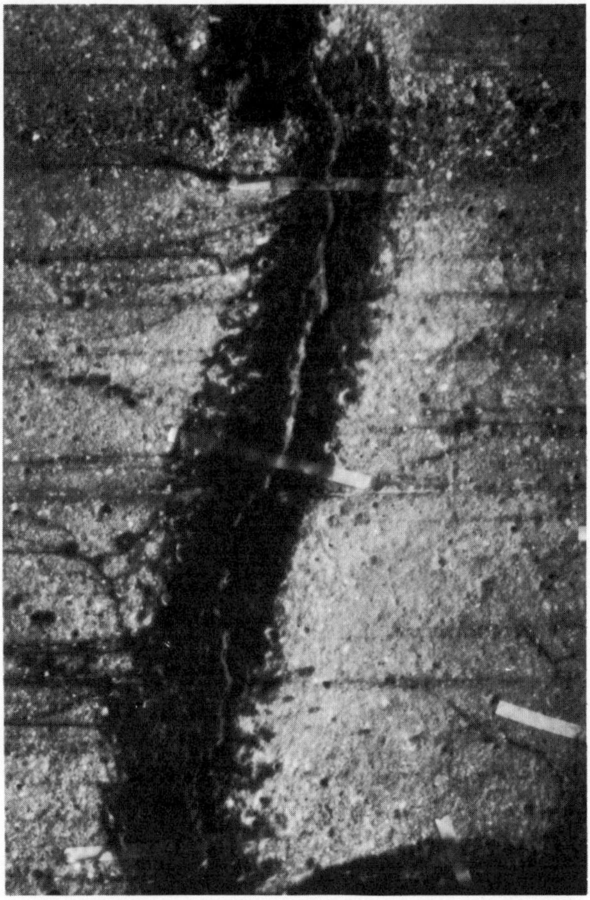

FIG. 17(b). Crack sealed with special thermoplastic sealant, ready for injection of resin. Injection points are located by means of adhesive tape. (Courtesy Cementation Chemicals Ltd.)

investigation. Additionally, resins possessing the good adhesion characteristics of epoxides but with improved heat distortion temperatures and creep characteristics would have advantage.

Resin materials having closely matched bulked thermal properties to concrete may also evolve. Low-viscosity materials based on the new anaerobics[80] and cyanoacrylates, are interesting possibilities, although what little work has been done so far has been disappointing. The anaerobics combine low viscosity with an indefinitely long 'pot life'. In theory, at least, these resins would remain unchanged until they were within the body of a crack and the air excluded, and under these conditions, they would then set. Resins having chemical reactivity towards a siliceous and silanol covered surface open up interesting areas of development relevant to glass-fibre reinforced concrete in addition to crack and concrete repair.

As it exists, resin injection brings together the use of what is after all a polymer adhesive and the need to reinstate fractured concrete, so offering an alternative to perhaps unsightly mechanical support and at worst, demolition.

RESIN BONDED ABRASION AND SKID RESISTANT SURFACES

In applications where finishes with very high resistance to abrasion are required, such as high speed water spillways in dams, and hydroelectric schemes, in heavy duty floors and also non-skid roads, a high degree of abrasion resistance is achieved by the incorporation of a high concentration of abrasion resistant materials in the surface layer exposed to these conditions.

Abrasion Resistant Surfaces

The principal stresses of surface in water spillways are the forces of impact and cavitation due to the high speed water together with the highly abrasive effect of sands and other solid material carried in the water producing a wet grit-blasting effect at the surface. This requires an abrasive resistant material to be fixed on the surface with high performance adhesives or binders.

Epoxy resin systems have been widely used for such applications but their use in these areas is rather poorly recorded in the literature.[81-84] The simulation in the laboratory of highly-abrasive conditions encountered on high-speed water spillways, is exceedingly difficult. However, it has been found that modifications of tests such as Taber Abrader (ASTM No. D.1044) and BS 416 for the evaluation of the wear resistance of road aggregates do, in fact, give a good indication of abrasion resistance of bonded abrasive resistant surfaces.

In one series of laboratory tests[85] to find the optimum wear resistance of heavy-duty epoxy resin flooring compositions, a number of different abrasion resistant materials were evaluated using BS 416 and employing three different epoxy resin binders which themselves had significantly different chemical compositions and mechanical properties. The results of this work which was carried out under dry conditions are given in Table 9. As can be seen from the table, the selection of the abrasion resistant material and the resin matrix both influence the abrasion resistance of the system although the abrasive material incorporated appears to play a more crucial role.

TABLE 9
ABRASION VALUES (BY BS 416) OF TROWELLED EPOXY RESIN FLOORING COMPOSITIONS

Aggregate	*Epoxy binder compositions*	*Weight loss after abrasion (g)*
Sand I	A	4·10
Sand I	B	2·75
Sand I	C	2·85
Sand II	A	5·5
Sand III	A	5·7
Gritstone	A	0·95
Granite	A	1·45
Calcined bauxite	A	0·95
Basalt	A	1·5
Cast iron grit	A	0·45
Copper slag	A	2·25
Sand I/Gritstone (50/50 by wt)	A	1·35

In the case of high-speed water with suspended solids, the nature of the abrasion resistance required is more akin to laboratory tests carried out under dry conditions since the very small particles of very hard high abrasion-resistant materials that may be removed from the surfaces during the tests are carried away in the flowing water.

In wet abrasive conditions, such as heavy duty industrial flooring, a small quantity of highly abrasive resistant material will remain in contact with the surfaces and will increase the abrasive nature of the 'grinding paste' produced between the heavy-duty wheel and the surface, since the abrasive resistant material in the surface is generally much harder than any sand or grit carried into the factory on wheels.

Abrasive resistant tests under wet 'grinding paste' conditions, however, do indicate a similar order of resistance although the binder appears to play a more significant part. In applications where the abrasive resistant surface is immersed in water for long periods the resin binder plays a more important part since the strength of the adhesive bond between the particles of abrasion resistant materials can, if the wrong resin binder system is used, drop markedly under prolonged wet conditions. In formulating resins for abrasion resistant surfaces it would appear that the adhesive properties of the resin binder to bond the abrasion resistant particles together firmly is the most important factor when selecting a resin system. In the selection of systems for large scale applications, costs must also be considered and on this basis bauxite calcined under defined temperature conditions, has been most commonly used as the abrasion resistant material.

Non-skid Road Surfaces

Epoxy-resin adhesives have also been used extensively in the production of skid-resistant road surfaces, particularly in the United Kingdom. Here, although the main requirement is to bond a highly abrasive-resistant material on to road surfaces, there are a number of other requirements which are equally important. To achieve a skid resistant road surface a skid resistant material is required to be bonded to a sound road surface with minimum cost and minimum hindrance to traffic flow.

In the United Kingdom, the initial interest in improving skid resistance stems from statistics on accidents which show that a very high proportion of accidents involving injury and death occurred in urban areas within approximately 20 m of road junctions and pedestrian crossings and that a large proportion of these accidents involve skidding of vehicles.[86] Many of the road surfaces where such accidents have occurred, although perfectly sound, have poor skid resistance and the UK Transport and Road Research Laboratory have developed a portable pendulum skid resistance tester[87] and by using accident statistics and other information, they have established that below a certain level of skid resistance (as measured by this portable tester) the incidence of skidding accidents was remarkably increased.

Using this portable skid resistance tester, under defined test conditions skid resistance values from 0–100 can be obtained on an arbitrary scale, specifically relating to this pendulum test method. At figures below 65 on this scale, the statistics have shown that the possibility of skidding accidents occurring under typical urban traffic conditions is markedly increased. Such a degree of skid resistance can be achieved by resurfacing hot-rolled asphalt surfaces with 2–4 cm of additional material, and on concrete roads by profiling the surfaces during construction.

However, these techniques can normally only be considered as relatively short term remedies as it has been shown that under heavy duty trafficking, the surfaces tend to become polished and lose their skid resistance, mainly due to the fact that most of the aggregates which are freely available and sufficiently cheap to be used in road surfacing applications, do not have sufficiently high resistance to abrasion.

Other work carried out by the UK Transport and Road Research Laboratory[88] to find aggregates which were not prone to lose skid resistance due to polishing under heavy traffic had shown that certain aggregates, in particular specific grades of calcined bauxite were satisfactory. Calcined bauxite costs over ten times that of conventional road aggregates and is, therefore, uneconomic to use in conventional road surfacing techniques. However, it was considered that one of the possible ways of using its excellent skid resistance to produce a skid resistant road surface at an acceptable price, was to bond a single layer of calcined bauxite particles on to a sound road surface using a suitable adhesive. Small-scale trials were initiated in London in 1966, using a number of different adhesive compositions including epoxy resin based systems to determine whether it was practicable to produce a durable skid resistant road surface in this way.[85]

The technique used to apply the non-skid dressing was to apply a thin layer of adhesive composition on to the clean road surface at a uniform spreading rate of about 1 $kg\,m^{-2}$ and then dress with excess of calcined bauxite particle size 1–3 mm.

In the development of a fully practicable adhesive system for non-skid roads, several problems had to be overcome. Since most of the potential areas for the application of such non-skid surfaces are on very heavily trafficked roads, where hindrance to traffic during peak hours is not permissible, it is usually necessary for all application work to be carried out at night and the treatment must be sufficiently cured to permit reopening of the road for morning traffic without affecting the durability of the treatment. To achieve acceptable cure rates a rapid curing epoxy resin system was developed and to improve the cure rate further it was found beneficial to apply the adhesive layer hot—this enabled a system with a high viscosity at ordinary temperatures to be used and this helped overcome one major problem which had become apparent during the proving trials. In these trials it was found that if a low viscosity epoxy resin system was used, when the non-skid aggregate was applied in excess on to the wet adhesive layer a significant proportion of the resin adhesive tended to 'wick' by capillary action, up between the aggregate particles leaving insufficient adhesive in contact with the road surface. This problem can be readily overcome, either by adjusting the rheology of the epoxy resin system so that 'wicking' did not occur or by increasing the thickness of the adhesive layer but obviously the former was to be preferred on cost. The initial results of these small trials, with epoxy resin based adhesives were so promising that in 1967 the Greater London Council decided to go ahead with full scale trials on seven major road junctions where accident statistics had indicated that a large number of skidding accidents had occurred. These full scale trials confirmed both the durability and effectiveness of such non-skid road surfacings and that this method of treating asphaltic road surfaces to improve their skid resistance is now becoming a standard practice in the United Kingdom.

The total area of road surface to be treated with the non-skid dressing at any one site is generally of the order 400 m^2 for approaches to pedestrian crossings to 2000 m^2 for approaches to major road junctions. To cover these areas efficiently and quickly, particularly with an adhesive system requiring application hot, it was necessary to develop special metering, mixing and application equipment. Suitable machines are now available which are capable of applying the epoxy resin composition at a covering rate of up to about 80 $m^2\,min^{-1}$ (ref. 89).

REFERENCES

1. Anon., *J. Amer. Concrete Inst.*, Committee 403, September 1962, pp. 1121–42.
2. Anon., *Engineering News Record*, July 1962, p. 32.
3. Anon., Epoxy Resins for use in Civil Engineering Projects, *Technical Report No.* 6-521, August, 1959, US Army Engineer Waterways Experiment Station.
4. Buck, B. I., *British Plastics*, October 1959, pp. 475–8, 486.
5. Anon., Cement & Concrete Association, Advisory Note No. 12, London, 1968.
6. Allen, R. T. L., *The Repair of Concrete Structures*, Cement & Concrete Association, London, 1974.
7. Department of the Environment, Housing and Construction, *Advisory Leaflet* 77, London, 1974.
8. L'Hermite, R. U. and Bresson, J., *Symposium*, Paris, 1967, Réunion Intern. Laboratoires Essais Rech. Mat. Construct. (RILEM), Paris.
9. Bresson, J., *Ann. Inst. Technique Bâtiment Trav. Publics.*, Suppl. No. 278, 1971.
10. Hugenschmidt, F., Epoxy Adhesives for Concrete and Steel, 1*st Int. Congress Polymer Concretes*, 7 May 1975. *Proceedings*, '*Polymers in Concrete*', The Construction Press, London, 1976.
11. Letts, A., Epoxides in the Construction Industries, *Shell Symposium*, Shell Centre, 1969.
12. Hodgeson, N. E., Resin Emulsions—Their Use in Concrete, *Admixtures for Concrete*, Cement & Concrete Association, Fulmer, 1975.
13. Shakespeare, J. A., *GLC Dev. and Mat. Bull.*, **33**(3), 1970, pp. 1–9, and **52**(2), 1972, pp. 1–5.

14. GIBBINS, P. F., *Consulting Eng.*, **33**(3), 1969, pp. 26–9.
15. Ciba-Geigy Ltd, *Instruction Sheet* No. S.30B, Duxford, Cambridge, 1972.
16. HEWLETT, P. C., Cementation Research Ltd, Research Report R.1/71, 1971.
17. ROWLAND, J. *Concrete*, **4,** 1970, p. 21.
18. SIMS, F. A., The Application of Epoxy Resins in Bridge Construction with Particular Reference to Rawcliffe Bridge, *Symposium*, '*Resins and Concrete*', Plastics Institute/Inst. Civ. Eng., 1973.
19. O'BRIAN, T., Jointing Structural Precast Concrete Units with Resin Adhesives, *Symposium*, Paris, 1967, Réunion Intern. Laboratoires Essais Rech. Mat. Construct. (RILEM), Paris.
20. *Ann. Inst. Technique Bâtiment Trav. Publics.*, Suppl. 204, 1964.
21. TUPHOLME, C. H. S., *International Construction*, **9,** 1970, pp. 35–9.
22. SCHULTZ, R. J., *J. Amer. Conc. Inst.*, **73,** 1976, p. 155.
23. Shell International Chemicals, *Resin*, **2** (ed. Ingham, M. P.), 1970, pp. 9–12.
24. DAVIES, P., *Civ. Eng.*, August 1976, pp. 77–81.
25. ARUP, O. M. and ZUNZ, C. J., *Structural Engineer*, **47,** 1969, pp. 99–132.
26. FLEMING, C. J. and KING, G. E. M., *Symposium*, Paris, September 1967, Réunion Intern. Laboratoires Essais Rech. Mat. Construct. (RILEM), Paris.
27. MOUSTAFA, S. E., *J. Prestressed Concrete Inst.*, **19**(4), 1974, p. 54.
28. Anon., Shell International Chemical Co., *Notes for Epikote Formulators*, Res./F/601, London, 1969.
29. JOHNSON, R. P., *The Structural Engineer*, **48**(6), 1970, pp. 227–52.
30. MIKLOFSKY, A., GONSIOR, J. and SANTINI, J. J., *Further Studies of Epoxy Bonding Compounds*, Highway Research Board Publications, National Academy of Sciences, National Research Council, Washington, DC, 1965.
31. TOPALOFF, E., *Bauingenieur* (Berlin), **2,** 1964, pp. 50–64.
32. Eidgenossische Materialprufungs und Versuchsanstalt, EMPA No. 24 171/2. Unveroffentlichter Bericht, 1974.
33. IRWIN, C. A. K., The Strengthening of Concrete Beams by Bonded Steel Plates, *TRRL Supplementary Report No.* 16*OUC*, 1975.
34. SHAW, J. D. N., The Use of Epoxy Resin for the Restoration of Strength of Deteriorated Concrete Structures, *Symposium*, '*Advances in Concrete*', Birmingham, 1971. Concrete Society, London.
35. SHAW, J. D. N., Epoxy Resin Compositions—Materials for the Engineer, *Liquid Polymers Symposium*, University of Surrey, 1972.
36. BRESSON, J., *Ann. Inst. Technique Bâtiment Trav. Publics.*, Suppl. 297, 1972.
37. *Schweizerische Bauzeitung*, **19,** 9 May 1974, pp. 457–74.
38. CROSSLEY, N., Private communication, 1971.
39. Department of the Environment, *Building Research Establishment Digest*, **99** and **101.**
40. MANDER, R. F., Bonded External Reinforcement—A Method of Strengthening Structures, *Symposium*, '*Adhesives and Sealants in Building*', Plastics & Rubber Institute, London, 1977.
41. Roy. Inst. Brit. Architects, *Rescon—resin-based jointless flooring system*, London, 1972.
42. Ministry of Transport. *Tech. Memo. B.E.*6., 1967.
43. DIAMANT, R. M. E., *Architect & Building News*, **234,** 17 July 1968, pp. 104–5.
44. *Polymers in Concrete*, SP-40, Amer. Conc. Inst., Detroit, 1973.

45. HOP, T. and MIODYNSKI, Z., *Building Science*, **2**, 1967, pp. 147–63.
46. MCCURRICH, L. H. and KAY, W. M., Polyester and Epoxy Resin Concrete', Symposium 'Resins and Concrete', Plastics Institute/Inst. Civ. Eng., 1973.
47. BROCARD, J. and CIRODDE, R., *Symposium*, Paris, 1967, Réunion Intern. Laboratoires Essais Rech. Mat. Construct. (RILEM) Paris.
48. HURLEY, S. A., Private communication, Cementation Research Ltd, March, 1977.
49. LEVITT, M., *Plastics in Building Structures*, Pergamon Press, 1965.
50. GAMSKI, J. M., *Symposium*, Paris, 1967, Reunion Intern. Laboratoires Essais Rech. Mat. Construct. (RILEM), Paris.
51. NUTT, W. O., *Polymer Cements*, Blue Circle Group Development Division, The Cement Marketing Company, London, 1968.
52. BENSON, L. H., SHEARING, H. J. and BELLIS, R. G., Urethane Concretes in the Chemical Industry, *ACHEMA Conference*, Frankfurt, August 1973.
53. *The Fire Resistance of Resin Jointed Concrete*, ICI Ceramics Division—Trade literature relating to U-crete.
54. HEWLETT, P. C. and LETMAN, J., *Concrete*, **8**(1), 1974, pp. 30–4.
55. HEWLETT, P. C. and WILLS, A. J., A Fundamental Look at Structural Repair by Injection using Synthetic Resins, *Symposium*, '*Resins and Concrete*', Plastics Institute/Inst. Civ. Eng., 1973.
56. HEWLETT, P. C., HUTCHINSON, M. T. and SPENSLEY, R. H., The Repair of Cracked Structures by Means of Resin Injections, *Aspects of Adhesion*, ed. Alner, D. J. and Allen, K. W., 1973, Transcripta, London, p. 102.
57. GAUL, R. W. and SMITH, E. D., Effective and Practical Structural Repair of Cracked Concrete, SP-21, Amer. Conc. Inst., Detroit, 1966, pp. 29–36.
58. TANDON, M., *Indian Conc. J.*, **41**, 1968, pp. 302–13.
59. *A Guide for Pressure Grouting Cracked Concrete and Masonry Structures with Epoxy Resins AD*-755926. Army Construction Engineering Research Lab., Nat. Tech. Inf. Service, US Dept. of Commerce, 1973.
60. MATHEWS, P. R., Paint Research Association, private communication, 1966.
61. BOGUE, R. H., *The Chemistry of Portland Cement*, 1955, Reinhold, New York, 1955, p. 793.
62. TAYLOR, H. F. W., *The Chemistry of Cements*, Academic Press, London and New York, **1**, 1964, p. 460, **2**, 1964, p. 442.
63. Highway Research Board, *Symposium, Structure of Portland Cement Paste and Concrete*, Washington, DC, 1966, p. 492.
64. EITEL, W., *The Physical Chemistry of the Silicates*, 1954, University of Chicago Press, Chicago.
65. DEBYE, P. and VAN BEEK, L. I. H., *J. Chem. Phys.*, **31**, 1959, p. 1595.
66. ZISMAN, W. A., *Ind. Eng. Chem.*, **55**(10), 1963, pp. 18–38.
67. OLSEN, D. A. and OSTERAAS, A. J., *J. Phys. Chem.*, **68**(9), 1964, pp. 2730–2.
68. SHAFRIN, E. J. and ZISMAN, W. A., *J. Phys. Chem.*, **71**(5), 1967, pp. 75–82.
69. DE LOLLIS, N. J., *Rubber Chem. Technol.*, **43**, 1970, p. 229.
70. WEYL, W. A., Paper presented at a meeting of Committee D-14 on *Adhesives*, Buffalo, June, 1946.
71. TRIVISONNO, N. N., LEE, L. H. and SKINNER, S. M., *Ind. Eng. Chem.*, **50**, 1958, p. 915.

72. ERICSON, P. and SILVER, I., *8th Annual Technical and Management Conference*, Section 14, 1953, pp. 1–14.
73. RUTZLER, J. E., *Adhesives Age*, **2**(6), 1959, pp. 39–43, and **2**(7), 1959, pp. 28–32.
74. Union Carbide Corporation, 270 Park Avenue, New York, 10017.
75. ZISMAN, W. A., Constitutional Effects on Adhesion and Adhesives, in *Adhesion and Cohesion*, ed. Weiss, P., Elsevier, New York, 1962.
76. CHUNG, H. W., *J. Amer. Conc. Inst.*, May 1975, pp. 233–4.
77. State Testing Establishment for Building Materials (Staatlichen Versuchsanstalt fur Baustoffe), Official Report No. 376-1/73 *Making Good of Cracked Reinforced Concrete Beams by the Structural Bonding Process* (*SCB Process*), 1973.
78. LEE, H. and NEVILLE, K., *Handbook of Epoxy Resins*, 1967, McGraw-Hill, New York, pp. 14–24.
79. Balfour Beatty Power Construction Ltd, Publication TG D3, *Vacuum Impregnation—The Balvac Impregnation Process*, Liverpool.
80. The Borden Chemical Co. (UK), North Baddesley, Southampton, CASCO ML Series.
81. ERICKSON, H. B., Repair of Cavitation Damage in Concrete with Epoxy Resin Materials. SP21-8, Amer. Conc. Inst. 1968.
82. Anon., Dam Fights Erosion with Plastic Coat of Armour, *Engineering News Record*, 11 March, 1976, p. 22.
83. WIMBUSH, A. M., *Repairs to Concrete—Proprietary Materials*, Internal Report, Binnie & Partners, London, 1973.
84. Anon., *Repairs to Mangla Dam, Pakistan, using Epoxy Resins*, Internal Report, Binnie & Partners, London, 1975.
85. SHAW, J. D. N., Epoxy Resin Compositions for Construction—Recent Developments. Paper presented to Japanese Society of Materials, November 1971.
86. HATHERLEY, L., MAHAFFY, K. and TWEDDLE, A., The Skid Resistance of City Streets and Road Safety, *J. Inst. Highway Engineers*, **16**(4), 1969, pp. 7–14.
87. GILES, C. G., SABEY, B. E. and CARDEW, K. H. F., Development and Performance of the Portable Skid Resistance Tester. *TRRL Technical Paper, No.* 66, HMSO, London, 1964.
88. JAMES, J. G., Calcined Bauxite and other Artificial Polish Resistant Road Stones. *TRRL Technical Paper, No. LR*84, Crowthorne, 1967.
89. DURST, W. J., The Shellgrip Surfacing System, *Highways and Road Construction*, **42,** 1974, pp. 12–16.

Chapter 3

ADVANCES IN EPOXY ADHESIVE TECHNOLOGY

E. W. GARNISH

Ciba-Geigy (UK) Ltd, Duxford, UK

SUMMARY

Epoxy adhesives have existed for about three decades and the technology has entered a mature phase. In recent years much has been learnt about the structure of the resins, and the molecular and microscopic structure of the cured compositions. This knowledge is leading to a greater understanding and control of their performance as adhesives especially in the blends with polymers that give the products with highest performance. New resins, hardeners and modifiers continue to appear and wider usage brings new applications to the fore. Some of these topics are discussed using a small and partial selection of the still rapidly expanding literature on epoxy resin systems.

INTRODUCTION

The terms 'epoxy resin' and 'structural adhesive' are today almost synonymous. The resins came on to the market a little over thirty years ago. Now they find applications in the widest range of industries and sciences, from aerospace to zoology. Many of these uses depend on the adhesion of epoxies to surfaces.

Epoxy technology has entered a mature phase. Because of the wide permeation of epoxy adhesives, advances tend to be linked now with the progress of adhesion science and technology in general. However, it is interesting to consider how themes springing from earlier days of synthetic adhesives still attract attention albeit with a high level of theoretical and instrumental sophistication.

HISTORICAL NOTES

At the Swiss Industries Fair in 1946, CIBA Ltd introduced a new class of thermosetting materials: the epoxy resins. There was one casting system for use as electrical insulation and one adhesive system for joining impervious materials. This was Araldite† Type I. Bonded joints made of aluminium alloy were shown. These were cyclically tested under load.

The first successful metal bonding adhesive Redux† had been invented a few years earlier in the UK. It was based on a phenolic resole combined with a powdered polyvinyl formal resin. It was first used in the De Havilland 'Hornet' twin-engined fighter which came into service in 1944. This was the fastest combat aircraft that the RAF had received. Like the Mosquito design from which it was derived, the airframe was largely of wood construction. However, in the wing there was a unique feature where an adhesive had been used to bond wood to the aluminium spar. This was a load-bearing joint. It was the first cautious practical step towards structural bonding using metals.

The basic patent on epoxy adhesives was applied for in Switzerland in 1945,[1] that on the vinyl-phenolic in the UK in 1941.[2] The selfsame products introduced then are still on the market and in successful use.

The two products represent strands of innovation that continue. The simple epoxy system, a solid resin compounded with a latent catalyst was a 100% reactive formulation. It cured at 120–180 °C without evolving volatiles and with low shrinkage. No high pressures were needed to make the assembly. A large variety of materials could be bonded. The vinyl-phenolic system, a resole coating on the adherend with powdered polyvinyl formal sprinkled on it cured at 145 °C with partial solution and reaction between the components. The blend of thermoset and thermoplast combined the permanence of the one and the resilience and toughness of the other.

The two strands have continued largely separate until recently. On the one hand a large number of '100% solids' epoxy formulations exist which contain constituents of only modest molecular weight so that they might easily be spread on and flow in the bonded assembly. These are pastes or liquids or sharply-melting solids. On the other hand, there have been developed blends of epoxy resins with polymers or elastomers that have molecular weights of 20 000 or more. These blends are presented as flexible tapes or films. When cross-linked, the networks are tougher and more resilient than those formed by any of the low molecular weight systems.

† Araldite and Redux are registered trade marks of CIBA–GEIGY (UK) Ltd.

THE RESINS

The output of epoxy resins is small in relation to the total production of plastics. Thus in the USA in 1976, 110 000 tonnes of epoxies were produced in a total plastics market of 13·35 million tonnes.[3] The epoxy resins sold for adhesive use were estimated at 8000 tonnes. The hardeners, modifiers, fillers etc. would be additional to these figures. Similar patterns would be shown in the statistics of other markets.

The market economics mean that resins have to command more than just adhesive use to maintain a place in a manufacturer's list. The first commercial resins introduced were those based on bisphenol-A (BPA). Very many epoxies have been synthesised since, yet they continue to predominate. The formal representation of the structure is that shown in Fig. 1.

At a zero degree of polymerisation the diglycidyl ether is a crystalline solid. The modified liquid epoxy resins commercially available have $n \simeq 0{\cdot}2$ and show an epoxy value *ca.* 5 mol/kg. In this text they are referred to as 'liquid BPA resins'. With increasing values of n, more viscous and then solid resins are seen. The higher molecular weight resins play a minor part in epoxy adhesive technology.

Many other phenols have been used or suggested as starting points for epoxies. Monofunctional phenols provide reactive diluents and polyfunctional phenols provide cured resins with high heat distortion temperatures. For these novolaks are usually used. The resins have the structure shown in Fig. 2.

Commercial epoxy novolak resins have appeared recently where $m \simeq 0$. The material is then difunctional, akin to the conventional liquid resins, but viscosity is much lower with no tendency to crystallise. They are referred to as bisphenol F resins.[4]

Epichlorohydrin (ECH), the compound usually used to introduce the epoxy moiety when making an epoxy resin, reacts readily with active hydrogen attached to nitrogen. Aromatic amines have been used commercially to make useful resins but more recently heterocyclic amines have provided a fertile source of new structures.[5]

With the aid of specific catalysts, epichlorohydrin will react with aliphatic hydroxyl groups leading to a number of low viscosity reactive diluents and resins. These continue to be marketed.

Glycidyl esters are made and sold. These usually have a low viscosity.

Carbon-to-carbon double bonds can be converted to epoxide groups by peroxy compounds and this area has received much attention as a source of

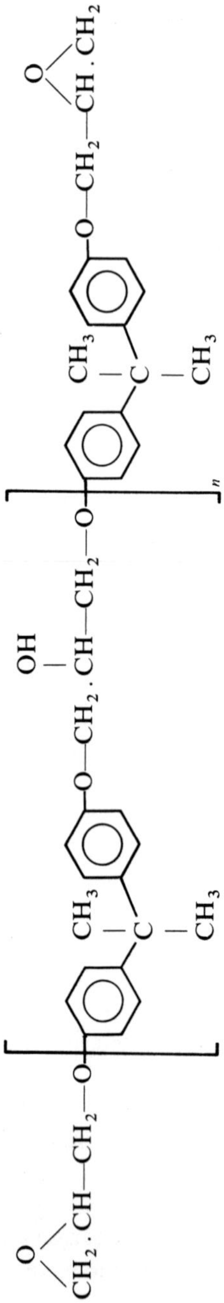

FIG. 1.

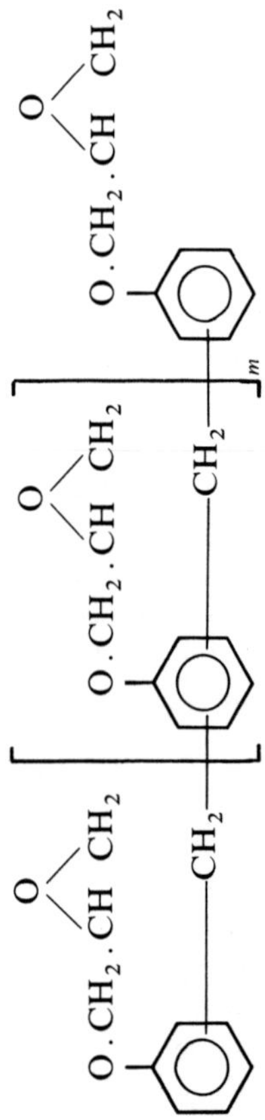

FIG. 2.

new resins. The route has production difficulties and has not produced the number of viable products that at one time seemed likely.

The actual—as opposed to the formal structure—of the resins based on BPA is now understood in some detail through the application of techniques such as thin-layer chromatography (TLC), gel permeation chromatography (GPC), and nuclear magnetic resonance (NMR). The reactive resins used in adhesive technology are the lower members of the homologous series of Fig. 1; liquids to moderate melting point glassy solids, epoxy values high to medium.

Direct interaction of epichlorohydrin with BPA in high ratio, with a base to give ring closure, leads to the liquid resins. At lower ratios, solid resins result. Making them by this method is called the 'taffy' process. Solid resins are also made by an 'advancement' process when BPA is reacted directly with low molecular weight liquid resin. The taffy process gives species as in Fig. 1 with $n = 0, 1, 2, 3 \ldots$; the advancement process gives $n = 0, 2, 4, 6 \ldots$.

TLC has been used[6] to study resins with average molecular weight below 1500. It was concluded that the method of preparation of the resins did not influence the results for resins of low molecular weight but was significant for resins of molecular weight greater than 600.

GPC is a more powerful technique to distinguish between the oligomers and hence between the 'taffy' and the 'advancement' processes.[7] Typical results from this technique are shown in Figs. 3 and 4.

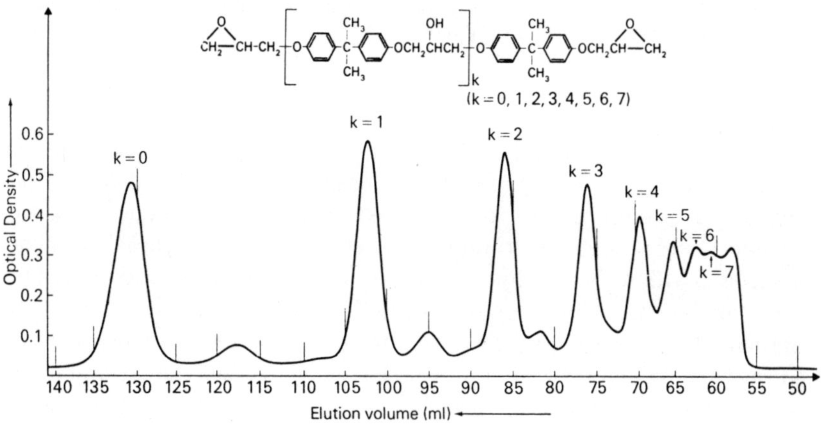

FIG. 3. High-resolution GPC curve (polystyrene gel Biorad SX2, with tetrahydrofuran as solvent) of a low molecular weight solid epoxide resin made by the 'taffy' process. From reference 7. (Courtesy *J. appl. Polym. Sci.*)

GPC showed that the measured molecular weight distributions for the taffy process matched calculated values based on Flory statistics only approximately. This could be due to the presence of impurities, side reactions and branching. The molecular weight distribution from the advancement process was found to be complex and not easily deduced theoretically. However, a further study[8] suggested that the extent of branching is probably small even in resins with molecular weight, $\bar{M}_n$, up to 4000.

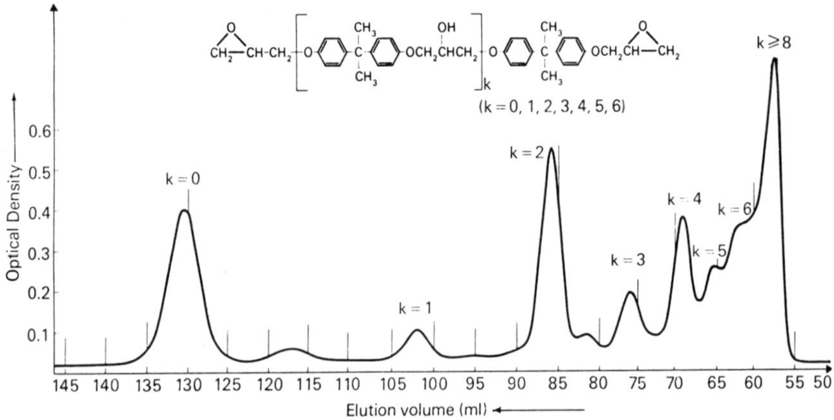

FIG. 4. High-resolution GPC curve (polystyrene gel Biorad SX2, with tetrahydrofuran as solvent) of a low molecular weight solid epoxide resin made by the 'advancement' process. From reference 7. (Courtesy *J. appl. Polym. Sci.*)

NMR spectroscopy is a technique that can show slight discrepancies from the expected –OH and –CH proton levels due to hydroxy and chlorohydrin terminations to molecules.[9] The GPC technique shows up small molecular weight distribution differences between batches that can be associated with resin flow differences observed in workshop conditions. The same analytical techniques can be used to follow the advancement of hardenable compositions, e.g. one-part pastes or adhesive prepregs in storage. These modern techniques are leading a detailed understanding of the composition of the resins as well as to a higher level of quality assurance in application practices.

Besides the attention paid to the BPA resins that dominate the market, new examples of the other reaction possibilities by which epoxy resins may be made continue to come forward. In the patent literature any new epoxy resin *ab initio* is usually regarded as a possible adhesive. Sometimes the claim is specific. The range covered is indicated in the examples that follow.

The diglycidyl ether of bisphenol-S (dihydroxydiphenylsulphone) may be advanced with further bisphenol-S, with catalyst, to give linear resins which can be used as adhesives.[10]

Cyano substituted phenols or aromatic amines are said to be advantageous starting materials from which to make epoxy resins used as adhesives.[11]

Aromatic hydrocarbon resins can be used as a basis for epoxy resins by successive reaction with polyvalent phenols and ECH to give resins used as adhesives.[14]

Hydrogen atoms attached to nitrogen in amine[15] or urethane[16] groups are sufficiently active to react with ECH and lead to epoxy resins suggested as adhesives.

Terephthalamide reacted with ECH in basic conditions gives an epoxy resin. Cured with 6% piperidine for 30 min/150 °C shear strength on aluminium is 46 mPa with self-extinguishing properties.[17]

Glycidyl esters from sulphamoylbenzoic acid and ECH have been claimed specifically as adhesives.[18]

1,7-Bis(epoxybutyl)neocarborane is claimed as an adhesive.[19] This is made by the peracid route. With diaminodiphenylsulphone it will cure at 100 °C; shear strength is 14 mPa at room temperature.

Methylol groups have been placed on the phenolic rings to give a dual functionality to the final epoxy resin. BPA-based resins modified in this way have been offered commercially[12] and a resorcinol-based material revealed in a patent.[13]

p-Isopropenyl phenylglycidyl ether with or without the 2-methyl isomer[20] can be copolymerised with styrene to give a resin readily cross-linked with conventional hardeners to give an adhesive for a variety of surfaces. Copolymerisation of *p*-vinyl phenylglycidyl ether with an ethylenically unsaturated tertiary amine[21] produces an adhesive to bond metals.

Polyglycol diglycidyl ethers are often used as modifiers to the conventional bisphenol resins. They reduce viscosity but are said to have other attributes in adhesive bonding. On the one hand they are said to give oil-penetrating adhesives[22] with hot or cold cure yet on the other hand impart hydrophilic properties to give good adhesive properties to wet wood.[23]

These aliphatic ethers also give rubbery properties to improve adhesive strength by 30%.[24]

Glycidyl esters of long-chain polycarboxylic acids may be used to give flexibilising modifiers for the conventional resins.[25] The long chain may be a rubber-like oligomer.[26]

Resins specifically directed to extending the useful temperature range of epoxies continue to come forward. Bis(epoxyalkyl)carboranes have been suggested.[27] Hardened with DDS, steel-to-steel joints useful over a wide temperature range are produced:[28]

Temperature	*Lap shear strength*
– 196 °C	27·8 mPa
260 °C	18·1 mPa

In another approach a view was taken of where the weak links were in the conventional structure. Bis-epoxides from phenylene oxides were chosen to give better results.[29] A structure is shown in Fig. 5.

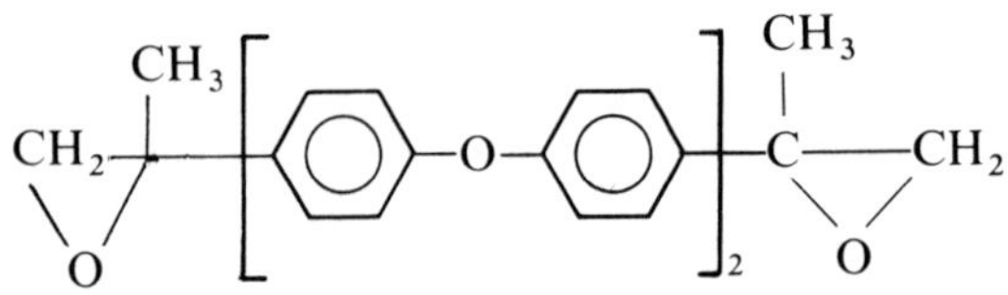

FIG. 5.

These materials gave good adhering resins, a laminate resisting 260 °C for 500 h.

1,5-Diglycidyl naphthalene from the allyl precursor[30] is a crystalline material and has been put forward. However, the transition temperature of 250 °C when cured with diaminodiphenylmethane is only a little higher than the 240 °C epoxy novolaks. A more conventional approach is to increase the complexity of the aromatic core of a resin, for example, a glycidyl ether of tris(4-hydroxyphenyl)methane can be prepared in good yield.[31]

Nevertheless, the thermal endurance of a cured epoxy adhesive is also dependent upon the hardener used. Anhydride hardeners are not commonly used in the adhesives but it has been demonstrated that benzophenone tetracarboxylic acid di-anhydride used with liquid BPA resin gives a cured adhesive joint with a life of at least 1000 h at 260 °C.[32]

The same compound in combination with maleic anhydride and a metal acetylacetonate with mixtures of epoxy resins[33] is also suggested as an adhesive.

As known resins such as epoxy novolak resins or selected hardeners with conventional resins can be shown to give good thermal endurance, it is difficult for new specialist epoxy resins to find a foothold.

THE HARDENERS

The first epoxy adhesive was compounded with a latent hardener requiring a hot cure.

Dicyandiamide is the latent hardener still widely used in one-part epoxy adhesives. It is a high-melting solid and requires rather severe curing conditions. Despite a long history of use the mechanism of reaction has remained obscure. Evidence has been presented[34] suggesting that the powdered compound dissolves in BPA resins during cure, with melamine being formed. Amide links are formed with further cure. Other work has suggested that contaminating water can play a part in the reaction[35] or in interactions at the interface.[36]

The quality control of the compositions, including water content, has been discussed.[37]

Substituted ureas can be used as accelerators for dicyandiamide. Monuron is one such compound. A mixture of the two triggered a reaction of a liquid BPA resin at 135 °C in a differential scanning calorimetry (DSC) test whereas dicyandiamide alone triggered the reaction at 200 °C. It was suggested that dimethylamine was the active accelerating species.[38]

However, although these investigations have been published much still remains to be explained about the action of dicyandiamide and its accelerators.

Accelerators to reduce the severity of the cure regime continue to be suggested for dicyandiamide since the compound gives adhesives with good cured properties. Recent examples include phosphonium or sulphonium salts,[39] tetraalkylacylguanidines,[40] and dimethyldithiocarbamic acid.[41]

Reaction products of dicyandiamide are sometimes suggested for stable one-part compositions curing at modest temperatures. Adducts with tetrahydrofurfuryl alcohol[42] and with phthalic anhydride polyamine condensates[43] are examples.

Cyanamide is a precursor of dicyandiamide and this with an accelerator such as benzyl dimethylamine gives a reasonable B-stage composition[44] curing to a high-strength adhesive.

The bifunctional substituted ureas may be used as hardeners with catalysts such as a cyanoalkylguanidine.[45]

Various types of imidazoles have also been tried as latent catalysts for epoxy adhesives.[46] For example, metal imidazole complexes have been screened.[47] Imidazolium salts can be used to accelerate dicyandiamide.[48]

Aromatic disulphonamides[49] and phthalamic acid[50] derivatives have also been suggested as latent hardeners.

Aliphatic amines were being used as hardeners for epoxy resins in the early 1950s. They continue as an important group now. With careful formulation they are effective curing agents at room temperature though warm cures give improved results. A type especially useful is the polyaminoamide made by reaction of polyamines with dimer acids. These were introduced a few years later than the simple amines. They have rather high amine equivalent weights and so are used in high proportion with the resins. They have low irritation potentials and give the cured compositions a certain flexibility and toughness.

Despite the rather long history of these types of hardener new candidates continue to be put forward. They tend to seek improved combinations of low irritation potential, thorough cure, and toughness.

Polyaminoamide variants occur, for example, methyl esters of unsaturated dicarboxylic acids have been reacted with polyamines.[51] Mixtures of monomeric and polymeric fatty acid amino amines have been suggested.[52]

In other ways, products leading to higher amine equivalent weights are explored. For example, low molecular weight poly(aminoalkylol) resins from polyamines and ethylene carbonate[53] have been suggested.

The kinetics of cure of polyaminoamides with liquid BPA resins have been examined.[54] It has been suggested that one reaction occurs at room temperature, two at 40 °C and three at 100 °C.

The relation of bond strength to cure regime, a function of time and temperature, is of practical importance and can also be discussed in thermodynamic terms. Amine[55] catalytic and anhydride[56] cure of conventional resins in lap joints have been examined in this way. Activation energies about 10 kcal mol^{-1} were suggested.

Liquid polysulphide rubbers have been used as hardeners for epoxy resins for some years. Recently there has been a new upsurge of interest in polymercaptans as hardeners for epoxies because of the fast setting speeds obtained. This is evidenced in the number of papers and patents that are issued. It has been suggested that with basic catalysts compositions cure in thin layers even at $-20\,^{\circ}C$.[57] Certain cycloaliphatic polythiols[58] are claimed to give epoxy compositions to bond GRP without extensive cleaning of the surface. Reaction products of multi-aziridino compounds with hydrogen sulphide can be used as hardeners.[59] Higher lap shear strengths[60] or more extended temperature profiles[61] represent the incremental developments claimed with fast cure.

Mercaptans can have very nauseating odours and attention has to be paid to this aspect if the materials are to be acceptable.

THE CURED COMPOSITIONS

There are no simple relationships between the bulk mechanical properties of a cured epoxy resin and the properties as an adhesive. For example, seven resins with *m*-phenylene diamine hardener have showed[62] that stoichiometry and cure regime had an influence on cured adhesive properties but there was no correlation with bulk tensile, flexural or compressive strength. These types of relationship have recently been reviewed.[63]

Deeper relationships between the chemical structure, the morphology, and adhesive properties of epoxy resins are now studied using powerful and sophisticated analytical techniques. Dynamic characterisations over wide temperature ranges are increasingly used methods. Torsion pendulum, DSC, thermal mechanical analysis (TMA), all are used over wide temperature ranges extending from 80 °K to over 500 °K to characterise adhesive materials.

It has been found that the morphology of poly-addition hardened epoxy resins is influenced not only by the segments between the cross-links but also by the cross-linking fragments formed during the reactions.[64] With suitable choice of resins and hardeners short range orientation can occur, and crystallisation or globular structures are possible.[65]

Thermal analysis shows that cured epoxides can show on annealing separation into zones of less cross-linked and high cross-linked networks that increase the flexural strength.[66]

Nodular structures have been detected in cured liquid BPA resins with polyamine and with anhydride hardeners.[67] Electron microscopy of free surfaces, fracture surfaces and interfaces was used. DSC has been used to compare quenched and annealed specimens of a liquid BPA resin cured with diaminodiphenylsulphone against a commercial adhesive where this system had been modified by a linear polymeric additive.[68] The T_g of the modified system was broader than that of the control. Annealing narrowed the range for both control and commercial adhesive.

Flexibilising modifiers or co-reactants in epoxy systems enhance the ability of the network to respond as a viscoelastic material.[69] Variation of test temperature and test rate on the peel strength showed there were viscoelastic effects below and above T_g with both urethane and nitrile rubber modified epoxies.

TOUGHENED COMPOSITIONS

The first toughened structural adhesive was the polyvinyl formal modified phenolic system. Phenolic resin systems require rather high pressures during the bonding process and are restricted to fairly narrow bond widths because of water evolved during cure. Blends of epoxies with thermoplasts or elastomers were introduced in the mid 1960s to avoid these difficulties and to develop high hot strengths and peel strengths. Nitrile rubber with epoxy is one such blend that has wide usage. The gains in properties due to the modification are large (Table 1).[70]

Amorphous nylon with epoxy is a thermoplast blend. It also was developed in the 1960s and gave a combination of lap shear and peel strength that has hardly been surpassed. However, this blend in joints is susceptible to deterioration due to moisture.[71]

Advantageous blends with epoxies continue to be actively investigated. Some examples are cited below:

Variants with solid carboxy functional nitrile rubbers still appear. They may be cured with dicyandiamide and monuron,[72] for example. Tacky forms have been made by including an acrylic acid copolymer.[73]

Other rubbers that have been used with epoxies are epichlorohydrin-alkylene oxide copolymers,[74] and 1,4-polybutadienes.[76]

Thiol terminated polychloroprene with epoxy resin plus xylene formaldehyde cured with a polyaminoamide is said to bond metals coated

TABLE 1

ADHESIVES MODIFIED WITH NITRILE RUBBER: CURED PROPERTIES

Epoxy	*Lap shear strength (mPa)*	*T-peel strength* (Nmm^{-1})
Unmodified epoxy	15·2	0·7
Epoxy with nitrile rubber	24·8	3·9

with oil.[76] A hydroxy-terminated polychloroprene has also been used in epoxy adhesives.[77] An unbranched polychloroprene used 3–9% in an epoxy polyaminoamide system[78] improved peel strength from almost nil to 5·5 N mm^{-1}.

Blends of a liquid BPA epoxy resin with a phenoxy resin cured with an aziridine are said to give high peel.[79] A phenoxy resin with an epoxy-tipped ester glycol copolymer gave good peel strength when cured with an anhydride.[80] A phenoxy resin with ethylene olefin ester copolymer also makes an advantageous blend.[81] Polysulphones can also be used in epoxy blends.[82]

Blends with high molecular weight materials give adhesive compositions that have to be presented as films or solids. A growing number of methods of toughening epoxy compositions are being found where modifiers with high molecular weight are built up in the resin during cure. This approach brings the two strands of innovation in unmodified and polymeric toughened epoxies together.

For example, the structure of a cured epoxy-urethane composition is a network of flexible and rigid units.[83] Polyurethane oligomers are compatible with epoxies. They may be made in the epoxy resin and an accelerated dicyandiamide cure used to give a high peel adhesives.[84] The *in situ* reaction can give an adhesive with a wide temperature range.[85]

There are several clear ways in urethane epoxy blends whereby chemical reaction occurs between the constituents. It is increasingly realised that a principle of all the successful blends is that there should be some degree of reaction between the resin and the thermoplast or rubber. It is also being appreciated that the cured composition may not be homogeneous but have a structure on the macromolecular or microscopic level. As a consequence, developments of new blends now proceed in a less arbitrary fashion.

The purposeful development can be seen in the growing body of literature upon blends of liquid acrylonitrile butadiene rubbers with epoxies. These rubbers have molecular weights of 3000–5000 and are available with functional groups that can react with the epoxy systems. Because they are liquids the blends do not have to be presented only as films or solids. For this feature and for the understanding of blends in general it is worth surveying the work done on this topic.

The early work started in 1966 when McGarry and coworkers[86] found that 5–10% additions of the liquid rubbers to epoxy resin/alicylic amine systems produced a cured resin containing a second phase of discrete, well-dispersed rubber particles. The cured resins had much improved crack resistance and impact strength.

The composition of the rubbers could be varied. The effect was tested by 5% additions of carboxyl-terminated liquid rubbers with a range of acrylonitrile contents[87] to liquid BPA resin and piperidine. The experiments showed a strong dependence on the acrylonitrile content; the best results were obtained at 10–18%. The commercially available grade of rubber contained 18% and with this additive a fifteen-fold increase in fracture energy was found in the cured resin.

Other terminal functional groups tested in these rubbers have been phenolic, methylol and thiol.[88] The latter two were far less effective in giving a high fracture energy to the cured liquid BPA and piperidine system.

The carboxyl functional liquid rubbers have received the most detailed attention. The toughening effect has been shown to occur with liquid BPA resin and a variety of curing agents; aliphatic polyamines, aromatic amines, piperidine, a Mannich base, a boron trifluoride amine complex and with anhydrides.[89] With the anhydrides there was no heterophase even though a large toughening effect was found (Table 2).

Elsewhere it has been shown that a number of the liquid elastomers (MW *ca.* 3000) at 10 phr doubled the impact resistance of a cycloaliphatic epoxy resin cured with an anhydride; ERL 4221† and hexahydrophthalic anhydride.[90] The commercial carboxyl functional rubber was adjudged best because tensile strength and elongation were improved as well as impact strength. In this epoxy system a second phase formed. Precipitation would occur well before the gelation point.[91] Increasing the liquid rubber progressively to 80 phr gave only small gains in impact strength with progressively greater adverse effects on transition temperature, tensile strength, and modulus. Phase inversion occurred at 50–60 phr.

Phase inversion has also been shown to occur with a liquid BPA resin and piperidine at a volume fraction of 0·5 liquid rubber.[92]

A cycloaliphatic epoxide with aromatic amine, ERLA 4617† plus *m*-phenylene diamine also gave a two phase cured resin using the liquid

TABLE 2

TOUGHENING EFFECT OF CARBOXYL FUNCTIONAL RUBBER

Rubber addition (%)	*Glass transition temperature* (°C)	*Fracture energy* (Jm^{-2})
nil	143	27
5·01	140	140

† Resins marketed by Union Carbide Corp.

elastomers in 10–15 % additions. There was a four-fold increase in fracture toughness.[93]

These various studies showed that the liquid rubbers, soluble in the uncured resins, often gave large gains in toughness usually, but not invariably, associated with the separation of a separate rubbery phase in the cured material. Impact strength might be increased fifteen-fold. The rubbers had functional groups that took part in the epoxy cross-linking reactions during cure.

In a development of the principle it was found that chain extensions of the carboxy functional elastomer with epoxide resin and with BPA gave an even greater gain in toughness.[94] This reaction could be carried out prior to or during the normal cure. Fracture energy was now forty times the value of the simple resin-hardener system. It was observed that a bimodal size distribution of precipitated particles occurred. Failure occurred by shear crazing around the small particles and crazing around the large particles.[95]

All these investigations were carried out on cast resin specimens. The advantages were not so clear cut when the principles were applied to the resins within the adhesive bond line. W. D. Bascom and his co-workers have examined this problem. Using a liquid BPA resin with piperidine and modified with the liquid rubber it was found[96] that there was a maximum fracture energy of about 3·2 kJ m^{-2} in both bulk and adhesive specimens at about 17 wt % modification with the carboxy functional elastomer. The unmodified resin gave a value of 0·12 kJ m^{-2}. However, the modified resin in the adhesive specimens showed arrest behaviour with energy 0·45 kJ m^{-2}, only three to four times that of the control. Furthermore, the adhesive fracture energy was dependent on bond thickness with the maximum occurring when resin thickness was equal to the plastic zone size at the tip of the propagating crack. In thin bond lines there was a restraint upon the development of the plastic zone and fracture energy was reduced.

In further investigations a 45° scarf joint was used where both opening and shear modes of failure were combined.[97] Fracture energy for the mixed mode could be estimated. Liquid BPA resins cured with amine or anhydride hardeners gave mixed mode values similar to opening and shear mode estimates measured separately. The amine cured adhesive, when modified with the liquid rubber, gave higher mixed mode results but the gain was much less than would have been expected from the very high opening mode results. Some commercial high peel adhesives showed similar behaviour. It was later shown[98] that when the scarf angle was 30° or 60° all advantage was lost in fracture toughness estimated by the mixed mode method.

However, suppliers have brought forward tough, peel resistant adhesives based on the precipitated second phase concept[99] and patents are published revealing compositions using the liquid rubbers. Anhydrides,[100,101] a boron trifluoride complex[102] and imidazoles[103] are examples of hardeners. Solid nitrile rubbers may be used in the compositions as well.[104] To obtain better high temperature shear strength a dihydroxynaphthalene[105] or bisphenol-S[106] was suggested as a co-reactant in place of BPA. The same advantage was claimed if the rubber resin adduct was reacted with a vulcanising agent.[107]

The techniques used in studying the liquid rubber blends with epoxies will undoubtedly be used in other aspects of adhesive development. Fracture mechanics has been much used in the investigations.

SOME APPLICATIONS

Where hardeners do not have sufficient life with the resins recourse to physical separation can still produce a one-part adhesive. Thus a carrier fabric may be coated with alternative resin and hardener stripes to make a film adhesive cured under heat and pressure.[108]

Microencapsulation is another separation technique. Anhydride hardeners for hot cure and a stannic chloride complex for ambient cure have both been tried as encapsulated hardeners.[109] Encapsulated polyamines can be used but have limited life.[110] They can be encapsulated in gelatine if they be water insoluble.[111]

An alternative is to encapsulate the resin. Superficial reaction with a boron trifluoride complex can produce the capsules.[112] A threaded fastener adhesive has been made with resin capsules and an amine hardener as the free phase[113] and a stannic chloride complex has also been used.[114]

Reliable lifetimes and reproducible conditions for capsule fracture and mixing of components may be limitations upon the technique.

Epoxy compositions that cure without the application of heat are often required to work in rather adverse conditions, in combinations of low temperatures and high humidity.

The effectiveness of resin-amine compositions can be much reduced in exterior conditions; there will be reduced activity due to lower temperatures and also absorption of water and of carbon dioxide. With a simple aliphatic polyamine crystals of the bicarbonate can be formed in these conditions. The absorption reduces the extent of reaction and of the adhesive strength. This shows in the breaking loads in a torsional shear test (Table 3).

Particular combinations of materials are claimed to give a synergistic

TABLE 3

EFFECTS OF EXPOSURE TO AIR ON AN EPOXY/POLYAMINE ADHESIVE

Composition	*Condition*	*Breaking load after cure (kN)*
Liquid BPA resin plus diethylene triamine	Minimal air exposure	8·5 kN
	45-min air exposure	0·58 kN

effect to overcome combinations of adverse conditions. They can be quite complex. Thus a hardener mixture of an ethylene polyamine, dicyandiamide, an imidazole and N-aminoethyl piperizine is claimed for exterior uses.[116]

Related problems will affect adhesives to be applied and used underwater.[117] These are a type of product that can be expected to be of increasing industrial application. Carbon dioxide absorption will not be a factor and in seawater moisture absorption may be less. High molecular weight amine hardeners have been made for this application. Adducts of aliphatic polyamines with glycidyl compounds[118] are one approach.

A polymercaptan plus blown fish oil[119] is a suggestion with some aptness for this application.

Such formulations will need to be rather reactive because they have to work in rather low temperatures. Accelerating additives will often be included.

Although temperatures will be low and massive displacement of water from the surface to be bonded is necessary, answers to the problems are not impossible.

Highly-fluorinated trimellitate anhydrides have been suggested to give good wetting out of surfaces but hydrophobicity when used as a curing agent.[120]

Dependability is a most important attribute required of an adhesive bond; the service life must be sufficient for the envisaged purpose. The stresses of humidity, loading and heat all play their part in determining the service life. Epoxy adhesives are closely examined in these respects.

The durability of the bonded joint is a topic reviewed in another chapter. Contributing to that durability are changes that may occur in the adhesive layer since it is clear that stresses have effects on the resin systems. The torsion pendulum is a useful technique now used to examine the ability of a system to absorb energy under stress.

It is found that exposure of a cured epoxy-polyaminoamide resin to high humidity results in a substantial decrease in the complex dynamic modulus[121] and a 40 °C reduction in thermal transitions. These effects are reversed by dehydration. Similar effects are seen in epoxy resin blended with thermoplasts as well.[122] Durability of lap-shear joints made with epoxy polyaminoamide adhesive is found to depend on the degree of cure, both excess epoxy or amine groups being deleterious.[123]

At elevated temperatures polymers become more susceptible to creep. Lap joints of nylon-epoxide adhesive show the classic three stages of creep in these conditions with the stress threshold approximately the same as the precision elastic limit.[124]

A different approach is seen in attempts to relate the durability of adhesives with their fracture properties. Precisely made double cantilever beam methods may be used[125] or recourse taken to simpler tests such as a standardised wedge driven between bonded uniform metal strips.[126] Exposure of a stressed crack to an adverse environment, e.g. high humidity for as little as 24 h is believed to give a relation to service experience.[127] It is increasingly realised that in bonded specimens the performance of the crack at the interface is an important factor.

Predictions of durability can be attempted by considering work of adhesion changes when a liquid such as water intervenes at an interface.[128]

CONCLUDING REMARKS

Commercial epoxy resins now have a history of over thirty years use in industry. Although the technology has entered a mature phase advances in their applications as adhesives continue. The topics discussed above have referred to a very narrow selection of the literature that has appeared on these systems in the past few years.

In that time new resins and new hardeners have been put forward as well as useful modifiers and toughening agents. Of perhaps greater significance is the new knowledge that has appeared on macromolecular structure and microscopic morphology related to properties. These advances have been greatly dependent on new techniques in chromatography, microscopy, spectroscopy and thermal analysis as well as theoretical advances as in fracture mechanics.

These new principles will no doubt lead to future advances in epoxy adhesive technology.

REFERENCES

1. Ciba Ltd, GB 629 111, Priority CH 13.7.46.
2. De Bruyne, N. A., GB 577 823, Priority CH 4.5.42.
3. Anon., *Modern Plastics International*, **7**(1) 1977, p. 43.
4. Chiao, T. T., Jessop, E. S. and Penn, L., *SAMPE, 7th Nat. Tech. Conf.*, 1975, p. 167.
5. Habermeier, J., *Angew. Makromol. Chem.*, **35,** 1974, p. 9.
6. Wiesner, I., *Angew. Makromol. Chem.*, **36,** 1974, p. 153.
7. Batzer, H. and Zahir, S. A., *J. appl. Poly. Sci.*, **19,** 1975, p. 585.
8. Batzer, H. and Zahir, S. A., *J. appl. Poly. Sci.*, **19,** 1975, p. 601.
9. Hussain, A., Paper 3, *Plast. Rubb. Inst. 4th Conf. Res. Prog. Resin Plast.*, 1976.
10. Witco Chem. Corp., US 3 733 305, Priority US 3.2.71.
11. Shell Oil Co., US 3 723 361, Priority US 24.9.70.
12. Apogee Chemical Inc., Tech. Bulletin, 1972.
13. US Secretary of the Army, US 3 925 315, Priority US 20.9.74.
14. Mitsubishi Gas Chem. Co., JA 48 60795, Priority JA 2.12.71.
15. Ciba-Geigy AG., DT 2 148 409, Priority CH 29.9.70.
16. Ciba-Geigy AG., DT 2 257 110, Priority CH 24.11.71.
17. Nissho Chem. Industry Co. Ltd, JA 71 27816, Priority JA 3.6.68.
18. Nippon Soda Co. Ltd, JA 72 42828, Priority JA 21.4.69.
19. Singer Co., DT 1 966 316, Priority DT 25.9.69.
20. Ismailov, R. G., Aliev, S. M., Bairamov, M. R., Aliev, S. A. and Alieva, S. G., *Azerb. Neft. Khoz.*, **3,** 1971, p. 34.
21. Aliev, S. M., Bairamov, M. R., Aliev, S. A. and Allakhverdiev, I. K., *Vysokomol. Soedin.*, Ser. A., **16**(10), 1974, p. 2227.
22. Fuller H. B. Co., US 3 551 249, Priority US 18.7.66.
23. Sumitomo Bakelite Co. Ltd, JA 75 54627, Priority JA 17.9.73.
24. Hiroto, M. and Inagaki, M., *Nippon Secchaku Kyokai Shi*, **9**(2), 1973, p. 68.
25. Shin Nippon Rika Co. Ltd, JA 72 141147, Priority JA 19.12.68.
26. Ciba-Geigy AG., DT 2 220 549, Priority CH 29.4.71.
27. Singer-General Precision, Inc., DT 1 948 598, Priority US 25.9.68.
28. Barnes, R., *NTIS Rept. N* 71-16730, 1970.
29. Neville, R. G., Mahoney, J. W. and McDowall, K. R., *J. appl. Poly. Sci.*, **12,** 1968, p. 607.
30. Shell Oil Co., US 3 635 843, Priority US 17.7.69.
31. Dow Chemical Co., NL 72 11359, Priority US 19.8.71.
32. Barie, W. P. and Franke, N. W., *Adhesives Age*, **14**(10), 1971, p. 37.
33. Gulf Research & Dev. Co., US 3 932 343, Priority US 9.5.74.
34. Sacher, E., *Polymer*, **14**(3), 1973, p. 91.
35. Higgs, D. A. and Manley, T. R., *Chem. & Ind.*, **23,** 1973, p. 1112.
36. Fountain, R., *Polym. Eng. Sci.*, **14**(9), 1974, p. 597.
37. Fountain, R. and Haas, T. W., *J. appl. Poly. Sci.*, **19,** 1975, p. 2059.
38. Son, P. N. and Weber, C. D., *J. appl. Poly. Sci.*, **17,** 1973, p. 1305.
39. Union Carbide Corp., US 3 691 133, Priority US 25.3.71.
40. B. F. Goodrich Co., US 3 732 286, Priority US 1.11.71.
41. US Steel Corp., US 3 903 048, Priority US 2.4.73.

42. SOROKOVOI, V. K., USSR 412 224, Priority US 22.11.72.
43. REN Plastics Inc., US 3 632 795, Priority US 9.1.69.
44. Schering A-G., DT 2 122 955, Priority US 10.5.71.
45. American Cyanamid Co., US 3 660 316, Priority US 20.11.70.
46. DOWBENKO, R., ANDERSON, A. A. and CHANG, W. H., *Ind. Engng. Chem. Prod. Res. Dev.*, **10**(3), 1971, p. 344.
47. Nippon Gosei Kako Co. Ltd, JA 47 25300, Priority JA 23.2.71.
48. PPG Ind. Inc., US 3 635 894, Priority US 6.11.69.
49. Dow Chemical Co., US 3 849 375, Priority US 2.4.73.
50. Ciba Corp., US 3 627 730, Priority US 28.11.69.
51. NAKAO, K. and MITO, M., JA 75 160 399, Priority JA 18.6.74.
52. General Mills Chemicals Inc., US 3 666 695, Priority US 6.10.70.
53. CANEVA, C., GUSMANO, G., MAURA, G. and RINALDI, G., *Chim. Ind.* (*Milan*) **53**(7), 1971, p. 669.
54. SACHER, E., *40th Ann. Conf. on Electrical Insulation and Dielectric Phenomena*, 1971, p. 249.
55. YANO, S., KOSUGE, N. and KANAMARU, K., *Kogyo Kagaku Zasshi*, **73**(7), 1970, 1476.
56. MOTOKI, I., TOKIMITSU, F. and TANIGUCHI, R., *J. Adh. Soc. Jap.*, **1**(4), 1965, p. 20.
57. GRAVER, R. B., *J. Paint. Tech.*, **42**(540), 1970, 37.
58. Dow Chemical Co., US 3 873 502, Priority US 18.4.72.
59. General Tire & Rubber Co., US 3 725 460, Priority US 10.5.67.
60. Phillips Petroleum Co., US 3 803 089, Priority US 29.12.69.
61. Phillips Petroleum Co., US 3 821 166, Priority US 1.9.70.
62. MIKA, T. F. and PARADIS, D. L., *SAMPE, 2nd Nat. Tech. Conf.*, **2,** (1970), p. 233.
63. PETKE, F. D., *Amer. Chem. Soc., Polymer Preprints*, **16**(1), 1975, p. 532.
64. BATZER, H., LOHSE, F. and SCHMID, R., *Angew. Makromol Chem.*, **29/30,** 1973, p. 349.
65. LOHSE, F. and SCHMID, R., *Chimia*, **28**(9), 1974, 576.
66. KREIBICH, U. T. and SCHMID, R., *J. Polym. Sci.*, **53,** 1975, 177.
67. RACICH, J. L. and KOUTSKY, J. A., *J. appl. Polym. Sci.*, **20,** 1976, p. 211.
68. KAELBLE, D. H., CIRLIN, E. H. and DYNES, P. J., *SAMPE, 19th Nat. Symp.*, 1974, p. 469.
69. CLARKE, J. A., *Amer. Chem. Soc. Div. Org. Coat. Plast. Chem. Papers*, **31**(2), 1971, p. 105.
70. LEWIS, A. F. and SAXON, R., Chapter 10, *Epoxy Resins*, ed. Kakuichi, H., Dekker, New York, 1969.
71. BODNOR, M. J. and WEGMAN, R. F., *SAMPE J.*, **5**(5), 1969, p. 51.
72. Minnesota Mining & Mfg. Co., US 3 894 113, Priority US 15.9.66.
73. Avery Products Corp., US 3 639 500, Priority US 9.5.68.
74. American Cyanamid Co., US 3 830 874, Priority US 5.8.71.
75. GOLDSCHMIDT AG. Th., DT 2 158 878, Priority DT 27.11.71.
76. Aica Kogya Co. Ltd., JA 75 158 624, Priority JA 13.6.74.
77. Denki Kagaku Kogyo K. K., JA 75 142 633, Priority JA 8.5.74.
78. Goldschmidt AG. Th., DT 2 216 786, Priority DT 7.4.72.
79. Sony Corp., US 3 873 637, Priority US 9.8.71.

80. Minnesota Mining & Mfg. Co., US 3 904 813, Priority US 18.3.74.
81. Ciba-Geigy (UK) Ltd, GB 1 240 709, Priority GB 18.7.68.
82. Ciba-Geigy (UK) Ltd, GB 1 169 613, Priority GB 21.2.67.
83. KOROLEVA, V. M., KOSTOCHKO, A. V. and NIKOLAEV, V. N., *Internat. Polym. Sci. Tech.*, **1**(3), 1974, p. 82.
84. Sekisui Chem. Co. Ltd, JA 74 97 952, Priority JA 21.12.72.
85. Dow Chemical Co., US 3 525 779, Priority US 15.11.68.
86. MCGARRY, F. J. and WILLNER, A. M., *Amer. Chem. Soc. Div. Org. Coat. Plast. Chem. Papers*, **28**(1), 1968, p. 512.
87. ROWE, E. H., SIEBERT, A. R. and DRAKE, R. S., *Modern Plastics*, **47**(8), 1970, p. 110.
88. ROWE, E. H., SIEBERT, A. R. and LIPIEC, J. M., *SPI 27th Ann. Tech. Conf.*, 1972, Paper 2-D.
89. MEEKS, A. C., *Polymer*, **15,** 1974, p. 675.
90. SOLDATOS, A. C. and BURHAMS, A. S., *Ind. Engng. Chem. Prod. Res. Dev.*, **9**(3), 1970, p. 296.
91. VISCONTI, S. and MARCHESSAULT, R. H., *Macromolecules*, **7**(6), 1974, p. 913.
92. KALFOGLOU, N. K. and WILLIAMS, H. L., *J. appl. Poly. Sci.*, **17,** 1973, p. 1377.
93. LAIBLE, R. C., SULTAN, J. N. and MCGARRY, F. J., *SPE 32nd Ann. Tech. Conf.*, **32,** 1974, p. 113.
94. ROWE, E. H. and RIEW, C. K., *SPE 32nd Ann. Tech. Conf.*, **32,** 1974, p. 663.
95. SULTAN, J. N. and MCGARRY, F. J., *Polym. Engng. Sci.*, **13**(1), 1973, p. 29.
96. BASCOM, W. D., COTTINGTON, R. L., JONES, R. L. and PEYSER, P., J. appl. Poly. Sci., **19,** 1975, p. 2545.
97. BASCOM, W. D., and TIMMONS, C. O., *Amer. Chem. Soc. Div. Org. Coat. Plast. Chem. Papers*, **35**(1), 1975, p. 53.
98. BASCOM, W. D., COTTINGHAM, R. L. and TIMMONS, C. O., *Naval Engng. J.*, **88**(4), 1976, p. 73.
99. WEBER, C. D., GROSS, M. E. and GOODRICH, B. F., *Adhesives Age*, **17**(2), 1974, p. 18.
100. Union Carbide Corp., DT 2 060 201, Priority US 19.12.69.
101. Gulf. Res. Dev. Co., US 3 948 849, Priority US 1.11.74.
102. Philips' Gloeilampenfabrieken NV., NL 70 18093, Priority NL 11.12.70.
103. Matsushita Elec. Wks., JA 51 020998, Priority JA 13.8.74.
104. Minnesota Mining and Mfg. Co., US 3 707 583, Priority US 18.10.68.
105. Dexter Corp., US 3 678 130, Priority US 10.9.70.
106. Dexter Corp., US 3 678 131, Priority US 10.9.70.
107. Dow Chemical Co., US 3 947 522, Priority US 12.3.73.
108. Nitto Electrical Industrial Co. Ltd, JA 75 112 433, Priority JA 16.2.74.
109. SCHAAB, C. K., HILBELINK, R. D., DAVIS, T. R. and HANNY, J. F., *SAMPE, 2nd Nat. Tech. Conf. Proc.*, **2,** 1970, p. 695.
110. BANK, M., BAYLESS, R., BOTHAM, R. and SHANK, P., *Amer. Chem. Soc. Polym. Preprints*, **13**(2), 1972, p. 1260.
111. Interchemical Corp., US 3 391 095, Priority US 20.9.65.
112. Dunlop Co. Ltd, GB 1 158 662, Priority GB 1.2.66.
113. Minnesota Mining and Mfg. Co., US 3 746 068, Priority US 15.5.67.
114. Dunlop Holdings Ltd, GB 1 253 843, Priority GB 9.3.68.
115. BELL, J. P., REFFNER, J. A. and PETRIE, S., *J. appl. Poly. Sci.*, **21,** 1977, p. 1095.

116. Shell Internat. Res. Mij. DT 1 951 600, Priority US 14.10.68.
117. Drisko, R. W., *Paint Varn. Prod.*, **58**(7), 1968, p. 31.
118. Mitsubishi Petrochemical Co. Ltd, JA 75 43200, Priority JA 22.8.73.
119. US Secretary of the Navy, US 3 891 583, Priority US 25.2.74.
120. US Secretary of the Navy, US 3 711 514, Priority US 12.2.70.
121. Butt, R. I. and Cotter, J. L., *J. Adhesion*, **8,** 1976, p. 11.
122. Althof, W. and Brockmann, W., *SAMPE, 21st Nat. Symp.*, 1976, p. 581.
123. Mika, T. F., *SAMPE, 5th Nat. Tech. Conf.*, 1973, p. 394.
124. Shen, H. K. and Rutherford, J. L., *Mat. Sci. Engng.*, **9**(6), 1972, p. 323.
125. Mostovoy, S. and Ripling, E. J., *Polym. Sci. Tech.* (*Adhesion Science and Technology*), **9,** 1975, p. 513.
126. Bethune, A. W., *SAMPE J.*, **11**(3), 1975, p. 4.
127. Marceau, J. A., Moji, Y. and McMillan, J. C., *SAMPE, 21st Nat. Symp.*, 1976, p. 332.
128. Gledhill, R. A. and Kinloch, A. J., *J. Adhesion*, **6,** 1974, p. 315.

Chapter 4

TACK AND MORPHOLOGY OF PRESSURE-SENSITIVE ADHESIVES

P. J. C. COUNSELL and R. S. WHITEHOUSE
Evode Holdings Ltd, Stafford, UK

SUMMARY

This chapter is concerned with macroscopic effects observed with the tack of pressure-sensitive adhesives and the microscopic structure of such adhesive films. A description of tack test methods is followed by a discussion of the effect of external parameters on tack results such as duration of contact, temperature and rate of testing, and it is demonstrated that for viscoelastic adhesives the WLF rate–temperature transformation can be applied to tack data. A technique of examining the chemical composition of the surface layers of pressure-sensitive adhesives is described and used to show that a non-homogeneous film is obtained with a number of rubber-resin adhesives studied. Scanning and transmission electron microscopy has given definite evidence of microscopic two phase structure in rubber-resin adhesive films within certain composition limits and this is compared briefly with the tack behaviour of such systems.

INTRODUCTION

Tack is an important characteristic of many different types of adhesive material. In its broadest sense, tack is the ability of the adhesive to bond rapidly under conditions of light contact pressure and short contact time. The presence of tack in adhesive systems means that complicated jigging systems are not needed to hold the components of an assembly together while the adhesive sets, and in many manufacturing operations this factor may be crucial in deciding whether to use an adhesive system instead of an alternative method such as mechanical fastening.

It is important at this early stage of the discussion of the subject to distinguish between the different types of tack, viz.:

Auto-tack (or *autohesion*)—in which the film adheres to a similar film on contact. This phenomenon is exhibited by contact adhesives and also in the plying together of strips of unvulcanised rubber as for example in the manufacture of tyres.

Pressure-sensitive tack—in which the film adheres tenaciously to any surface with which it comes into contact. The strength of the bond is influenced by the contact pressure—hence the term 'pressure-sensitive adhesive'.

Wet tack—this condition applies to adhesive systems in which the bond is made while the adhesive still contains a considerable quantity of carrier vehicle. The adequacy or otherwise of the wet tack is a function of the rheology of the total system.

This chapter is wholly concerned with the second type of tack, namely that exhibited by pressure-sensitive adhesives. It presents the modern view of its nature substantiated by recent work, particularly that of the authors and their colleagues.[1,2]

The general principles of formulating polymers for use as pressure-sensitive adhesives are discussed elsewhere in the book.

EXPERIMENTAL METHODS OF EXAMINING TACK AND MORPHOLOGY

Tack

The characteristic property of tack is the rapid development of bond strength and this in turn means that in any test procedure to measure tack the accurate control of the time sequence between making of the bond and breaking it is very important. Furthermore, as has already been stated the bond strength is influenced by the contact pressure during the bond making process and this parameter also has to be closely controlled. Other factors such as the rate of breaking (and making) of the bond, the chemical and physical nature of the substrate and the ambient temperature all affect the measured result and have to be clearly defined if reproducible data are to be obtained.

The various tack test procedures can be broadly classified into three general types:

(a) Peel test methods
(b) Rolling ball or cylinder methods
(c) Probe test methods.

In all cases it is necessary for the adhesive to be in a usable form which in practice means a film of precisely-defined thickness (or mass/unit area) on a suitable substrate. Such films are obtained by spreading the fluid adhesive, often with a doctor blade, followed by removal of volatiles in the case of solution or dispersion type adhesives or cooling in the case of hot-melt systems. It is generally experimentally convenient to spread the adhesive initially on to a silicone-coated release paper and subsequently to transfer the film to the chosen substrate by dry lamination and removal of the paper. At all stages precautions must be taken to avoid contamination by dust, grease or water.

Peel Test Methods

These methods involve the application of the adhesive, usually in the form of a tape on a flexible carrier substrate, to a defined surface under known conditions of time and light pressure (often just the weight of the tape alone). The peeling force required for debonding is measured under defined geometry, rate of separation and temperature conditions.

The application of this technique has been described in the literature by several workers.[5,6]

The disadvantage of all peel test methods used to assess adhesion is that the measured peel force is dependent on several factors such as modulus and thickness of substrate and adhesive layer as well as the inherent strength of the adhesive bond. This means that the peel test for assessment of tack is only really suitable for comparative purposes, as for example in quality control work.

Rolling Ball or Cylinder Methods

The principle of these methods is that a flat plane is inclined at a certain angle to the horizontal and the adhesive film is placed on this surface, tacky surface uppermost. A small ball or roller is allowed to roll down the plane for a pre-determined distance before rolling on to the tacky surface where its movement is arrested. The distance the ball or roller moves on the tacky surface is an inverse measure of the tack of that surface. This type of test was used by Dow[7] and Condon.[8]

The mathematics of this test are fairly complex since it was assumed when it was first developed that the ball moves at a constant velocity on the tacky surface until it is instantaneously arrested. More recent work has shown that the velocity of the ball decreases linearly with time.[9]

Although this test is complex in its interpretation it requires only simple apparatus and it can be quite useful when problems arise with pressure sensitive adhesives where very short contact times are employed. An example of such an application is in the splicing of webs of newsprint on printing presses while the latter are running at speeds of several hundred metres/minute.

It has been pointed out that the rolling ball/cylinder tack test is really a low angle peel test. A modified procedure devised by Bull *et al.*,[10] in effect reversed the layout of this type of test by placing the adhesive film, tacky side outwards around the periphery of a large rotating drum. The main problem with this type of test is that the dwell time and rate of test are inextricably linked and the profile of the wheel in contact with the drum also affects the result.

Probe Test Methods

In this type of test a probe of precisely defined surface composition and dimensions is brought into contact with the adhesive film and allowed to rest on the film for a known time (the dwell time) under fixed-load conditions, during which the tacky film deforms and 'wets' the surface of the probe with which it is in contact. The probe is then pulled off the film at a defined rate and the force is recorded. Alternatively, the work done during the separation of the probe from the film can be recorded by employing a suitable force × distance integrating device connected to the load-cell transducer. As with the other tack tests the film thickness and ambient temperature parameters must be accurately controlled. This type of test has been used by a number of workers, for example, Wetzel[3] and Hammond.[11] Its strength lies in the fact that it is theoretically sound and that all of the variables affecting the tack of the adhesive may be studied independently.

In the form used at Evode Ltd, the test utilised a bullet-shaped probe with a hemispherical tip of 9·4-mm diameter which is mounted in a holder in which it could slide freely upwards. Its movement in the downwards direction was restrained by a collar at the upper end of the probe. The holder was connected via a load cell transducer to the crosshead of a sensitive tensile test machine. The adhesive film was mounted on a glass plate which in turn rested on a levelling table.

During operation of the test the crosshead of the tensile test machine was lowered at a known rate causing the probe to contact the adhesive film. The crosshead was then stopped for the duration of the dwell time and then raised, again at a known rate separating the probe from the film. The force on the probe was measured by the transducer and gave a readout of tack force on the recorder. The pressure of the probe on the film could be altered by placing appropriate additional weights on the probe.

To obtain a statistically valid result for the tack force on a given sample, ten replicate determinations were carried out, moving the film slightly between each one so that a fresh area of film was contacted with the probe each time. The mean and standard deviations for the tack force (or tack energy) were then calculated by usual procedures. In normal practice cleaning of the probe with a suitable solvent was carried out between series of tests on different films, and not between individual tack measurements. No tendency for a residue of adhesive to build up on the probes was noticed nor was there a general upward or downward trend of the tack results within a given series of tests.

EFFECT OF VARIOUS PARAMETERS ON TACK AND DEDUCTIONS REGARDING THE NATURE OF PRESSURE SENSITIVE ADHESIVES

Tack and Dwell Times

When a pressure-sensitive adhesive comes into contact with another surface it immediately deforms and begins to 'wet' the surface so that intimate contact between the molecules of the adhesive and the surface is achieved on the molecular scale. Since the adhesive behaves as a pseudo-solid with a definite resistance to deformation and flow the wetting process is not instantaneous as it is for most low-viscosity liquids. Because of this, the tack of an adhesive measured by the probe technique is very dependent on the dwell time of the probe on the adhesive prior to separation, and increases with dwell time in the manner shown in Fig. 1. The factors affecting flow and wetting vary for the different films—and hence a different tack–dwell time curve is obtained.

The three adhesives are:

A. Natural rubber/partially hydrogenated wood rosin/toluene.
B. Polyvinyl isobutyl ether composition in petroleum solvent.
C. A hot melt applied system based on E/VA copolymers.

The curves tend to flatten at the longer dwell times and ultimately the

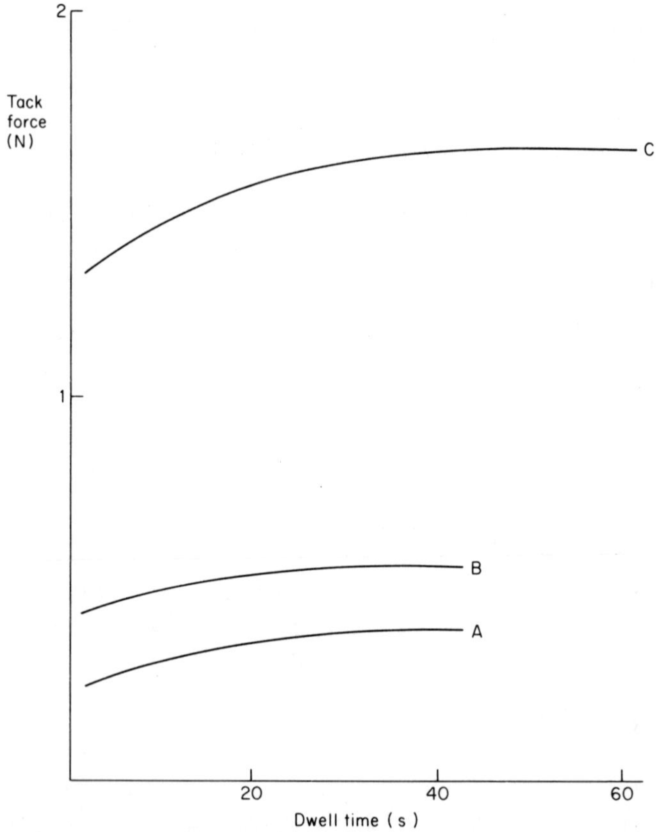

FIG. 1. Tack force as a function of dwell time. (A) Natural rubber/partially hydrogenated wood rosin/toluene; (B) polyvinyl isobutyl ether composition in petroleum solvent; (C) a hot melt applied system based on E/VA copolymers.

progressive increase in tack with dwell time ceases. In practice, this needs dwell times of the order of days to achieve and hence it is not experimentally convenient to investigate in detail. Dwell times as short as 0·4 sec are easily attainable with probe type apparatus.

Tack and Film Weight per Unit Area

The tack measured by the probe technique increases with the thickness of the adhesive film up to a limiting value above which tack is unaffected. This value varies from one adhesive-probe combination to another.

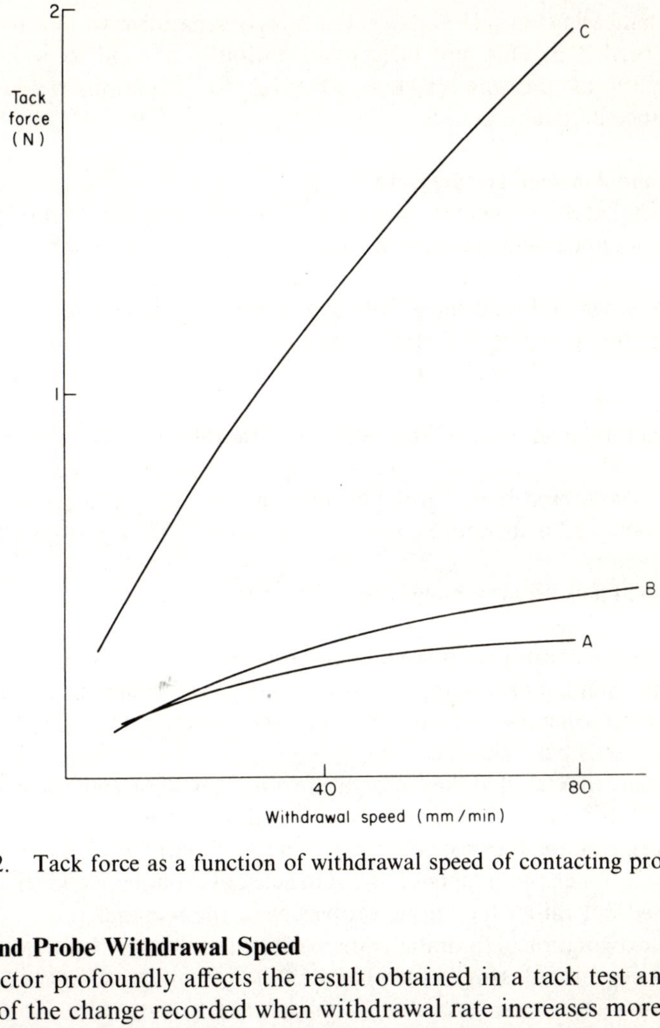

FIG. 2. Tack force as a function of withdrawal speed of contacting probe.

Tack and Probe Withdrawal Speed

This factor profoundly affects the result obtained in a tack test and the extent of the change recorded when withdrawal rate increases more than tenfold is shown in Fig. 2 for the same three adhesive films tending to a limit at the highest speeds for A and B and presumably also for C but at speeds beyond the response time of the recording system.

This observation that the rate of testing has an important bearing on the tack force result is paralleled by almost all other adhesion strength tests carried out on adhesive materials. It is well known that pressure-sensitive adhesives show a rather poor resistance to light loads sustained for long periods (i.e. creep resistance) and the shape of the tack/withdrawal speed curve indicating low tack forces at low rates of separation would enable this to be predicted. This type of test can obviously be used to evaluate the suitability of pressure-sensitive adhesives for situations where creep resistance is required.

Tack and Ambient Temperature

The fact that the ambient temperature can have an important bearing on the result obtained when joining two surfaces together with adhesives is well known. As was described earlier, in order to achieve a bond the adhesive must deform and 'wet' the adherend surfaces, and since viscous flow is a temperature dependent phenomenon so also is adhesive bond formation.

In the case of the tack test the situation is complicated by the fact that a bond is being made and subsequently destroyed after a short time interval. The factors such as increased adhesive viscosity at lower temperatures which tend to prevent bond destruction, thus giving a higher tack force result, also inhibit bond formation by retarding the wetting process. The result obtained is therefore a compromise between the opposing effects of these factors.

Typical tack versus temperature behaviour of three adhesive systems, A, B and C is illustrated in Fig. 3.

As the temperature falls the tack force at first rises reaching a maximum at some characteristic temperature. At still lower temperatures the tack force diminishes rapidly to zero, and the adhesive no longer exhibits pressure sensitive behaviour.

On the other hand, if the temperature is increased the tack force tends to decrease and in most cases the adhesive film softens so that probe–film interfacial failure does not occur, and cohesive failure within the adhesive is observed. When this happens it is not the tack of the adhesive which is being recorded but rather the tensile properties of the material.

The explanation of the initially increasing tack at lower temperatures lies in the rising modulus (or stiffness) of the adhesive, which causes the probe to encounter an increasing resistance to separation from the film. However, at progressively lower temperatures the increase in modulus prevents the film deforming sufficiently to allow wetting to occur and a bond is no longer formed, causing the tack to fall rapidly to zero.

Each adhesive system shows a different tack temperature relationship due to the varying response of the film moduli to ambient temperature. The temperature at which the tack–temperature curve meets the tack = 0 axis is closely related to the glass transition temperature of the adhesive system. Experiments have been carried out with a torsion pendulum apparatus to

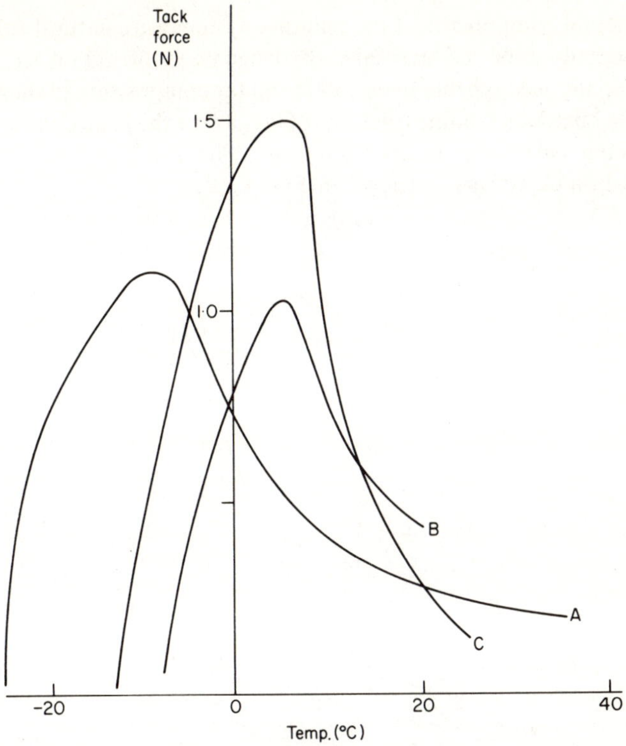

FIG. 3. Tack force as a function of ambient temperature.

examine the changing modulus behaviour of adhesives with temperature and these confirm the views stated above. A full description of the torsion pendulum work is beyond the scope of this chapter.

Tack and Adhesive Composition

With the type of pressure-sensitive adhesive with which this chapter is concerned, there are always at least two polymeric materials present, natural rubber and a tackifying resin. With such adhesives it is possible to vary the concentration of the basic polymer components over broad limits and still obtain tacky products, although the level of tack alters as the composition varies. The pioneering work of Wetzel[3] in examining such materials by the probe technique has already been mentioned. Our own experiments were carried out to complement studies on other effects observed with such rubber/resin adhesives which are described elsewhere in this chapter.

The actual components of the adhesive system were natural rubber and the glycerol ester of partially hydrogenated wood rosin (known commercially as Staybelite Ester 10®)† and the composition of the adhesives was varied between volume fractions of 0·1–0·9 of the resin component, the films being cast from toluene solution. The form of the tack versus composition curve was as shown in Fig. 4.

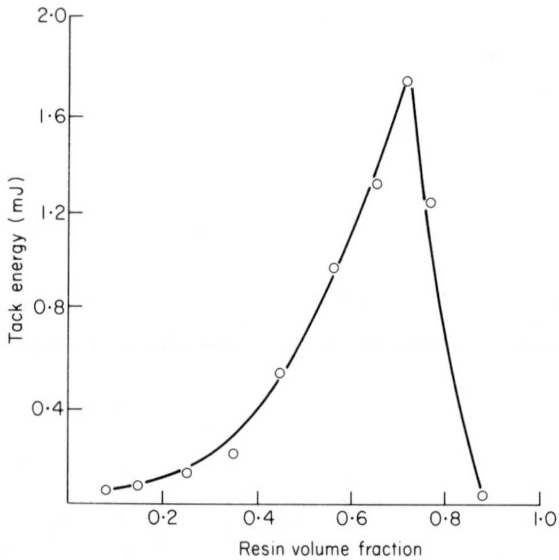

FIG. 4. Tack energy as a function of the volume fraction of resin in the adhesive.

In this particular case the tack energy (i.e. work done in separating the probe from the film) was measured. The results showed a progressive increase in tack energy as the resin volume fraction increased up to 0·7, with a subsequent sharp fall to zero tack as the resin increased still further. Other workers have observed very similar behaviour in such adhesive systems.

It is interesting to compare these results with those obtained in the previous section when tack–temperature behaviour was being examined. The rapid fall in tack from a maximum to zero with increasing resin content closely parallels the change of tack with falling ambient temperature. It is tempting to assume that both effects derive from the same basic mechanism, that of the increasing modulus of the adhesive with increasing resin content

† Staybelite® is the Registered Trade Name owned by The Hercules Powder Co. Ltd.

or decreasing temperature, causing the composition to pass through its glass transition point. Absolute experimental proof of this theory has not been obtained.

Nature of the Probe Material and Its Effect on Tack

It is well known in the technology of adhesives that some materials are more difficult to bond than others. Quite apart from the problems which arise from dirty, contaminated and rough surfaces, certain of the well-known plastics materials such as polyethylene, polypropylene and polytetrafluorethylene present a considerable problem with most types of contact and structural adhesive. Only pressure-sensitive types and hot melt systems are capable of giving reasonable results unless elaborate pretreatments are given to the plastics surfaces before application of the adhesive. The tack behaviour of pressure-sensitive adhesives to probes made out of some of these difficult-to-bond plastics is therefore of interest. Figure 5 shows the effect on measured tack of varying the nature of the probe surface, the results referring to constant conditions of loading. The results are presented as a graphical plot of tack force versus the critical surface tension (γ_c) value of the probe material, determined from the literature. This concept was originated by Zisman[12] and is the surface tension of the liquid which just wets the material. γ_c is accepted as closely related to the surface free energy of the material. The lower it is the more difficult the wetting of the material by an adhesive.

The adhesives are the same as before with the addition of D which was an acrylic solvent based composition. The relationship between tack force and probe γ_c value is linear although the slopes of the lines obtained with the different adhesives varied with adhesive composition and film thickness. This behaviour emphasises the relationship between the wettability of the probe by the adhesive and the tack result and shows that the probe with the lowest γ_c value (PTFE) gives similar tack results with all adhesives, emphasising the difficulty of finding formulations to bond to it.

Relationship Between Tack and Rate of Test

It has been established by many workers that when mechanically testing polymeric materials the rate at which the test is carried out has an important bearing on the result obtained, and it was shown by the theoretical analysis of Williams, Landel and Ferry[13] that the rate of test and the temperature at which the tests were carried out were connected, providing that the temperature was above the glass transition temperature of the polymer. It was possible to carry out tests at various rates and temperatures and to plot

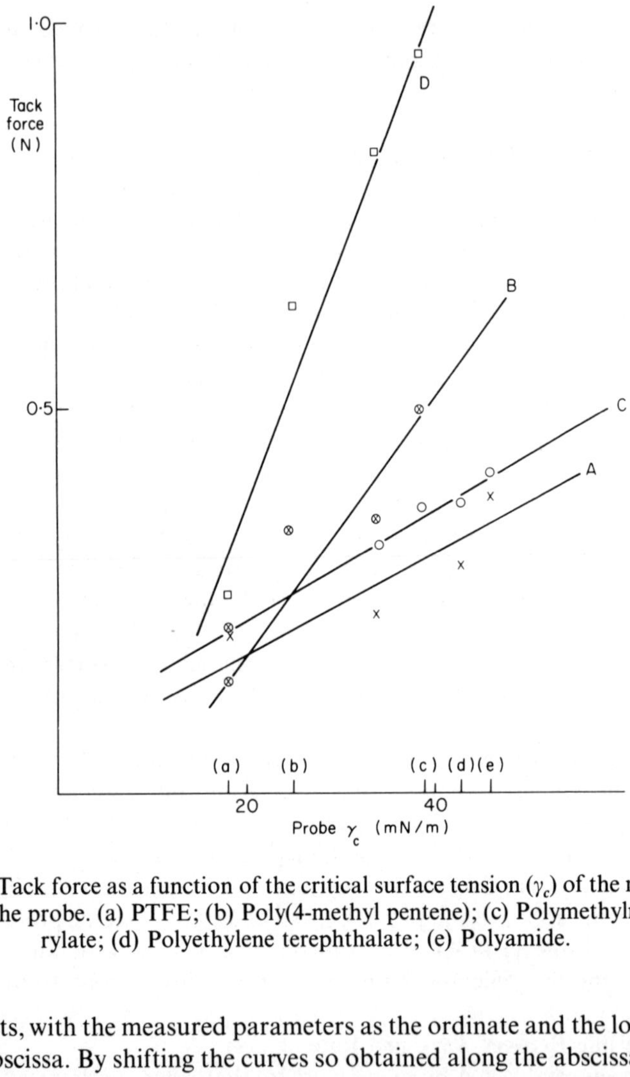

FIG. 5. Tack force as a function of the critical surface tension (γ_c) of the material forming the probe. (a) PTFE; (b) Poly(4-methyl pentene); (c) Polymethylmethacrylate; (d) Polyethylene terephthalate; (e) Polyamide.

the results, with the measured parameters as the ordinate and the log (rate) as the abscissa. By shifting the curves so obtained along the abscissa it was possible to combine all such curves into one continuous curve of measured parameter versus reduced rate. The actual shift applied is given by the WLF equation[1] and relates to the temperature at which the tests were actually carried out and the glass transition temperature of the polymer.

$$\log_{10} a_T = \frac{-C_1(T - T_g)}{C_2 + (T - T_g)} \qquad (1)$$

where: a_T = shift factor at temperature T
T_g = glass transition temperature
C_1, C_2 = constants which are the same for a wide range of viscoelastic polymers of different chemical compositions.

This relationship between rate of test and temperature enables the data obtained with various polymer systems to be expressed in a convenient and readily comprehensible form.

It has been shown by Kaelble[14] and also by Gent *et al.*,[15-17] and is discussed by D. W. Aubrey in Chapter 5, that the WLF procedure could be applied to adhesion tests both in peel and in other modes of debonding. It therefore appeared likely that this type of rate–temperature transformation could be applied to the probe tack test, since, as has already been stated, the tack result is strongly dependent on the modulus (or stiffness) of the adhesive.

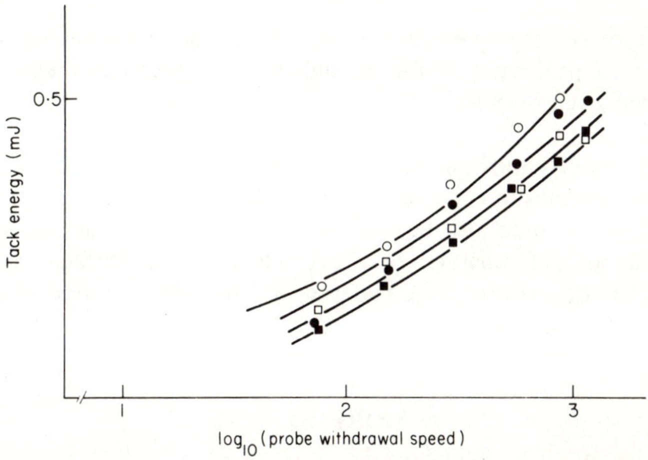

FIG. 6. Tack energy as a function of widrawal speed of contacting probe at different temperatures. ○ = 282 K; ● = 304 K; □ = 297 K; ■ = 309 K.

The effect of varying the test temperature on the tack energy was examined for a polyvinyl isobutyl ether based adhesive system using a polymethylmethacrylate probe under standardised film thickness, probe load and dwell time conditions, but at varying probe withdrawal speeds. The results are indicated in Fig. 6.

Using the WLF transform procedure all these results were reduced to a single smooth curve as in Fig. 7. The rate–temperature transposition

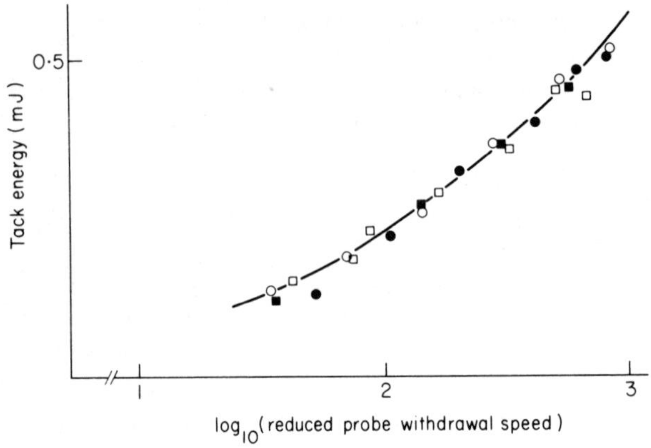

FIG. 7. WLF master curve per tack energy and withdrawal speed. ○ = 282 K; ● = 304 K; □ = 297 K; ■ = 309 K.

procedure is thus applicable to tack test data in common with other mechanical properties of adhesive polymers. A fuller account of this work has been given by Bates.[1,18]

This result confirms the view that the tack of the adhesive is a modulus related property and indicates that in order to possess tack that the adhesive composition must possess viscoelasticity. It is important to note that apart from this it should also be capable of wetting the surface of the probe material and in general this is dependent on the surface free energy characteristics of the probe material and the adhesive, and also on the relative roughness of the probe and adhesive surfaces.

MORPHOLOGY

Modern adhesive systems of the pressure-sensitive type are based on mixtures of several components, the exact formulations generally being derived empirically. The compatibility of the components is largely unknown and the phase structure of such systems has received comparatively little study. In addition, many of these systems are solvent based, i.e. the adhesive components are dissolved in a solvent which is used as a carrier to assist in the application of a thin film of the adhesive to the substrate. Although the system may be a homogeneous solution at normal solvent concentrations, during the evaporation stage any incompatibility within the mixture will become evident, and the nature of the solvent and evaporation conditions might be expected to influence the final structure of the adhesive film. The systems examined consist of natural rubber and Staybelite Ester 10®† dissolved in a variety of solvents to provide a pressure-sensitive adhesive mass.

Attenuated total reflectance (ATR) infrared spectroscopy can be used to assess the chemical composition of the surface of adhesive films. By relating this information to the known composition of the total film, some knowledge of the degree of heterogeneity of the adhesive film may be determined. Electron microscopy can also be used to provide information relating to the surface topography and bulk morphology of these films. These techniques may also be used to assess the effects of various film forming parameters on the heterogeneity of these adhesive films.

Spectroscopic Analysis of the Surface of Adhesive Films

Spectroscopic examination of the surface of adhesive films may be carried out using a spectrophotometer fitted with a multiple internal reflectance attachment. In this case a Perkin Elmer Model 457 grating infrared spectrophotometer, and a Wilks Model 9 multiple internal reflectance attachment was mounted in the sampling beam of the spectrophotometer. The crystal chosen for these studies was a 45° entrance and exit face KRS-5 crystal (eutectic mixture of thallous bromide and iodide), which could give up to a maximum of twenty-five reflections.

The depth of penetration (dp) of the evanescent wave into the adhesive film, perpendicular to the interface has been shown by Harrick[19] to be given by:

$$\mathrm{dp} = \frac{\lambda}{2\pi n_1}(\sin\theta^2 - n_{21}^2)^{-(1/2)} \tag{2}$$

where λ = wavelength of the radiation in air
θ = angle of incidence of the radiation
n_{21} = ratio of the refractive indices of the adhesive (n_2) to that of the crystal (n_1).

Using eqn. (2) with absorptions at 1730 cm^{-1} and 1230 cm^{-1} resin concentrations at depths of 1·33 μm and 1·87 μm are measured and show, as in Fig. 8, that there is indeed a surface excess of resin which decreases as the depth of penetration of the radiation increases. The same concentration gradient is obtained at both interfaces of the film. These surface thicknesses

† Staybelite Ester 10® is a Registered Trade Name owned by The Hercules Powder Co., and is a partially hydrogenated wood rosin ester.

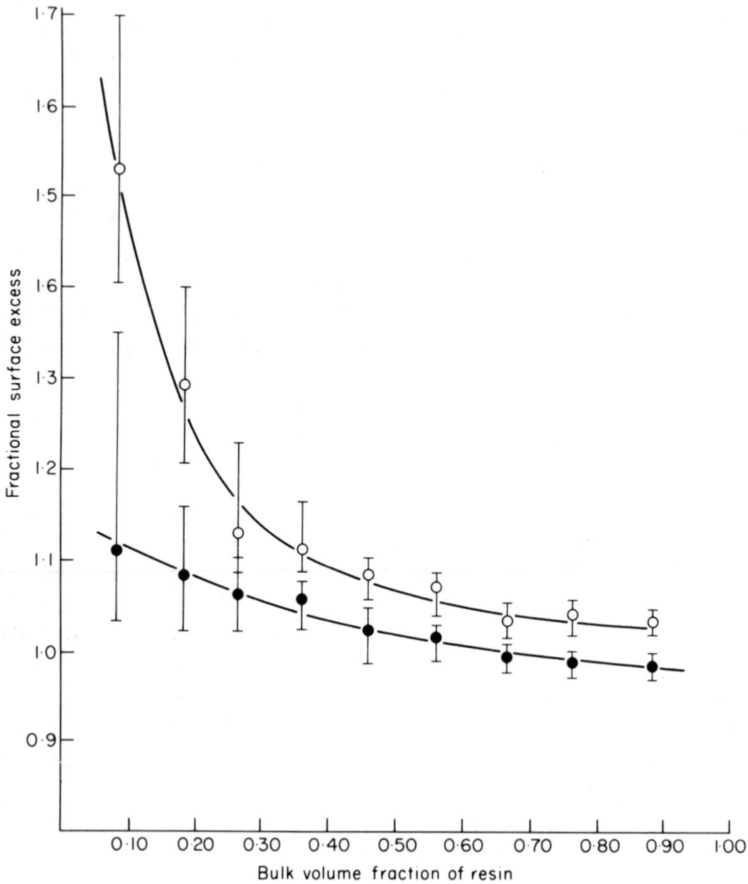

FIG. 8. Fractional surface excess of resin for 5·2 μm films cast from toluene solutions per two depths of penetration: 1·33 μm (○), 1·87 (●).

are not negligible in relation to the total film thickness and the composition of the film must therefore vary throughout its thickness with a minimum of resin in the centre and a maximum at the interfaces. Detailed results[20] have shown that the concentration changes can be represented by a parabola:

$$y = ax^2 + b \tag{3}$$

where y = volume fraction of resin
x = distance from centre of film
a and b are constants

Values for y_{max} (concentration at the surface) and y_{min} (concentration at the centre of the film) may be obtained by solving the integral of eqn. (3) for films of known mean composition (y_{av}). Table 1 shows these calculated results for a range of solvents. There is a strong indication that the surface of some films will consist entirely of resin.

However the occurrence of a surface separation of a pure phase does not represent an equilibrium situation but only that obtained during the process of solvent evaporation. Equilibration (say 500 days) restores a more or less uniform composition from surface to bulk.

The following mechanism of film formation by solvent evaporation is proposed which would account for the observed concentration profile. The initial solution, containing solvent, rubber and resin may be assumed to be homogeneous. As the solvent is lost, the concentration of rubber and resin increases. As the resin has a higher cohesive energy density ($4{\cdot}16 \times 10^8$ Jm^{-3} compared with $2{\cdot}94 \times 10^8$ Jm^{-3} for the rubber) it is likely that the rubber and resin are incompatible to some extent. Hence the resin will act as a non-solvent for the rubber. Under these conditions, the rubber-rich phase is likely to precipitate out of solution first, as a consequence of the lower contribution to the free energy of mixing with the solvent for a given volume of high molar mass rubber compared with an equal volume of low molar mass resin. Since the density of the natural rubber is less than that of the resin, any tendency to settling out will favour movement of the resin plus solvent phase downwards towards the substrate interface, leading to an excess of resin at this interface. However, the movement of the rubber-rich phase is likely to be restricted in view of the high viscosity of this phase. The solvent evaporation process involves transport of solvent from the bulk of the drying film to the air/adhesive interface by spontaneous convection currents and subsequent loss of solvent to the atmosphere. Since the solvent is resin-rich, this process will transport the mobile resin molecules to the air/adhesive interface, where solvent molecules are lost leaving an excess concentration of resin at the adhesive/air interface. Loss of solvent from the evaporating film is fast enough to prevent the resin diffusing back into the bulk of the film; essentially freezing the concentration profile for the film with surface excess of resin at both interfaces. The lower fractional surface excess of resin which is observed for thicker films is consistent with the longer drying time. This leads to more diffusion of resin-rich phase back into the bulk, thus producing a more homogeneous film.

The qualitative mechanism outlined, suggests that the surface excess of resin depends on the solubility characteristics of the ternary rubber–resin–solvent system and the film drying time. The latter variable

TABLE 1

SURFACE (y_{max}) AND CENTRE (y_{min}) VOLUME FRACTIONS OF RESIN FOR VARIOUS MEAN COMPOSITIONS (y_{av}) RELATED TO SOLVENTS AND THEIR PROPERTIES

Solvent	*Vapour pressure (mmHg)*	*Cohesive energy density* ($\times 10^8$ Jm^{-3})		y_{av}				
				0·086	0·266	0·458	0·664	0·884
Dichloromethane	437	406	y(max)	0·192	0·475	0·743	0·929	1·119
			y(min)	0·026	0·191	0·332	0·550	0·638
Chloroform	194	349	y(max)	0·154	0·437	0·685	0·998	1·083
			y(min)	0·036	0·151	0·312	0·400	0·736
Carbon tetrachloride	115	308	y(max)	0·130	0·374	0·584	0·815	1·102
			y(min)	0·051	0·153	0·382	0·550	0·704
Benzene	95	349	y(max)	0·167	0·461	0·652	0·966	1·304
			y(min)	0·040	0·170	0·317	0·455	0·554
Toluene	29	332	y(max)	0·207	0·334	0·601	0·746	1·048
			y(min)	0·019	0·236	0·377	0·606	0·801
Xylene	8	324	y(max)	0·140	0·410	0·708	1·060	
			y(min)	0·046	0·185	0·287	0·360	
Chlorobenzene	12	376	y(max)	0·149	0·389	0·620	0·992	1·386
			y(min)	0·051	0·192	0·363	0·422	0·458
Trichloroethylene	76	352	y(max)	0·117	0·365	0·615	0·925	1·220
			y(min)	0·068	0·213	0·355	0·464	0·614
1,2 dichloroethane	82	400	y(max)	0·238	0·498	0·685	0·906	1·308
			y(min)	0·009	0·150	0·338	0·505	0·540
1,1,1 trichloroethane	125	384	y(max)	0·173	0·489	0·683	0·911	1·264
			y(min)	0·035	0·149	0·332	0·498	0·604
1,1,2,2 tetrachloroethane	8	394	y(max)	0·148	0·373	0·540	0·963	1·018
			y(min)	0·039	0·206	0·409	0·461	0·777

may also affect the temperature of the drying film and hence the solubility characteristics.

The cohesive energy densities of the three components may be compared as a guide to the likely solubility characteristics for the system and the vapour pressures of the solvents may be compared as an indication of the rate of film formation. It is clear from a consideration of the data in Table 1 that no single overall correlation exists between the surface concentrations of resin and the cohesive energy densities and vapour pressures of the solvents over the whole concentration range of adhesive formulations. In general, however, the highest surface concentrations of resin tend to be shown by solvent systems having cohesive energy densities which are furthest removed from that of the rubber, and having high vapour pressures. These are the systems which would show minimum mobility for the rubber, due to fast drying times. A general trend can be discerned. If the heterogeneity of the films measured as $y_{max}-y_{min}$ (at 0·086 overall volume fraction of resin, where the resin excess is most pronounced) are plotted against cohesive energy density of the solvent, the heterogeneity appears to increase linearly as the cohesive energy density of the solvent increases from a value corresponding to that of the natural rubber up to a value corresponding to that of the resin. This is consistent with the view that maximum mobility of the resin plus solvent phase and minimum mobility of the rubber-rich phase leads to highest values for the resin concentration at the interfaces of the adhesive film.

The heterogeneous nature of the rubber plus resin adhesive films, when cast from solvents, and the dependence of the degree of heterogeneity on the solvent, thickness of film and time for equilibration demonstrate that the study of adhesive properties of such films must take those parameters into account, since the surface properties of such adhesives are usually thought to be of vital importance in the adhesive bonding process.

TOPOGRAPHY OF ADHESIVE FILMS

The appearance of films comprising of mixtures of natural rubber and Staybelite Ester 10 (cast from solvent) under the scanning electron microscope is that in Fig. 9A, the surface structure showing uniform ripples, randomly orientated. This structure is similar to that observed for films without tackifier and is indicative of a single phase structure. If a piece of grit is placed on the substrate surface prior to casting the film, the ripples appear to spread radially from the centre of the contamination. The

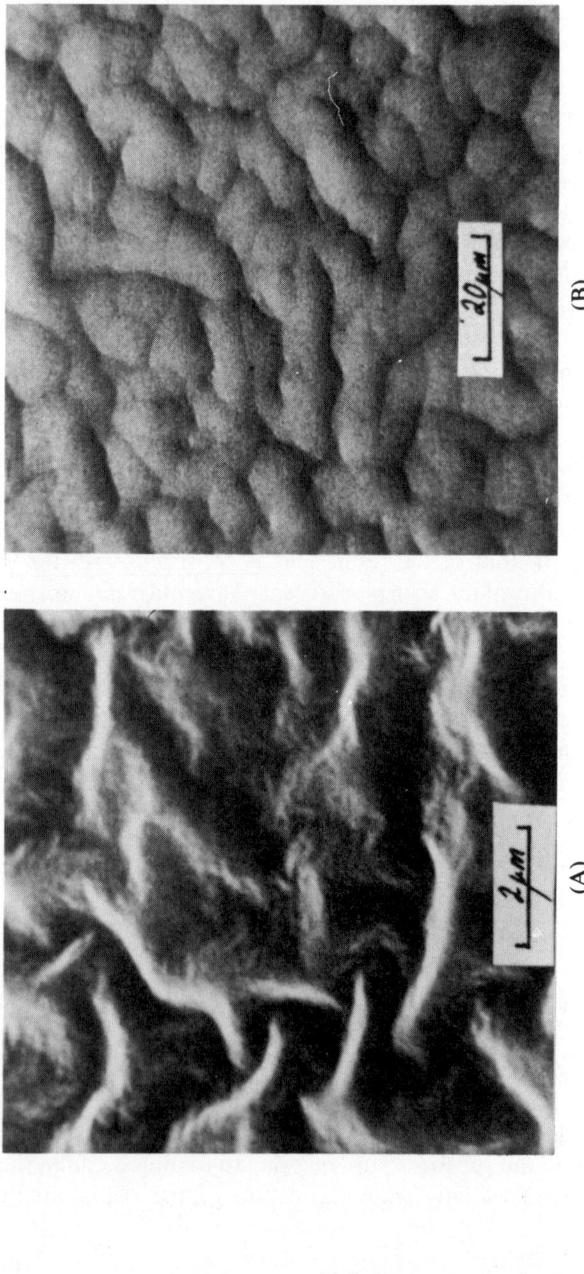

FIG. 9. Photomicrographs of surfaces of rubber–resin mixtures. (A) Natural rubber, Staybelite Ester 10; (B) Natural rubber, phenolic resin; (C) Nitrile rubber, Staybelite Ester 10; (D) Polychloroprene, Staybelite Ester 10.

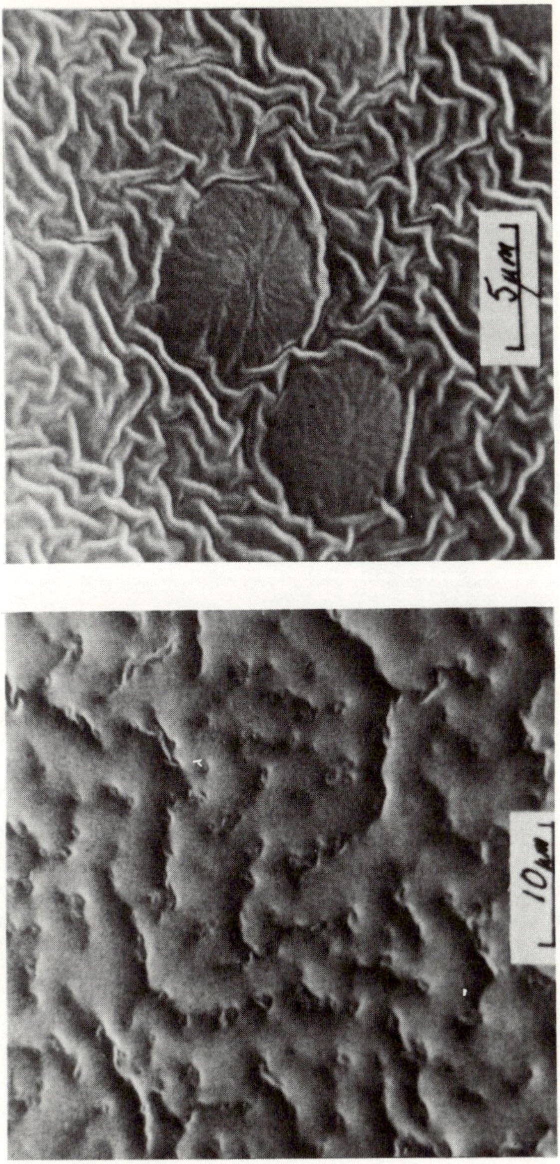

(D)

(C)

FIG. 9—*contd.*

orientation of the ripple structure disappeared within 5 μm of the point of contamination.

Further examination of these films indicated that the wavelength of this ripple structure was controlled by the dry film thickness, the rubber/resin composition and the solvent from which the film was prepared. Figures 10 and 11 illustrate how the magnitude of the ripple structure is altered by changes in film composition and the cohesive energy density of solvent. This

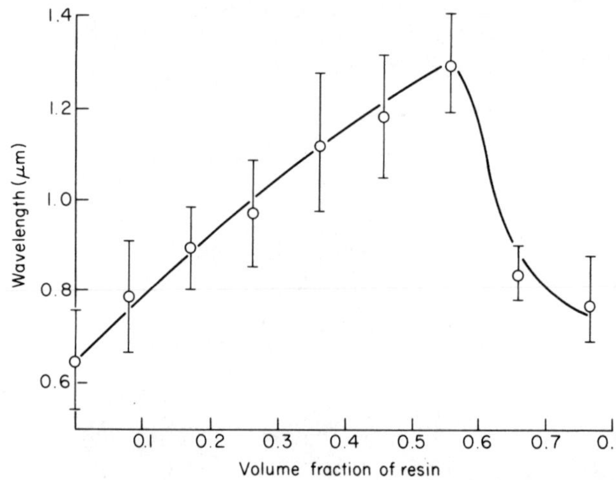

FIG. 10. Variation of ripple wavelength with volume fraction of resin.

surface ripple structure is produced during film formation. A possible mechanism is that during the evaporation process, solvent is lost from the air/adhesive interface and convection currents are set up within the adhesive solution. Loss of solvent at the interface results in a lowering of the surface temperature; thus forming a skin. The initial contour of the surface remains until it collapses due to a significant reduction in the volume of the adhesive. For the surface skin to be accommodated in the reduced area (due to solvent evaporation), it must fold and buckle, thus forming a ripple structure.

This type of surface structure has been observed by Princen[27] when he examined the films formed from various aqueous dispersions which included linseed oil. Hock examined similar rubber/resin systems to those used by the authors using a replication technique and produced entirely different micrographs showing a two phase structure. However, more recent work[2,27] has shown that Hock's technique of using an aqueous gelatin solution may result in distortion of the surface during the initial stages of the replication process.

Various other rubber/resin systems have been examined to see if the ripple structure observed for natural rubber/Staybelite Ester 10 films is obtained with such systems. It was observed that natural rubber/resin systems had very similar ripple structures providing the resins were of

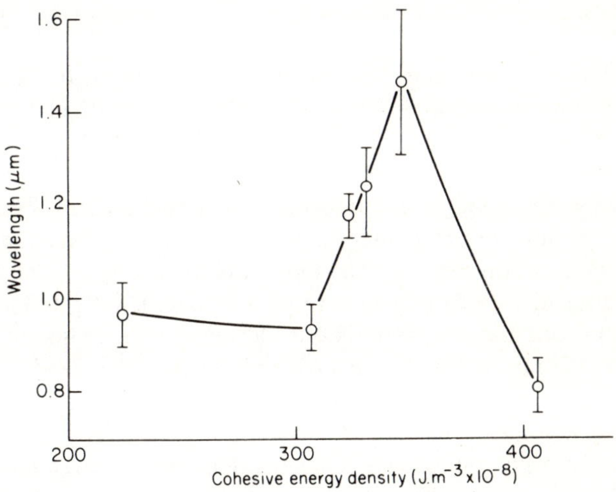

FIG. 11. Variation of ripple wavelength with cohesive energy density of solvent.

the same chemical type, i.e. resin derivatives but changed significantly with different types of resin (see Fig. 9B).

If the rubber system is changed, then completely different surface topographies may be observed. Figure 9C illustrates the topography observed with an acrylonitrile-butadiene rubber (Breon 1001) and Staybelite ester 10 (75% rubber, 25% resin); a distinctly non-uniform structure is observed. The smooth areas between the fibrils appear to be associated with a resin-rich phase, whereas the fibrils are associated with a rubber-rich phase. This concept is supported by the decrease in the magnitude of the fibrils with increased resin concentration. Figure 9D shows the surface structure observed with a film of a crystallising grade of polychloroprene (Neoprene AC medium) and Staybelite Ester 10. The resin and rubber are partially compatible forming areas indicated by a randomly orientated ripple structure (non-amorphous), the remaining rubber being present as spherilitic domains of crystalline rubber.

BULK STRUCTURE USING TRANSMISSION ELECTRON MICROSCOPY

The bulk structure can be investigated by electron microscopy by using sections cut by ultramicrotomy from a bulk sample held at a temperature below the glass transition of the mixture. The actual choice of sectioning temperature is very critical in producing specimens sufficiently thin to be transparent in the electron microscope.[21–25] To improve the contrast in the electron microscope the sections are stained with osmium tetroxide.[26]

Films having a volume fraction of resin, less than 0·174 exhibited a uniform appearance with no areas indicating higher concentration of rubber or resin. When these films were examined at higher magnifications, the uniform distribution of resin-rich phase in a rubber-rich matrix could be observed as slightly lighter areas on a darker background. Films having a volume fraction of resin greater than 0·75 were also one phase, but this time high magnification micrographs indicated small domains of rubber-rich phase dispersed uniformly in a resin-rich matrix. At intermediate resin concentrations there was definite evidence of a macroscopic two-phase structure with particles of a resin-rich phase dispersed randomly within a rubber-rich matrix. This is illustrated in Fig. 12 for a film which has a volume fraction of resin of 0·366.

The resinous phase consists of spherical inclusions of random size which was independent of the composition within the boundaries of the two phase system. Since it is relatively easy to distinguish between the two phases it may be possible to quantify the phase composition of each rubber/resin system. However, because the electron micrograph produces a pictorial account of the phase structure within a specimen of finite thickness (t) and not a plane surface, the projected area fraction (A_A) of the dispersed phase is not the true volume fraction of the phase (V_v).

Cahn and Nutting[29] have devised a means of estimating the true volume fraction of a dispersed phase within a specimen of finite thickness (t), and this may be obtained from:

$$V_v = A_A - \frac{S_v t}{4} \quad (4)$$

where S_v is defined as the particle surface area per unit volume and is equal to $2N_L$. N_L is the average number of intersections per unit length, of a random test line intersecting the dispersed phase. N_L and A_A may be obtained directly from micrographs and the true volume fraction calculated from the above equation.

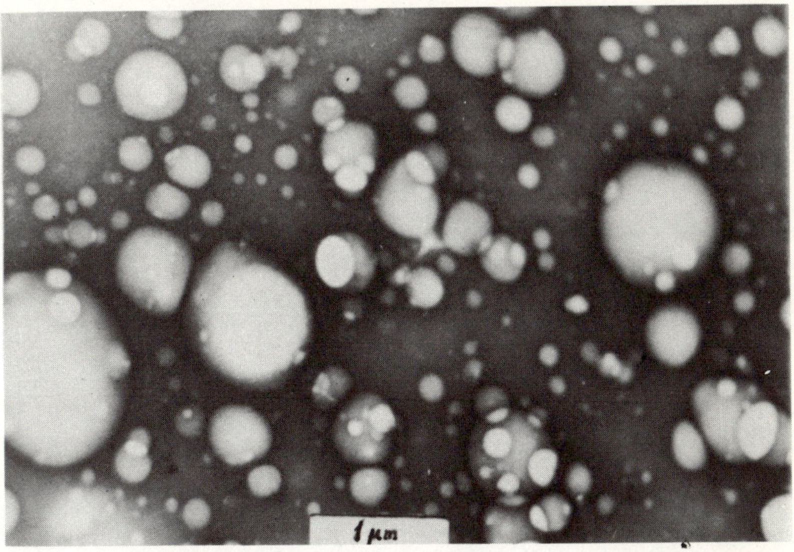

FIG. 12. Transmission electron micrograph of film, 40% resin, 60% rubber by weight; volume fraction of resin 0·366. Stained with osmium tetroxide. Original magnification 20 000 ×.

Figure 13 illustrates the variation of the true volume fraction of resin with volume fraction of resin. This was observed to be independent of the solvent from which the samples were prepared. If these results are compared with tack energy–composition relationship shown in Fig. 4 several general observations may be made:

(1) The maximum two phase character was observed at approximately 0·36 resin fraction and coincided with a rapid change in the rate of increase of tack energy with composition.
(2) The tack maximum at a resin fraction of 0·664 corresponds to a shoulder on the phase distribution curve.
(3) The tack energy falls to zero at a resin fraction of 0·789 which corresponds to the system reverting to a single phase.

Hock,[4,28] Kamagata[30] and Toyama[31,32] have previously claimed that this type of rubber/resin composition is two-phase. However, their micrographs of the same system were widely different and their interpretation was based on poorly defined images, with areas of slightly lighter intensity being claimed to be the dispersed phase. The technique described in this present

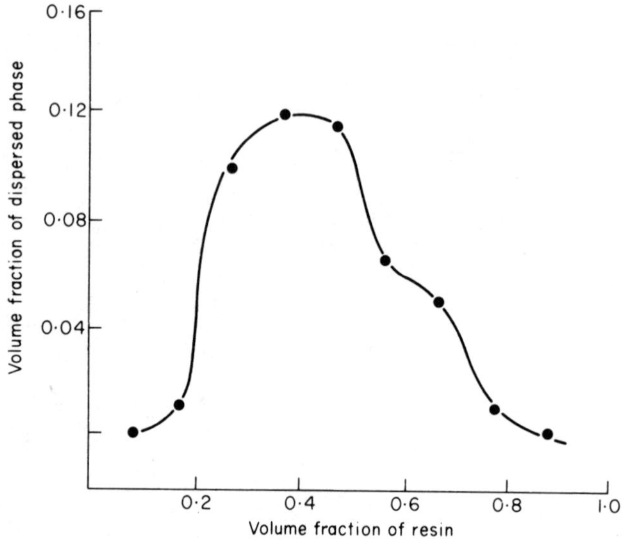

FIG. 13. Volume fraction of dispersed phase of resin a function of the total volume of resin present.

work has produced sharp detailed micrographs which show conclusively the presence of a two-phase structure.

Various other rubber/resin systems have been investigated using this technique and all exhibit a two phase structure similar to that observed for natural rubber and Staybelite Ester 10.

CONCLUSIONS

The tack of pressure-sensitive adhesives is a complex phenomenon profoundly influenced by the bulk mechanical properties of the adhesive material and therefore sensitive to how these are altered with the rate at which the tack test is performed. Tacky adhesives show viscoelasticity and as a consequence it is possible to apply the WLF rate–temperature transformation to data obtained from tack tests as well as from peeling experiments. Even the wetting phenomena associated with change of probe material can be accommodated by a WLF transform. The examination of tack enables information to be obtained quickly on the adhesive behaviour of compounded systems and this means that studying tack by a probe technique has applications beyond the sphere of pressure-sensitive adhesives, for example in following the weathering behaviour of some types of sealants.

These viscoelastic and surface properties are, of course, related to the microstructure of the films and a true mechanistic theory of tack must show not only how it operates but also why it changes with time and the solvent used for film casting. This is clearly related to the degree of separation of two phases and their eventual disappearance. A comprehensive understanding encompasses all these phenomena.

ACKNOWLEDGEMENTS

The authors express their thanks to Messrs Evode Ltd, for permission to publish and to colleagues whose work has been included. RSW acknowledges the Company support which enabled him to submit a thesis to the CNAA for the degree of Ph.D.

REFERENCES

1. BATES, R. Thesis. The City University London, 1975.
2. WHITEHOUSE, R. S. Thesis. Wolverhampton Polytechnic. CNAA, 1976.
3. WETZEL, F. M., *Rubber Age*, **82,** 1957, p. 291.
4. HOCK, C. W., *Rubber Age*, **82,** 1957, p. 471.
5. BUSSE, W. F., LAMBERT, J. M. and VERDERY, R. B., *J. Appl. Phys.*, **17,** 1946, p. 376.
6. CHANG, F. S. C., *Adhesives Age*, **1,** 1958, p. 32.
7. DOW, J., *Proc. Inst. Rubber Ind.*, **1,** 1954, p. 105.
8. CONDON, W. F., *et al.*, *S.P.T.* 14*th Annual Tech. Conf.*, 1959, Section 14, p. 1.
9. KAMAGATA, K., *J. Adhesion*, **2,** 1970, p. 279.
10. BULL, R. F., *et al.*, *Adhesive Age*, **11,** 1965, p. 20.
11. HAMMOND, F., *ASTM Spec. Tech. Publ. No.* 360 123.
12. ZISMAN, W. A., *Ind. Eng. Chem.*, **55,** 1963, p. 18.
13. WILLIAMS, M. J., LANDEL, R. F. and FERRY, J. D., *J. Am. Chem. Soc.*, **77,** 1955, p. 327.
14. KAELBLE, D. M., *J. Adhesion*, **1,** 1969, p. 102.
15. GENT, A. N. and PETRICH, R. P., *Proc. Roy. Soc. Lond.*, **A 310,** 1969, p. 433.
16. GENT, A. N. and KINLOCH, A. J., *J. Poly. Sci.*, A2 **9,** 1971, p. 659.
17. GENT, A. N. and SCHULTZ, J., *J. Adhesion*, **3,** 1972, p. 281.
18. BATES, R., *J. Appl. Poly. Sci.*, **20,** 1976, p. 2941.
19. HARRICK, N. J., *Internal Reflection Spectroscopy*, Interscience N.Y., 1967.
20. WHITEHOUSE, R. S., COUNSELL, P. J. C. and LEWIS, G., *Polymer*, **17,** 1976, p. 699.
21. PRICE, C., WATSON, A. G. and CHOW, M. T., *Polymer*, **13,** 1972, p. 333.

22. ANDREWS, E. H., *Proc. Roy. Soc.*, **A277,** 1963, p. 562.
23. COBBOLD, A. J. and MENDELSON, A. E., *Sci. Tools*, **18,** 1971, p. 1.
24. CRUDGINGTON, J., *J. Sci. Instr.*, **43,** 1966, p. 938.
25. ANDREWS, E. H. and STUBBS, J. M., *J. Roy. Microsc. Soc.*, **B2,** 1964, pp. 3, 221.
26. KATO, K., *Poly. Eng. Sci.*, **7,** 1967, p. 38.
27. PRINCEN, H., *Appl. Poly. Symp.*, **16,** 1971, p. 209.
28. HOCK, C. W., *J. Poly. Sci.*, **C3,** 1963, p. 139.
29. CAHN, J. W. and NUTTING, J., *Trans. AIME*, **215,** 1959, p. 526.
30. KAMAGATA, K., KOSAKO, H., HINO, K. and TOYAMA, M., *J. Appl. Poly. Sci.*, **15,** 1971, p. 483.
31. TOYAMA, M., ITO, K., HINO, K., NUKATSUKU, H. and ISHIKWA, T., *J. Appl. Poly. Sci.*, **17,** 1973, p. 3493.
32. TOYAMA, M., ITO, K., HINO, K., NUKATSUKU, H. and ISHIKWA, T., *Nippon Secchaka Kyokai Shi*, **9**(5), 1973, p. 241.

Chapter 5

PRESSURE-SENSITIVE ADHESIVES—PRINCIPLES OF FORMULATION

D. W. Aubrey
National College of Rubber Technology, London, UK

SUMMARY

Pressure-sensitive adhesives have traditionally been formulated using empirical principles based on intuition and experience; formulation has often been regarded more as an art than a science. In looking for a more rational basis for formulation, the more fundamental aspects of behaviour of any pressure-sensitive adhesive are first examined. These are (i) the ability of the adhesive to resist shearing loads, (ii) its ability to form bonds rapidly on contact and (iii) the ability of the bonds so formed to sustain high peeling loads.

Unfortunately, these aspects of behaviour conflict in their requirements as expressed in terms of the physical properties of the adhesive, and formulation therefore inevitably involves a compromise. The nature of this compromise is discussed and expressed in terms of viscoelastic behaviour, which in turn is related to aspects of chemical composition and structure of components of the adhesive. In this way some general (qualitative) principles may be established as guidelines for formulation. The formulator must also, of course, deal with additional, special, technical requirements (colour, ageing, heat-resistance, etc.) in particular adhesives.

The above aspects of formulation relate entirely to the achievement of certain technical *properties. Equally important, however, are the* economic *constraints which determine the feasibility of using a particular formulation in commercial manufacture. Thus, in deciding how to meet the technical requirements specified, the formulator must take into consideration the costs of raw materials and of the necessary processing in manufacture. The principal raw materials of interest are discussed, as well as the viability of processing them in the form of solutions, emulsions, hot-melts and reactive intermediates.*

INTRODUCTION

Technologists engaged in the design and formulation of pressure-sensitive products (tapes, sheets, labels, etc.) have an unenviable task, in that they are expected to solve a formidable and interacting set of problems, a full understanding of which would require expertise ranging over such disciplines as chemistry, physics, materials science, polymer science and technology and engineering. It seems hardly surprising that, over the course of many years, the various products have been developed largely by empirical technological advancements which have, in general, preceded a scientific understanding of their behaviour.

The complexity of the problem may be illustrated by considering that, for the production of a pressure-sensitive tape, the designer will have to select a suitable support or 'backing' member, a pressure-sensitive adhesive, a means of anchoring the adhesive to the backing and possibly a release agent to facilitate unwinding from a reel. For each of these an enormous range of materials are possible, and the selection will be based on compromise between chemical and physical properties, economic considerations and manufacturing possibilities. Further complications arise since the individual components of a tape are not mutually independent in their action, and it is found that a change in the thickness or stiffness of the backing can produce large changes in the apparent properties of the pressure-sensitive layer.

Nevertheless, since we are presently concerned only with formulation of the pressure-sensitive adhesive layer, we must consider it in isolation from the rest of the tape, but at the same time bear in mind that such properties as peel adhesion or tack (as normally measured) are properties of the whole tape rather than of the adhesive only. The approach used will be to consider first the main *technical* requirements which any pressure-sensitive adhesive must fulfil; these are to give adequate resistance to shear within the adhesive, to bond adequately by conforming to a surface within a sufficiently short time and to show an adequate resistance to subsequent separation by peeling. The ways in which specific materials can be adapted to give the best compromise between these requirements will then be considered, but here it is necessary to consider also certain *economic* constraints. It is clearly a prime duty of the formulator to secure adequate technical performance in his adhesive at a minimum of cost to the manufacturer.

These considerations are applicable to any pressure-sensitive adhesive product. It will be apparent, however, that specific products may require various special properties (clarity, colour, oil-resistance, ageing, low-temperature performance, etc.) which may impose further limitations on the choice of materials. Since there is virtually no limit to the combinations of such requirements, and since they may demand knowledge ranging over the whole field of polymer technology, it is not possible to deal with them except very superficially in the present review.

SHEAR STRENGTH IN PRESSURE-SENSITIVE ADHESIVES

It is appropriate to consider in the first place the behaviour of an adhesive under conditions of shear loading since it involves a relatively simple type of adhesive response and is not affected by interfacial conditions. The test for shear strength most relevant to service conditions would be a 'creep' experiment in which movement under constant load is monitored.

Shear strength of a tape joint is often assessed in the industry by a 'holding power' test, in which a weight (e.g., 1 kg) is hung on a tape bonded over a known area (e.g., 2 cm^2) to a test plate (Fig. 1(a)). The time taken (at a given temperature) for the tape either to move a certain distance *d* or to drop from the plate is taken as an indication of shear strength. Whilst this test may be perfectly adequate for routine testing of performance under

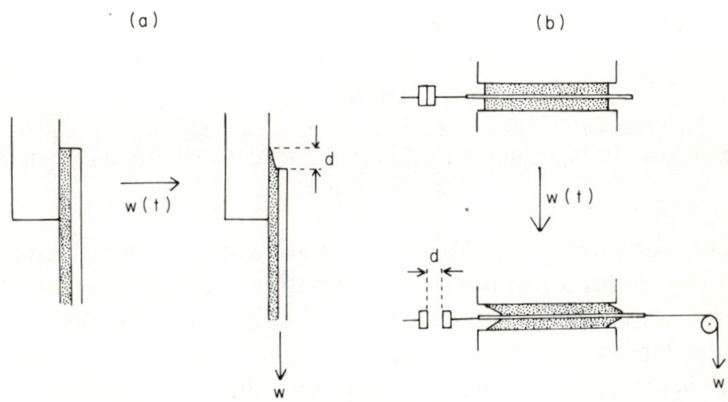

FIG. 1. Assessment of adhesive shear characteristics: (a) 'holding power' test, (b) shear sandwich creep test.

simulated service conditions, it does not as such give much information about any intrinsic property of the adhesive. This is because (a) the results may be strongly influenced by the thickness and flexibility of the backing, (b) the adhesive may separate, at least partly , by peeling from the upper end (a process not always obvious from visual inspection) and (c) measurement of distance moved is crude, and is usually taken at only one time interval.

A much more meaningful analysis of the shear behaviour of an adhesive may be obtained from a 'creep' test of the shear sandwich type described by Ferry.[1] In a suitable version of this test (Fig. 1.(b)), a thin inextensible metal sheet is sandwiched between two layers, of known thickness and area, of the adhesive. This assembly is sandwiched between two constant temperature platens, under a minimum of pressure. A fixed load is applied to the middle sheet and the creep displacement d is measured, by a contactless displacement transducer, at frequent time intervals. It is important to ensure that no slippage can occur at the interfaces in this test. The form of the curve of creep (displacement d) against time (Fig. 2, curve a) would then

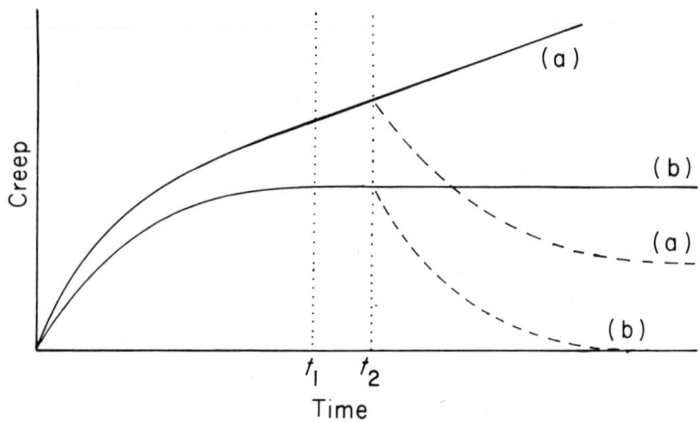

FIG. 2. Diagrammatic representation of creep– and creep recovery–time curves for (a) a viscoelastic liquid and (b) a viscoelastic solid. Creep recovery is shown by the dashed lines.

be interpreted as follows: at short times there is a large amount of creep due primarily to an elastic (recoverable) deformation of the adhesive; the elastic deformation approaches a limit at time t_1 (Fig. 2), and continuing creep due to 'steady-state' viscous flow is revealed (ideally) as a straight line, the slope of which is a measure of the capacity of the adhesive to undergo viscous flow. The presence of a continuous molecular network (i.e. some cross-linked polymer or 'gel') effectively eliminates the possibility of viscous flow, and zero slope is approached after time t_1. Materials exhibiting these two types of response are often designated 'viscoelastic liquid' (steady-state flow occurs) and 'viscoelastic solid' (no steady-state flow), respectively.

A moment's consideration shows that the axis scales of Fig. 2 will depend not only on the nature of the adhesive but also on the temperature, applied load and test-piece dimensions. A quantity dependent only on the properties of the adhesive (for a given temperature) is the *creep compliance* (= strain/stress), in which the dimensions are normalised such that the strain (= creep distance) represents that which occurs relative to opposite faces of a 1-m cube. Since stress is constant, the shape of the creep compliance–time curve will be the same as the creep–time curves of Fig. 2. Although pressure-sensitive adhesives can vary enormously in their creep compliance characteristics, an approximate indication of the scale units for room-temperature compliance of a good general-purpose adhesive would be a compliance range $0 \rightarrow 0{\cdot}15 \times 10^{-5}\,m^2\,N^{-1}$ over a time scale of 0 $\rightarrow$ 5000 sec. Although this is a short time scale compared with service conditions, it is usually sufficient to enter the steady-state region, and hence to predict with some confidence the behaviour of the adhesive over much longer time scales. The creep process can, of course, be speeded up if necessary by raising the temperature.

In order to distinguish more clearly between the elastic and flow components of creep it is common practice to remove the load (say, at time t_2 in Fig. 2), and obtain a creep recovery–time curve (shown as dashed lines in Fig. 2). Creep then eventually diminishes to zero for an adhesive containing network or gel (i.e. a viscoelastic solid), but assumes a finite value, representing the extent of viscous flow, for an adhesive containing no network (i.e. a viscoelastic liquid).

The quantitative significance of these curves is discussed fully by Ferry,[1] and the reader is advised to study his description before embarking on a detailed analysis. For the moment, however, we are more interested in a qualitative appraisal of the dependence of creep on structural features of the adhesive.

For the purposes of assessing 'holding power' of a tape, the elastic (recoverable) part of the creep is not normally of interest, since it is limited in extent and does not involve a 'failure' of the adhesive. Of much greater significance is the steady-state flow viscosity (proportional to the inverse slope of the straight lines in Fig. 2), since it is clear that, provided the slope is not zero, failure of the adhesive will occur ultimately whatever the temperature, applied load and adhesive thickness and area.

It would seem from these considerations that an adhesive containing a proportion of three-dimensional network or 'gel' cannot flow under any circumstances, and creep would reach a limiting value and then stop. This is true even for the case where an apparently insignificant amount (say, 1 %) of the adhesive is in the form of a network. Although the extent of elastic deformation may be quite large before the network reaches its limiting extension, no *flow* will occur. However, *failure* may occur since it will be appreciated that, at the limit of elastic deformation, the load will be supported by a small number of network covalent bonds. The adhesive may therefore fracture quite readily, with a resulting complex mode of shear failure. This situation rather resembles the rupture of a soft table jelly at room temperature; although it cannot flow it is too weak to sustain high shearing loads.

It becomes necessary therefore to increase the concentration of network or gel so that the adhesive can sustain high shearing forces. The concentration cannot be increased without limit, however, since it progressively reduces the extent of creep for a given time, and the ability to show a large deformation under low load is an essential requirement for making contact with surfaces (discussed in the next section). Thus, a compromise must be sought, and for a high shear-performance adhesive it is usually found perfectly adequate in practice to have about 30–50 % gel concentration.

If for economic or other reasons it is not possible to crosslink part of the adhesive, the steady-state flow viscosity must be made as high as possible. Thus, instead of having a proportion of infinite molecular weight (MW) material (i.e. network) present, there should be a proportion of 'near-infinite' (i.e. very high MW) material present. Again, the whole adhesive must not consist of high MW polymer or the bonding process will suffer due to the low short-time compliance. This is in effect saying that the steady-flow viscosity in creep is maximised in a satisfactory manner by having a broad *distribution* of molecular weights, in which the very high MW material is having the predominant effect in restricting flow. Such an effect is well known for its influence on the flow behaviour of elastomers in bulk but, to the author's knowledge, has not been systematically studied in pressure-sensitive adhesives.

Adhesives derived from natural rubber and tackifier resins will invariably show a broad distribution of molecular sizes, (the tackifier being regarded as low MW material) and shear behaviour is good provided that the rubber has not been over-masticated. Certain 'single polymer' adhesives such as polyacrylates may also show a broad distribution of sizes, arising from extensive branching via chain transfer reactions during their polymerisation; for the more demanding load-bearing applications, however, they need some network content. Linear, narrow-distribution polymers such as polyisobutylene or polyvinyl alkyl ethers will normally show poor shear strength, and it is common practice to blend very high with low MW versions of these materials to obtain the best results. Despite this improvement by blending they would not normally be chosen for load-bearing requirements. Of course, temperature has a profound influence on the response of an adhesive to shear deformation and, although adhesives containing no gel may function adequately at low temperature, some cross-linking is almost invariably necessary for performance at elevated temperatures. Most formulators operate intuitively in accordance with these principles although, in the absence of systematic scientific study, they can only be enunciated in a very qualitative way.

THE BONDING PROCESS

In making a bond to a surface with a pressure-sensitive adhesive, pressure is applied for a short period, at the most a few seconds, after which time it must show adequate resistance to separation. This short-time bonding/debonding operation is the basis of tests for the evaluation of 'tack', whether this be done by testing with the thumb, or with a 'probe' tack tester (Chapter 4), or by rolling a ball or cylinder down an inclined plane of adhesive, or by other methods. It is not proposed to discuss the phenomenon of tack in detail, since this is the subject of a separate contribution (Chapter 4).

It is useful to the formulator, however, to have some idea of the separate processes occurring during the 'bonding' time, which determines the extent of true molecular contact achieved over the apparent contact area, and the 'debonding' time, which involves separation by peeling of those parts in molecular contact. For the best 'tack' we clearly need a rapid 'contacting' process and a high peel adhesion under the conditions of debonding. Since the latter is primarily the subject of the next section, the bond formation process only will be considered here.

First, it is important to recognise that all surfaces are rough or irregular on various scales of magnitude, ranging from visible (e.g. paper or cloth) to sub-microscopic (highly-polished metals, glass etc.). Some idea of the degree of roughness left by various machining treatments for metals is conveyed by figures from Vohralik and Wake[2] (Table 1).

If the adhesive and substrate surfaces were perfectly smooth at all levels of magnitude, molecular contact (and hence adhesion) would occur

TABLE 1
HEIGHTS OF ASPERITIES ON MACHINED METAL SURFACES[a]

Treatment	*Average height* ($m \times 10^{-6}$)
Lapped or polished	0·02–0·25
Honed	0·10–0·50
Extruded	0·25–4
Ground	0·5–2·5
Drilled	2·5–5·0
Turned, shaped, milled	3–6

[a] Evode Copyright. See ref. 2.

instantaneously and completely. Generally speaking, the time taken for a given area of contact to be achieved in practice increases with the overall rugosity of the surface.

An essential prerequisite for good contact on practical surfaces is the establishment of a 'wetting' situation (contact angle θ of adhesive on substrate $< 90°$). For a non-wetting system ($\theta > 90°$) the equilibrium extent of contact will be small, irrespective of whether the adhesive is a simple liquid, a viscoelastic liquid or a viscoelastic solid. The extent of contact may be temporarily increased by application of pressure, but will subsequently be restored to its equilibrium value (although both processes will be time-dependent to an extent depending on viscoelastic properties); this is illustrated in Fig. 3(a).

The non-wetting situation is made use of in 'low-adhesion' or 'release' surfaces, where the reverse side of a tape backing or a temporary carrier or support for a pressure-sensitive adhesive is coated with a material of low surface energy. Macroscopically rough or embossed surfaces have also been used to limit contact in releasing liners, but are restricted in application to adhesives which undergo little or no cold flow on storage.

Since pressure-sensitive adhesives are almost invariably high in aliphatic hydrocarbon content they have low surface tension compared with most adherends, and the 'wetting' condition normally obtains. For the few occasions when it is required to stick to a very low energy surface, the formulator has little choice but to change from a largely hydrocarbon to a (much more expensive) silicone adhesive. Where a 'wetting' condition prevails, the surface energetics now favour an increase in molecular contact with time and, if the adhesive can flow, complete contact may be achieved eventually without the aid of external pressure (this requiring, in theory, a zero contact angle). If external pressure is applied, the degree of contact achieved in a given time will be determined not only by the interfacial tension and surface rugosity but also by the balance between elastic and viscous response in the adhesive. This situation is illustrated in Fig. 3(b), where at a given instant the surface energetics and applied pressure both

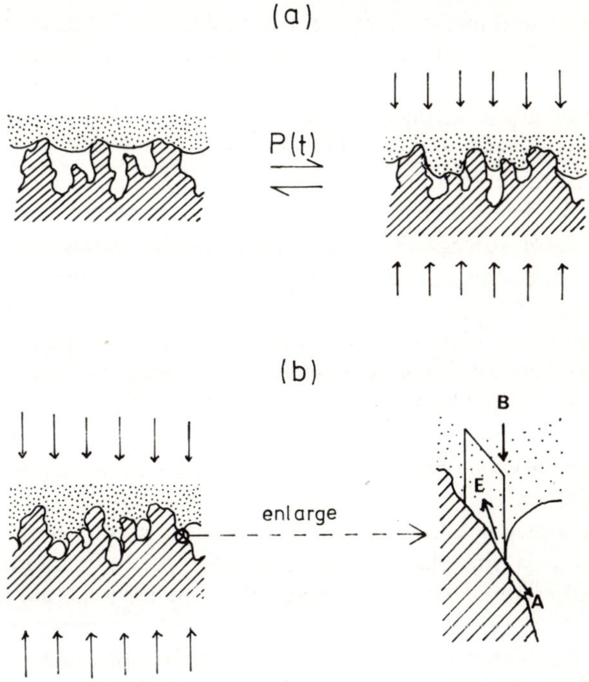

FIG. 3. Illustration of effect of applied pressure on contact with surface micro-irregularities for (a) non-wetting and (b) wetting conditions. B = Applied bonding force; E = Elastic recovery force; A = Adhesion force.

favour increased contact, the elastic recovery forces are tending to reduce contact and the viscous component of compliance determines the rate at which contacting occurs.

If the deformation of an element of adhesive near the contact boundary is considered (marked in Fig. 3(b)), it will be apparent that the type of deformation is fairly similar to that described in the preceding section, i.e. contacting occurs by creep under constant load (to a first approximation); in this case, however, the applied load is reinforced by an interfacial attractive force. From these considerations, one might expect a curve of 'contact' or 'contact-recovery' (after removal of applied pressure) to be of similar form to the 'creep' and 'creep-recovery' curves of Fig. 2. Since the interfacial energy component of the total load is not removed, however, there will be a residual equilibrium amount of contact for the case of a viscoelastic solid, determined by the balance between elastic recovery and interfacial forces.

Unfortunately, measurements of true molecular contact are notoriously difficult to make, and the author is unaware of any studies made along these lines, relevant to pressure-sensitive bonding. In principle, however, it should be possible to predict the degree of contact with time for some 'regular' rough surfaces (say, a regular fibrous grid).

It will be noted that, in a tack test, a (debonding) load in the opposite direction is applied immediately after removal of the bonding load. Thus, residual elastic forces will initially assist the debonding operation, and clearly the bonding and debonding operations are not mutually independent. If, however, a long time is allowed to elapse after the removal of the contacting load (as is usual when a tape is used in practice), elastic recovery will initially reduce contact but then it may continue to increase if the adhesive is capable of steady-state flow. If the substrate is a porous one such as paper, the adhesive would eventually 'strike-through', visual observation of which has been used as a semi-quantitative test of adhesive flow behaviour.

On the other hand, an adhesive containing a continuous molecular network, i.e. a viscoelastic solid, can never penetrate more than very superficially into a porous substrate, unless applied loads are sufficient to rupture network bonds. It is possible that transient networks, such as those provided by hydrogen-bond cross-links, may be disrupted and reformed with application and removal of a load, and will therefore show exceptional behaviour.

It is important that an adhesive formulator is fully cognisant of these complex interactions, despite the fact that they can be discussed only in a qualitative manner. The necessity for compromise between the high creep compliance requirement for rapid attainment of contact and constant or near constant long time compliance for high shear strength (holding power) is clear. Of course, the problem is largely eliminated if a chemical or physical change can be made to occur after bonding, as in chemically curing a liquid rubber or evaporation of solvent from a rubber cement. However, it is in the nature of a pressure-sensitive adhesive that no such transition or 'setting' can occur, and its performance depends on a very delicate balance of physical properties.

THE PEEL BEHAVIOUR OF PRESSURE-SENSITIVE ADHESIVES

Peel adhesion is expressed in terms of the force required to peel a standard width of tape from a rigid substrate with which it has been allowed to reach an equilibrium state of bonding. Industrial standards specify glass or metal substrates, prepared and cleaned by standard procedures, and specify pulling rates (e.g. 12 in/min), temperature, humidity and peel angle (either 90° or 180°), as well as conditions for preparing the joint. However, such tests are intended for routine control and comparative purposes and give little useful information about the behaviour of the adhesive. There have been many reported investigations of peel adhesion in which test variables (angle, rate, temperature) or joint variables (dimensions, materials, interfacial conditions) have been studied, and the overall picture would seem to be very complex. Nevertheless, an attempt will be made to select some results which throw light on the function and behaviour of the adhesive during peeling and hence, hopefully, indicate some principles of formulation for high peel strength. For this the author refers to results from recent studies at the National College of Rubber Technology, some of which have been reviewed in rather more detail.[3]

First, it is possible to obtain an indication of the effect of variation of pulling rate, whilst keeping all other factors constant. This is illustrated in Fig. 4 for the joint system: glass (substrate)–polybutyl acrylate (adhesive)–polyester film (backing).[4] The tape dimensions are fixed and typical of industrial ones, and peeling is at 90° angle. The whole 'master-curve' was obtained from results at various rates and temperatures by application of the method of time–temperature superposition, and it should be borne in mind that most of the pulling rates as shown have been 'shifted' from the actual pulling rates (at other temperatures) and the peel force values have been 'reduced' to their room temperature values from the actual values measured (at other temperatures). In view of the dubious validity of the latter reduction (which is based on the assumption that modulus is proportional to absolute temperature) too much reliance should not be placed on the actual magnitude of peel force; however, the whole curve may be accepted with fair confidence as showing the trends in peel force and changes in mode of peeling at room temperature as pulling rate is varied. Experimental data points have been omitted in the interests of clarity.

Figure 4 shows three regions of steady peeling, labelled A, B, and C. At the transition between regions A and B, there is a narrow zone (labelled

A/B) in which peel force oscillates regularly between well-defined limits (shown by dashed lines in Fig. 4). Between regions B and C, there is a broader zone of oscillatory peeling (labelled B/C) which embraces about four decades of rate. Such oscillatory peeling behaviour is commonly referred to as 'slip-stick' peeling.

Further study relating to this master-curve[3,4] shows that in region A, where force is strongly dependent on rate and separation involves much

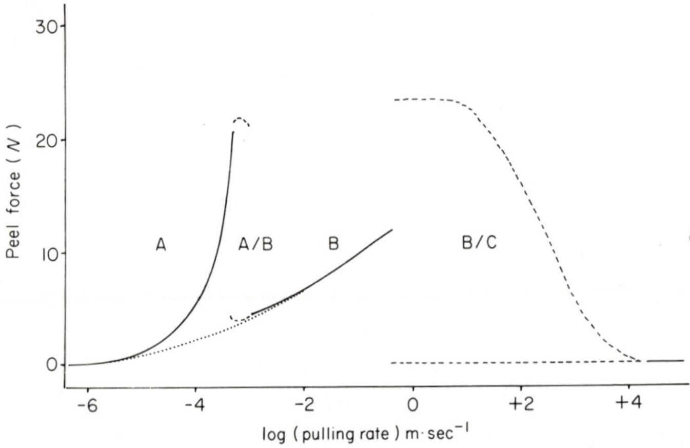

FIG. 4. Peel force versus (log)pulling rate at room temperature (296 °K) for the system glass substrate–polybutyl acrylate adhesive–polyester film backing (90 ° peel angle, joint width 25·4 mm, backing thickness 0·0025 mm, adhesive coating weight $26 \pm 1\,\mathrm{g\,m^{-2}}$). This graph results from time–temperature superposition (see text). The graph is adapted from results by Sheikh.[4]

filamentation within the adhesive, the predominant response of the adhesive is one of viscous flow. This behaviour is eliminated by light cross-linking of the adhesive, and region B then extends throughout the lower rates (shown by a dotted line in Fig. 4). This is analogous to the disappearance of steady-state shearing flow during creep when a network is introduced (discussed in an earlier section). In region B, where peel force is less rate-dependent and filamentation is of microscopic size, separation occurs at the adhesive–glass interface and the adhesive exhibits predominantly rubbery behaviour. This 'rubbery peeling' is that which is desirable and most often observed in the peeling of pressure-sensitive tapes. In region C, peel force is very low and independent of rate; there is no filamentation, separation is at the backing–adhesive interface and the adhesive is behaving as a glass. Slip-stick peeling behaviour (zones A/B and B/C) involves regular alternation in mode of peeling between those modes at immediately lower and higher rates, and is associated with the alternate storage and dissipation of elastic energy in the stretched backing member.

Thus, in peeling at a fixed temperature, the response of the adhesive changes from viscous through rubbery to vitreous with increase in pulling rate, and the mode of peeling changes accordingly. Conversely, if pulling

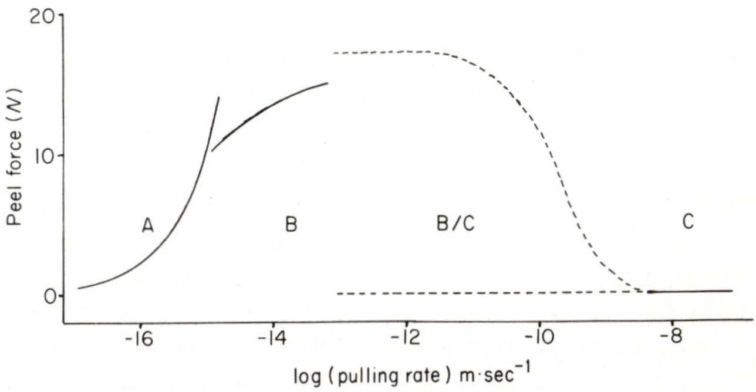

FIG. 5. Peel force versus (log)pulling rate at room temperature (296 °K) for the system glass substrate–polyvinyl acetate adhesive–polyester film backing (peeling conditions, joint dimensions and ancillary notes as for Fig. 4).

rate were fixed and temperature increased, the same changes would be observed (although in inverse order); such a graph could readily be constructed from the data for Fig. 4.

Now, referring particularly to region B as representing the desirable 'rubbery' or 'pressure-sensitive' type of peeling, it is necessary to examine the factors which determine the position of this region on the rate axis, and the magnitude of peel force for a given rate in the region.

By way of illustration, a complete change of polymer from polybutyl acrylate to polyvinyl acetate changes the master-curve to that shown in Fig. 5.[4] All other factors (dimensions, materials, peeling conditions) have been kept constant.

Reservations concerning the acceptability of Fig. 5 apply in the same way as those discussed earlier in connection with Fig. 4. However, if we may ignore minor differences between the two master-curves, the trend from viscous (region A) through rubbery (region B) to glassy (region C) peeling is again apparent. Examination of the abscissa of Fig. 5, however, shows that, compared to Fig. 4, this whole curve is displaced by some twelve decades to the left. The best reference point for determination of this displacement would be the rate for onset of 'glassy' peeling (since molecular weight variation will have minimal effect at this point), which will clearly correlate (inversely) with the glass transition temperature, T_g, of the polymer as normally determined. Since the 'slip-stick' peeling region B-C always (of the dozen or so systems examined) covers approximately 4·5 decades of rate, we may say that the position of the high-rate end of region B is determined approximately by the adhesive T_g. Thus, a rise in adhesive T_g will cause the region of desirable 'rubbery' peeling (region B) to move to the left and, conversely, a lowering of T_g will move it to the right. Clearly, polyvinyl acetate homopolymer cannot show useful pressure-sensitive behaviour (except at elevated temperature) because this displacement is too extreme. However, it will be appreciated that a slight lateral movement of region B (as resulting from a slight change in adhesive T_g) may produce a significant change in peel force for given pulling rate. In order to illustrate this, some peel force–rate curves for 'rubbery' peeling only have been drawn in Fig. 6; again, joint dimensions and composition are constant except for changes in the adhesive component.[3–5]

If we consider a pulling rate of 1 cm sec^{-1} (i.e. -2 on the log m sec^{-1} scale) as representing a typical speed of stripping of a tape from a surface, it is evident from inspection of Fig. 6 that peel force increases in the order NR → PBA → PVEE ≈ NR/Resin. The VA/BA copolymer adhesive shows 'slip-stick' behaviour, not 'rubbery' peeling at this rate. These differences are certainly due mainly to the differences in viscoelastic state, as revealed by T_g differences. It would appear, therefore, that in order to obtain a high peel force with 'rubbery' peeling at or near the 1 cm sec^{-1} rate, NR and PBA require to have their T_g values raised to shift the curves to the left. One effect of the addition of tackifier resin is to progressively raise T_g,[5] and it is seen that the presence of 40 % terpene resin in a NR adhesive is sufficient to give desirable peeling characteristics. The T_g of PBA might alternatively be raised by copolymerisation with, for example, a little acrylic acid or vinyl acetate, as in many typical commercial systems. In order to utilise a polymer which has too high a T_g (e.g. the VA/BA copolymer of Fig. 6), a plasticiser such as a dialkyl phthalate may be added to move the curve to the right; again, such systems are often used commercially as pressure-sensitive adhesives. PVEE apparently needs no modification, and indeed is often used as the simple homopolymer.

The lateral displacement of region B along the rate abscissa is not, however, the only factor which determines the magnitude of peel force for a given rate. Also of great importance will be the interfacial attractive forces

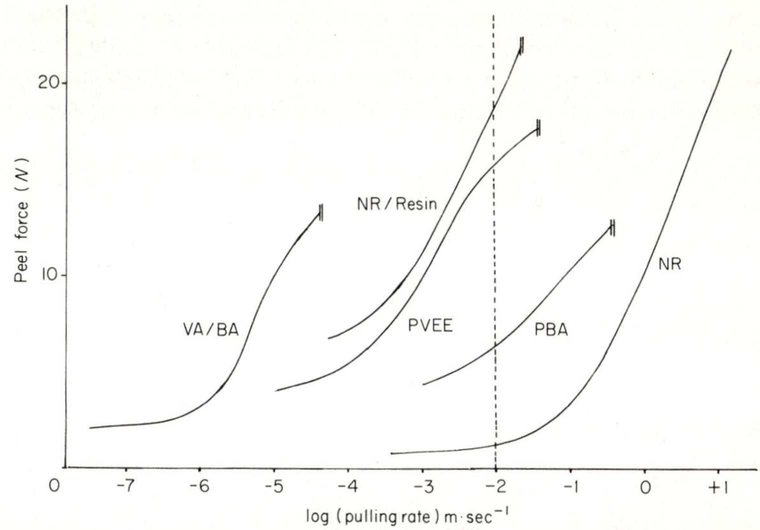

FIG. 6. Peel force versus (log)pulling rate at room temperature (296 °K) for various adhesives in the same joint system, and showing only the region of 'rubbery' peeling. ‖ = Region B ends. Peeling conditions and joint dimensions as for Fig. 4. Some of data for the curves are from time–temperature superposition. Drawn from results by Sheikh[4] and Sherriff.[5] VA/BA = 50/50 w/w vinyl acetate/butyl acrylate copolymer; NR/Resin = 60/40 w/w NR/®'Piccolyte S115'; PVEE = Polyvinyl ethyl ether; PBA = Polybutyl acrylate; NR = Natural rubber; ®'Piccolyte' is a registered trade name owned by Pennsylvania Industrial Chemical Corporation.

between the adhesive and the substrate, which will influence the vertical height of the curve, rather than its horizontal position. In order to isolate this effect, the nature of the *bulk* adhesive must be kept constant whilst only *interfacial* conditions are varied. This can be done by surface modification of either the substrate or the adhesive, and its effect is illustrated from some actual results in Fig. 7. Again, for Fig. 7 the joint materials and dimensions are kept constant (the same as for Fig. 4), and only interfacial conditions are changed.

It may be seen from Fig. 7 that the introduction of carboxylic acid groups in the surface only of a PBA adhesive progressively increases the force of peeling from glass for a given pulling rate in the region of 'rubbery' peeling (region B). Extensive testing by Ginosatis[6] showed that there was no lateral displacement of the curves (i.e. the surface carboxyl groups have not affected the bulk viscoelastic response of the adhesive), and the vertical shifting with carboxylation was attributed to increased interfacial attractions at the adhesive/glass junction. Ginosatis was further able to distinguish clearly between bulk effects (lateral shifting) and interfacial effects (vertical shifting) arising when a PBA adhesive was replaced entirely by a carboxylated PBA adhesive.

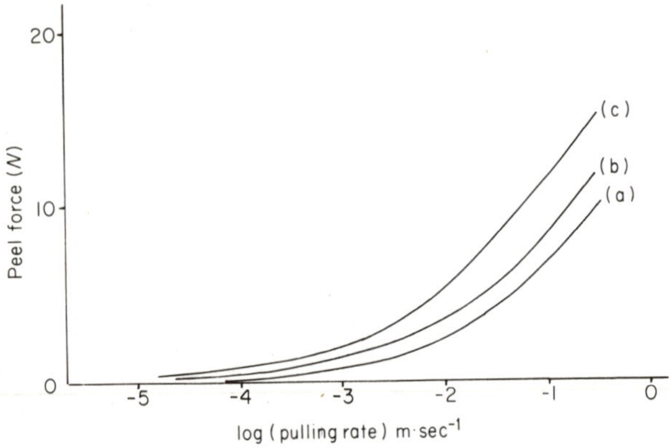

FIG. 7. Peel force versus (log)pulling rate at room temperature (296 °K) for a joint with adhesive comprising (a) PBA homopolymer (b) PBA homopolymer with a 1 % carboxylated surface and (c) PBA homopolymer with a 10 % carboxylated surface. The '1 % and 10 % carboxylated surfaces' were obtained by very thin supercoating of the PBA homopolymer with a 99/1 and a 90/10 w/w butyl acrylate/acrylic acid copolymer, respectively. Peeling conditions, joint composition and dimensions as for Fig. 4, except that the adhesive is lightly cross-linked. Some of the data for the curves are from time–temperature superposition. Drawn from results obtained by Ginosatis.[6]

It will be apparent from the foregoing considerations that, in formulating adhesives with peel adhesion in mind, there are two principal factors to be considered. One of these is that the adhesive must exhibit the correct bulk viscoelastic behaviour; to a first approximation this is determined by the T_g of the adhesive and, for the best results (judging by typical commercial tape formulations) this should be in the region of 230–235 °K. The formulator will often need to adjust viscoelastic properties by addition of tackifier or plasticiser, or by copolymerisation, as discussed earlier. (It is not intended to imply that the only function of added tackifier or plasticiser is to modify peel adhesion; in changing the distribution of molecular sizes they also have an important effect in increasing compliance at time scales appropriate to the bonding operation, as discussed in an earlier section.)

The second factor, that of securing high interfacial attractive forces, is much less amenable to control by the formulator, since he often has no knowledge of the substrates to be used in service. However, the introduction of a few strongly polar (e.g. –C≡N) or hydrogen-bonding (e.g. –COOH, –$CONH_2$) substituent groups can have a profound influence in promoting adhesion to many substrates. The precise mode of action of such groups will not normally be obvious, however, since as discussed earlier they also have an effect on bulk viscoelastic behaviour.

MATERIALS AND METHODS OF MANUFACTURE

It will have become apparent that, in formulating a pressure-sensitive adhesive, a formulator will need to strike a balance of viscoelastic properties which will give a suitable compromise between the various technical aspects of behaviour. The balance will involve consideration of viscoelastic response at very long times (for shear strength), at intermediate times (for a rapid bond-forming process) and at very short times (for peel adhesion). However, technical considerations cannot be isolated from economic ones; the latter are equally important, since it is clearly the formulator's main task to obtain the desired technical properties as cheaply as possible. Thus, the formulation will also be influenced by the relative prices of suitable materials and by the manufacturing process, insofar as the latter can be made more economic by adjustment of the formulation. The following sections are intended to illustrate some of the factors which need to be considered.

Materials

Little has been said so far about the chemical nature of the materials used for pressure-sensitive adhesives, although it will be apparent that the required viscoelastic properties impose some limitation on the range of materials which can be used.

The conventional approach has been to take an elastomer, particularly natural rubber (NR) and to adjust its viscoelastic behaviour by the addition of suitable amounts of a 'tackifier resin' and sometimes other materials. Such a system is probably still representative of the bulk of pressure-sensitive adhesives used today. The tackifier resins have the special property of a high T_g at very low molecular weight, and are usually cycloaliphatic materials of highly-condensed ring structure. They may be considered as falling into three main groups: rosin esters and modified rosin esters (e.g. the pentaerythritol ester of hydrogenated rosin, available as ®Pentalyn H); polymerised α- and β-pinenes (the latter unfortunately now in short supply, formerly available in quantity as, for example, ®Piccolyte S115); synthetic hydrocarbon resins obtained by polymerisation of unsaturated petroleum fractions (e.g. those available more recently under the names ®Hercules HC Resin, ®Wing-tack, ®Escorez, and ®Sta-Tac). As might be expected, performance as tackifiers varies with price and, quite often, mixtures of different resins are used to obtain the best technical/economic compromise. Adhesives based on blends of NR with such resins may be formulated to show an excellent combination of shear strength, tack and peel adhesion for room-temperature applications, without the necessity for cross-linking the rubber. Where 'rubbery' clean peeling at elevated temperature is required, as in making tapes for paint stoving applications, the rubber will normally be lightly cross-linked (using either a phenol-formaldehyde resin or a sulphur/accelerator curing system) by subjecting the tape to a brief curing cycle during manufacture. The effects and uses of such crosslinking have been discussed in detail by Butler.[7]

In general, however, the same useful combination of properties cannot be obtained by replacing the NR with other hydrocarbon rubbers (e.g. styrene–butadiene rubber (SBR), butadiene rubber (BR), ethylene–propylene rubber (EPM or EPDM), butyl rubber (IIR)); this leads one to suspect that there is some hitherto-unrecognised feature of NR which is responsible for its unique performance in adhesives. The latter could be accounted for by a momentary occurrence of strain-induced crystallisation during peeling, which would prevent the 'cohesive' (flow) mode of failure, commonly observed if synthetic rubbers are used; however, the author is not aware of any experimental evidence to support or refute this hypothesis.

Synthetic isoprene rubber (IR) may be used to replace part or all of the natural rubber; but the only advantage it appears to offer is one of lower coloration; it is certainly (from the author's experience) more difficult to

® Pentalyn is a Registered Trade Name owned by Hercules Powder Company.
® Piccolyte is a Registered Trade Name owned by Pennsylvania Industrial Chemical Corporation.
® Wing-tack is a Registered Trade Name owned by Goodyear Company, Chemicals Division.
® Escorez is a Registered Trade Name owned by Standard Oil Company.
® Sta-Tac is a Registered Trade Name owned by Reichold Chemicals, Incorporated.

obtain adequate shear strength from the currently-available grades of IR. Possible reasons are a narrower molecular weight distribution in IR compared with NR, and/or a reduced propensity for strain-induced crystallisation in the IR, due to its lower *cis*-1,4 content.

The principal disadvantage of NR-based adhesives and, indeed, of all adhesives based on highly-unsaturated rubbers is the poor ageing behaviour; this is so despite the presence of the commonly used non-discolouring, phenolic antioxidants. Although a reel of tape may show adequate storage stability, in service applications the adhesive may be exposed much more directly to oxygen, heat and sunlight, all of which will accelerate oxidation with resultant softening and discolouration of the adhesive.

Where a long useful service life and absence of coloration are essential, a manufacturer must look to some of the synthetic saturated, elastomeric materials. Blends of different molecular weight grades of polyisobutylenes or of polyvinyl alkyl ethers have been used, and are still used to a limited extent. Despite their good ageing behaviour however, these materials are of very linear structure and tend to give poor shear strength; moreover, it is not possible to cross-link them by any of the conventional methods. Their versatility is further restricted by the fact that they can be copolymerised with relatively few other monomers. Undoubtedly, the greatest success has been achieved using polymers derived from esters of acrylic acid, particularly *n*-butyl- and isooctyl- (e.g. 2-ethyl-hexyl-) acrylates. Such polymers, being made by free-radical polymerisation, may be varied in composition by copolymerisation with a wide range of other monomers; they show extensive branching and hence better shear strength, and offer easy cross-linking by several different mechanisms. However, since this type of polymer is specially prepared for use in pressure-sensitive adhesives, formulation is usually out of the hands of the tape or product manufacturer and is the responsibility of the polymer producer.

Reference to Fig. 6 and the associated discussion shows that the simple homopolymer of *n*-butyl acrylate requires chemical modification (in a form which raises its T_g) in order that peel strength at normal test rates ($0{\cdot}1$–$1{\cdot}0\ \mathrm{cm\ sec^{-1}}$) is reasonably high. The homopolymer of 2-ethyl-hexyl acrylate would also require modification in the same direction. Modification is most economically accomplished by copolymerisation with vinyl acetate, (of course, one could equally say that polyvinyl acetate has been modified in a way which reduces its T_g, but for the present purpose we are regarding the polymers as being based on the acrylate polymer); there are currently available many copolymers of alkyl (usually octyl) acrylates and vinyl acetate. Another common method of modification is by copolymerisation of the acrylate with a small amount (say 5 %) of a carboxylic acid monomer (e.g. acrylic, methacrylic, maleic, itaconic acids); this system has been used extensively by at least one major manufacturer, and offers certain specific advantages (discussed earlier). Quite commonly, however, commercial adhesives contain acrylate, vinyl acetate and carboxylic monomer.

Although the performance of uncross-linked acrylate-based polymers may be perfectly adequate for some applications (e.g. decorative or protective sheet products), cross-linking becomes necessary to improve shear strength in the more demanding applications requiring some degree of load-bearing ability. There are an almost overwhelming number of methods of cross-linking these polymers, described in the patent literature. Cross-linking of simple acrylate-based polymers may be accomplished very easily using free radical reagents; thus, gelation occurs in a few minutes by heating with a little (say 2 %) benzoyl peroxide at 100 °C; alternatively, a short exposure to u.v. light, in the presence of a sensitiser such as benzophenone, will often suffice. Such processes, however, require a special curing operation, generally frowned upon by manufacturers. Much effort has been devoted to the development of more rapid low-temperature curing systems, initiated at the time of coating, and often allowed to proceed to completion *in situ* within a few days storage of the product at room temperature. Such curing systems almost invariably require the incorporation of small amounts of special functional groups into the polymer; among the types of functionality which have been introduced for this purpose are:

(i) Acid or anhydride groups (e.g. from acrylic acid (I) or maleic anhydride (II)). These may be used in reactions with bases to give ionic cross-links, in reactions (suitably-catalysed) with polyalcohols to give ester cross-links, or in catalysed reactions with epoxy resins to give ester cross-links.

(ii) Amide groups (e.g. from acrylamide (III) or methylolacrylamide (IV)). These would normally be used for cross-linking by interaction with formaldehyde donors (urea or melamine–formaldehyde condensates). With suitable acid catalysis these reactions can be made to occur quite rapidly at room temperature.

(iii) Hydroxyl groups (e.g. from 2-hydroxyethyl methacrylate (V)). The presence of reactive primary hydroxyl groups enables rapid cross-linking to occur via their reaction with a polyisocyanate (e.g. 4,4′-diphenylmethane diisocyanate), in the presence of tin compounds or tertiary amines as catalysts. Hydroxyl groups may also be reactive in systems involving formaldehyde donors ((ii) above).

(iv) Epoxy groups (usually from glycidyl methacrylate (VI)). Primary and secondary aliphatic polyamines crosslink such polymers rapidly (as in epoxy-resin technology); anhydrides or carboxylic acids may also be used, but are slower and heat may be required. It is sometimes convenient to incorporate carboxylic and epoxy groups in the same polymer which, after suitable catalysis, is then self cross-linking.

I

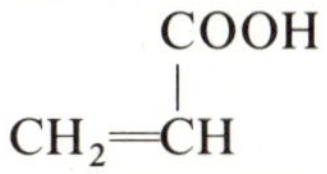

acrylic acid

II

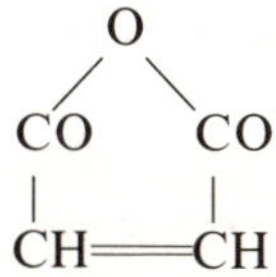

maleic anhydride

III

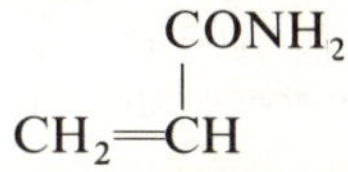

acrylamide

IV

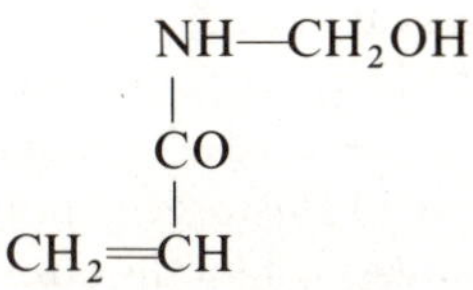

N-methylol acrylamide

V

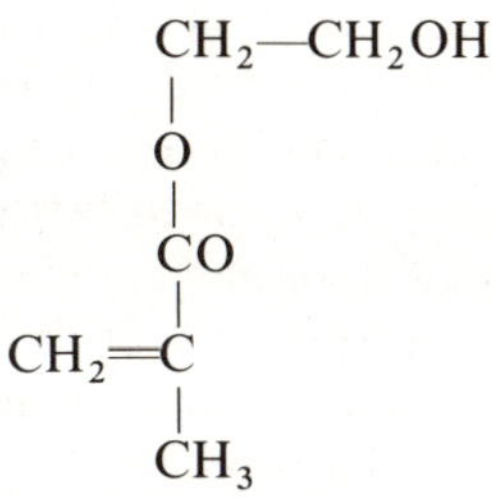

2-hydroxyethyl methacrylate

VI

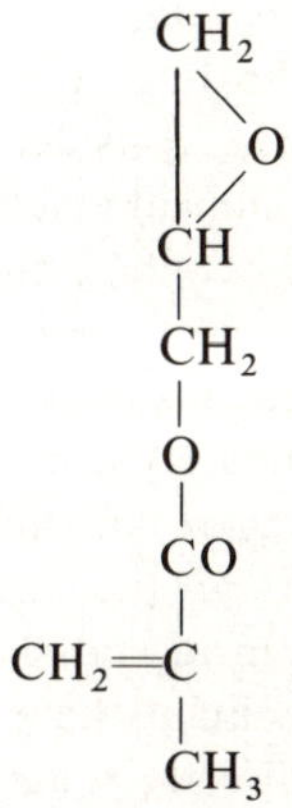

glycidyl methacrylate

A severe limitation of these low-temperature curing systems is that cure rate reduces very gradually as the concentration of functional groups diminishes, and crosslinking will still be progressing during quite long periods after manufacture. This can produce noticeable changes in properties (tack, peel adhesion) over periods of weeks or even months in the final product.

Although acrylate-based polymers show excellent ageing behaviour, and their resistance to degradation at high temperature equals that of almost any other organic polymer, there are occasions when the heat stability of the polyacrylates is inadequate. Thus in adhesives for continuous exposure to temperatures of 150–200 °C (e.g. in high-temperature electrical tapes), there is little alternative but to select a (semi-inorganic) silicone adhesive. Very little is disclosed by the manufacturers about the chemical nature of these adhesives but, like the polyacrylates, they would normally be cross-linked after coating. They are said to consist essentially of mixtures of silicone rubbers and silicone resins.

Solution versus Emulsion Processing

The choice of method of processing to be used is important to the formulator in that it may impose restrictions on the range of materials which can be used.

The most successful method of manufacture, over a period of many years, has been to coat the backing or support member with a solution of the adhesive in a volatile organic solvent. The solvent is evaporated off under moderate heating and ventilation and, if justified by economic or environmental considerations, may be recovered and re-used. Although excellent technical results may be achieved by the use of this process, the disadvantages of high cost of lost solvent (or of its recovery), flame hazards, and environmental pollution are obvious. The deposition of adhesive films from an aqueous dispersion would clearly eliminate these disadvantages, but unfortunately it also introduces others.

Whereas there is no difficulty in obtaining rubbers such as NR and SBR in the form of latex (i.e. aqueous dispersion), the dispersion of tackifier resins into the form of stable aqueous emulsions presents problems. The more desirable tackifiers have melting points $> 100\,°C$ and cannot therefore be emulsified directly by agitation with soaps unless suitable pressure vessels are available. Tackifiers softened with solvents may be emulsified, but this to some extent defeats the object of the process. Dispersion by colloid-milling of solid tackifier resin is a difficult and not economically-viable process. Even when a tackifier has been apparently well dispersed and suitably blended with the rubber latex, it must be appreciated that, in order to obtain a uniform homogeneous film, the tackifier and rubber particles must mutually dissolve each other during drying, a process which is by no means easy to achieve. Because of this, there are difficulties in obtaining reproducible pressure-sensitive behaviour, and difficulties in eliminating haze in cast films. Other difficulties arise because: water softens the (commonly-used) cellulose backing film; adhesive properties are adversely affected by excessive use of soaps and other dispersing aids; latices need to be protected against coagulation; and water, despite its cheapness, consumes considerable thermal energy and time for its elimination.

Of the synthetic polymers in general, only those made by free-radical polymerisation can be obtained economically in the form of latex or 'emulsion'. Thus, emulsion polymerisation is a very practicable proposition for the acrylate-based polymers and, since (in principle) no further treatment of the emulsion is necessary prior to coating, the use of such emulsions offers considerable potential economic advantages in manufacture. Emulsions of this type clearly overcome many of the disadvantages associated with rubber/resin systems (above). Simple emulsions of acrylate/vinyl acetate copolymers are indeed extensively used in non-critical applications such as labels and decorative vinyl sheet material. However, subsequent cross-linking of emulsion polymers cannot be achieved as readily as solution ones. Since it is necessary to allow time for full coalescence of particles before cross-linking begins, it is, of course, no good cross-linking the polymer whilst still in emulsion form, because subsequent particle coalescence cannot then occur. The main symptom of lack of particle coalescence is a low shear strength. A further disadvantage of emulsion adhesives arises from the lack of wetting of siliconised paper (on which they need to be coated uniformly for manufacture by the 'transfer' process); this is usually overcome by a drastic increase in viscosity (by adding a thickener such as polyacrylic acid), so that the rate of 'de-wetting' of the initially-formed film is less than the rate of coagulation by drying of the emulsion.

The emulsion method of processing must be used for the coating of paper with a so-called 'delayed-tack' adhesive. A mixture of emulsions of the polymer (e.g. polybutyl methacrylate or butylacrylate–vinyl acetate copolymer) and a crystalline plasticiser (usually dicyclohexyl phthalate) is evaporated at low temperature, such that discrete particles of the polymer and crystalline plasticiser are deposited on the paper to form a coating completely free of tack. In this condition the paper may be printed and cut into the form of labels. The adhesive is 'activated' in preparation for bonding by heating to about 65 °C, when the plasticiser melts and mixes with the polymer to give 'pressure-sensitive' behaviour. After cooling, the plasticiser gradually separates from the polymer and crystallises, with the disappearance of tack (although the morphology of the film is not the same as in its original condition).

Solventless Reactive Systems

There are clearly even greater economic advantages to be achieved if the use of both solvent and water as coating media can be dispensed with. This is the aim of much intensive development work currently proceeding in the pressure-sensitive products industries; the approaches used are to apply the adhesive to the substrate either as a reactive liquid or as a hot melt (dealt with in the next section).

It is convenient to divide the approaches to convertible reactive systems into two types; first, systems involving the polymerisation of monomers *in situ* on the backing member (i.e. the conversion involves a chain reaction) and, secondly, those where molecules of a low molecular weight intermediate polymer are linked together (this usually involving 'stepwise' reactions). The latter process is often referred to as 'chain-extension' to distinguish it from polymerisation.

Some of the difficulties emerge clearly as soon as one attempts to polymerise a monomer (say 2-ethyl-hexyl acrylate) supported in the form of a thin layer on a backing film. Apart from the difficulties in metering and maintaining the monomer as a thin film, it is necessary to provide an oxygen-free atmosphere to avoid inhibition of polymerisation. The generation of free radicals by a thermal process (e.g. decomposition of a peroxide) is undesirable due to the appreciable volatility of the monomer, and may necessitate the recovery of volatilised monomer. The low temperature generation of free radicals by u.v. or high-energy radiation is a more promising proposition, both types of system having been extensively covered by patents. It will be appreciated that both bulk polymerisation and cross-linking reactions will be proceeding simultaneously in such systems, the former predominating in the early stages (when there is much monomer and little polymer) and the latter predominating in the later stages (when there is much polymer and little monomer). The possibility of close control over the structures produced in such processes is clearly very limited. There will probably always be some free residual monomer (unless it is specially removed) which, apart from its effect on properties, will leave an unacceptable odour. Monomer/polymer syrups would normally be used for their ease of handling, but since the polymer will probably have been isolated from a separate emulsion or solution polymerisation, some of the advantage is lost. Some of the disadvantages due to the high volatility of the monomer may be overcome by chemically combining polymerisable vinyl or isopropenyl groups with a low molecular weight (liquid) polymer, but, as yet, such products have not been made cheaply enough to justify their extensive use in this application.

A more promising approach (in the opinion of the author) is to take a low molecular weight (viscous liquid) polymer containing functional groups, and link together molecules, as necessary, by a stepwise (non-chain) process. This type of reaction, referred to as 'chain-extension' if functional groups are terminal or 'cross-linking' if they are lateral, is much more amenable to control in a bulk process.

It is clear that it does not differ fundamentally from the process of cross-linking after conventional solution processing (discussed earlier), the only change being that, since the polymer now has much lower molecular weight to start with, more functional groups need to be used. Thus, all of the types of functionality (formulae I → VI) described for use in crosslinking the polyacrylates can, in theory at least, be used for the conversion, *in situ*, of a liquid polyacrylate into a pressure-sensitive adhesive. Again, such systems have been the subject of numerous patents but, so far, have not been developed commercially on a large scale, presumably because of the expense of specially preparing the low MW functional prepolymers. Liquid functional polymer intermediates are manufactured on a large scale, however, for other applications, notably polyurethanes. Particularly attractive from an economic point of view are the liquid polyether polyols, which may readily be chain-extended by treatment with a diisocyanate. With suitable attention to the requirements (molecular weight distribution, T_g, etc.) specified earlier in this review, there seems to be no obvious reason why good pressure-sensitive behaviour should not be obtained from chain-extended polyols based on poly(propylene oxide) or poly(tetrahydrofuran). Indeed, judging by the recent voluminous patent literature on such compositions, the possibility of further development must be at least promising.

Hot Melt Systems

Ideally, a 'hot-melt' pressure-sensitive adhesive is one which, when heated, becomes a sufficiently mobile liquid to allow spreading on to a backing by conventional techniques (e.g. by a 'doctor' blade or 'curtain'-coating process). However, the term is often extended to cover the extrusion of heated adhesives under moderate pressure; clearly, most uncross-linked adhesives would be described as 'hot-melt' if extrusion conditions could be varied without limit. In order to make the process potentially viable as an alternative to solution or emulsion processing, however, it is desirable to have a high degree of fluidity in the heated adhesive. Thus, the adhesive is required to show pressure-sensitive behaviour, with high shear strength, at room temperature, changing to a fluid which flows readily at processing temperature (say, 150–200 °C).

These requirements certainly cannot be met by a conventional simple, homogeneous pressure-sensitive adhesive. A typical general purpose rubber-resin or polyacrylate adhesive is far from being sufficiently fluid at hot-melt processing temperatures; if the molecular weight is reduced sufficiently to provide high temperature fluidity, shear strength at room temperature becomes completely inadequate. In order to accommodate both the low and the high temperature requirements in an adhesive, it is necessary to pass through some kind of physical *transition* at intermediate temperatures. Below the transition temperature range, part of the adhesive should behave as if it were of very high or infinite molecular weight, but above the transition it should behave as low molecular weight material. In making hot-melt pressure-sensitive adhesives, both the crystalline-amorphous transition (melting temperature T_m) and the glass transition (T_g) have been used; in these cases, a separate crystalline or glassy dispersed phase serves to strengthen the viscoelastic continuous phase at low temperature, but not at high temperatures. The principles of action of the two types of hot-melt adhesive are quite similar, as will be apparent from the following description.

Hot-melt adhesives working on the crystallising principle almost always require long regular sequences of methylene groups which separate into a waxy crystalline phase at low temperature, but melt when heated, to allow for easy flow; as such they may be regarded as developments of the familiar paraffinic waxes. Materials composed entirely of methylene groups, however, like microcrystalline waxes or polythene, display too high a level of (rigid) crystallinity and leave no room for the desirable pressure-sensitive viscoelastic behaviour; the proportion of crystalline material, which has no useful function in the pressure-sensitive sense, must therefore be reduced to a minimum. An idealised approach would be to chemically combine some methylene sequences (sufficiently long to crystallise) with an amorphous, low MW viscoelastic material. The crystallites formed at room temperature would then act to increase the effective molecular weight, as may be illustrated diagrammatically in Fig. 8.

If the overall proportion of crystalline material were sufficiently small, the adhesive could show useful pressure-sensitive behaviour; on heating, its effective molecular weight would be drastically reduced and it would become a relatively mobile liquid. Adhesives acting in this way have been made by copolymerising acrylates and related polymers with monomers having long aliphatic sequences (e.g. vinyl stearate, vinyl laurate) and at least some of these have been used on a commercial scale.

More common however, are the copolymers of ethylene in which the crystallising sequences are part of the main chain. Copolymerisation of ethylene with monomers such as vinyl acetate or ethyl acrylate interferes with structural regularity and reduces the number and lengths of

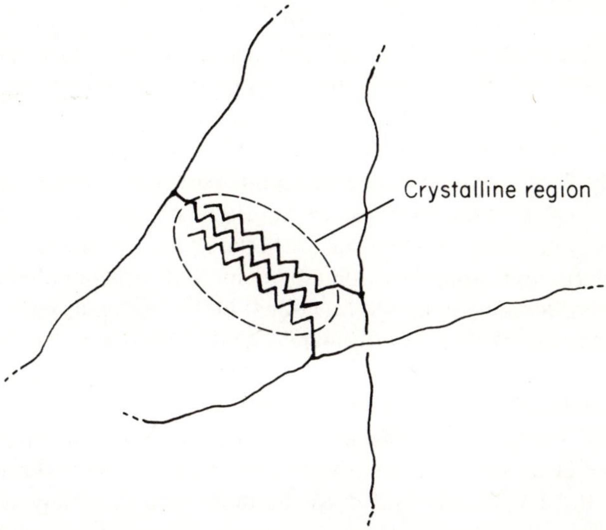

FIG. 8. Diagrammatic illustration of influence of crystallinity on physically-effective molecular weight in adhesives.

crystallising sequences. Thus, both the degree of crystallinity and the T_m range are reduced with increasing comonomer concentration. With this possibility of various monomer ratios, and also the possible variation in overall molecular weight, there are a great number of polymers of this type from which a formulator may choose. Since the acrylate or vinyl ester comonomer will not enter the crystalline regions, they will be important in determining pressure-sensitive behaviour. Adjustment of molecular weight and T_g of the non-crystallising phase will usually be necessary, by the addition of oils, tackifier resins and other materials. Many recommended formulae are to be found in manufacturers' literature, but these hot-melt polymers have not, in general, enjoyed the same success in pressure-sensitive adhesives which they have gained in other types of hot-melt adhesives. It may well be that the amorphous parts are too stiff to display acceptable pressure-sensitive behaviour; however, the possible emergence on a large scale of elastomeric block copolymers of ethylene with propylene (or other monomers) may hold out hope for more successful developments along these lines.

Somewhat similar in their mode of action are the non-crystallising 'thermoplastic' elastomers which have enjoyed rapid commercial development during the last few years. These are block copolymers which exist in the form of two phases at all temperatures; the dispersed phase is a glass at room temperature and softens to a rubber at 'thermoplastic' processing temperatures. Most appropriate for consideration in the present instance are the styrene–butadiene block copolymers, in which the styrene comprise about 25% by weight of the polymer. Provided that block lengths are sufficiently long, styrene sequences and butadiene sequences behave essentially as homopolymers; since polystyrene and polybutadiene are immiscible, phase separation occurs, with the styrene blocks being confined to sub-microscopic spherical domains dispersed in a butadiene matrix. These glassy domains fulfil two functions: first, they act as linkage points for the butadiene chains, thereby increasing the effective molecular weight; secondly, they are of such size and rigidity to act in the capacity of a reinforcing 'filler'.

Since an enormous volume of literature has built up around such materials during recent years, the reader intending to make a serious study of these is recommended to the more comprehensive reviews by Dawkins.[8]

These styrene–butadiene block copolymers are discussed in detail in Chapter 9. It will suffice here to draw attention to the simple two-block (SB) copolymers such as ®Solprene 1205† in addition to the SBS and radial, teleblock polymers mentioned in that chapter.

It will be clear that the formation of styrene domains in the 2-block copolymer cannot have the effect of network formation in the butadiene matrix, since one end of each butadiene sequence will always be free. Flow will therefore occur very readily in the butadiene matrix, even at ordinary temperatures; at elevated temperatures flow still occurs almost entirely in the butadiene phase, even through the polystyrene domains will have softened to a leathery or rubbery state. Easy flow at low temperature, however, is incompatible with the shear requirements for a pressure-sensitive adhesive.

† ®Solprene is a Registered Trade Name owned by Phillips Petroleum Company.

An SBS or radial block copolymer, however, will normally exist as a complete three-dimensional network at ordinary temperatures, since all the ends of the polybutadiene sequences will be attached to polystyrene domains. Thus, these materials behave as non-tacky elastic solids at room temperature; on heating to temperatures above the T_g of polystyrene ($\sim 368\,°K$) flow may occur, but in this case it must involve disruption of the polystyrene domains. For this reason the melt viscosity is much higher than that of the 2-block polymers (of the same molecular weight); the teleblock polymers have the advantage of a somewhat reduced melt viscosity compared with the 3-block SBS polymer.

Neither of the above two extremes of behaviour correspond with that required for a pressure-sensitive adhesive. In terms of molecular sizes and arrangements we clearly need something between the 2-block and 3-block (or teleblock) morphology; thus, for rapid bond-forming (and for 'tack') we require the high short-time compliance characteristic of the 2-block material, whereas for shear strength the presence of a tenuous network structure provided by 3-block material would be very desirable.

In addition to the blending of such materials so as to modify the distribution of molecular sizes and arrangements, the formulator must further consider the question of adjustment of the composition for satisfactory peel adhesion. Since the styrene blocks are confined to domains which (probably) do not suffer deformation during peeling, peeling behaviour will be determined mainly by the viscoelastic response of the ('filled') polybutadiene matrix. Since this is of low T_g ($\sim 183\,°K$), it will require considerable addition of tackifier resin or other material to produce a satisfactory viscoelastic response (*see* p. 140).

It will be apparent from such considerations that the formulae of hot-melt pressure-sensitive adhesives based on thermo-plastic elastomers will often be complex mixtures of polymers, resins, oils etc., and the precise function of each additive may sometimes be obscure. In addition to the polymer types mentioned above, other hydrocarbon rubbers (e.g. 'random' SBR, styrene-isoprene block copolymers) are often used.

Finally, thermoplastic elastomers are often used to economise on solvent consumption, rather than to eliminate it completely. Since molecular weights are very low compared with conventional rubbers, homogeneous solutions of high concentration may be used (providing that the solvent dissolves both blocks in the molecule); the heterophase morphology is then restored when solvent is evaporated. For this reason, thermoplastic elastomers are receiving increasing use as components of the more conventional rubber-resin solution-applied adhesives.

REFERENCES

1. Ferry, J. D. *Viscoelastic Properties of Polymers* 2nd edn., 1970, John Wiley and Sons, New York, pp. 11, 19–22, 147.
2. Vohralik, V. and Wake, W. C., Chapter 2 in: *Industrial Adhesives and Sealants*, ed. Jackson, B. S., 1976, Hutchinson–Benham, London.
3. Aubrey, D. W., *De Nederlandse Rubberindustrie* **2,** 1975, p. 1 (in English).
4. Sheikh, A. A., Ph.D. Thesis, (C.N.A.A.), 1973, National College of Rubber Technology, London.
5. Sherriff, M., Ph.D. Thesis, (C.N.A.A.), 1973, National College of Rubber Technology, London.
6. Ginosatis, S., M.Sc. Thesis, 1977, The City University, London.
7. Butler, G. L., *Aspects of Adhesion*—4, ed. Alner, D. J., 1968, University of London Press, London, p. 82.
8. Dawkins, J. V., chapters 8A and 8B in: *Block Copolymers* ed. Allport, D. C. and Janes, W. H., 1973, Applied Science Publishers, London.

Chapter 6

ACRYLIC ADHESIVES

F. R. Martin
Loctite Ltd, Tallaght, Eire

SUMMARY

Acrylic chemistry has been used for about 50 *years in the coatings industry. Recently this technology has been applied to develop pressure-sensitive adhesives with uniquely controlled properties such as tack, cohesive strength and water solubility. Growth in this new area has been partially responsible for an overall sales growth in acrylic adhesives and sealants of about* 20 % *per year from* 1963 *to* 1973.

The most recent developments in acrylic adhesives are the cyanoacrylates and the liquid acrylic adhesives (100 % *solids reactive fluids*). *Cyanoacrylates are almost ideal adhesives, forming strong bonds literally in seconds to a wide range of substrates, including rubber and plastics. Only limitations of cost, peel strength and moisture resistance have prevented these materials being used as general purpose structural adhesives. Liquid acrylic adhesives require activators but provide impact and peel strength and durable bonds to a variety of substrates. Speed of cure and ease of use make these new adhesives attractive alternatives to older adhesive technologies such as epoxies as well as mechanical fastening systems.*

INTRODUCTION

The versatility and scope of acrylic chemistry has long been exploited in the formulation of coatings and, more recently, of sealants and adhesives. According to Brief and Skeist[1] the use of acrylic resins in sealants and adhesives grew at a rate of 20% per year from 1963 to 1973. This growth is attributed by them primarily to new applications areas such as pressure sensitive adhesives and non-woven textiles. Similarly, the large growth rates in the use of phenolic and amino resins was primarily due to their use respectively in softwood plywood and particle board. During the same period the use of natural products such as dextrins, casein, and animal glue showed little or no increase. The substantial growth in the adhesives industry has therefore been in synthetic materials, where improved performance more than outweighs the cost of the materials used.

Present developments in acrylic adhesives are aimed at meeting the requirements of future applications. To define such targets better, studies such as the DuPont Delphi Forecast[2] have been carried out. While that study is essentially an opinion poll of people in the adhesives manufacturing and user industries, it does focus on one major new requirement or growth opportunity for new adhesive systems. In the next decade solvent based adhesives are expected to lose a large percentage of the total market they now enjoy, primarily because of increasingly severe legislation regulating the use of solvents. Water and air pollution and industrial safety, with the additional factor of increasing solvent costs make the search for alternative systems very worthwhile. Among the new systems possible are water based adhesives, hot melts, and the so-called 100% solids reactive fluids. Opportunities for new adhesives lie not only in replacing old systems such as natural products or solvent-based adhesives, but also in finding applications where adhesives have not previously been used. The driving force for the use of adhesives in such cases is essentially economic, whether the saving is in weight for aircraft or in a reduced reject rate for an assembly operation. Among the most important requirements that adhesives must meet to gain acceptance are:

Strong, durable bonds. Bond strength and durability sufficient to guarantee the bonded article for its useful lifetime are obvious prerequisites of any adhesive system. This is particularly important in new application areas where the user is unfamiliar with adhesives. Not all bonding applications, however, require the strength and durability of adhesive joints in aircraft and some compromise in bond strength to obtain improvements in other properties is usually desirable. With the increased use of plastics in manufactured goods the ability of an adhesive to bond dissimilar materials such as metals, ceramics, and plastics can also be more important than bond strength.

Ease of use. This includes simple storage and handling operations, the capability of being applied quickly and efficiently, and of being used effectively with little or no training or particular skills.

Speed of cure. In an assembly line operation speed of cure can be equivalent to valuable factory space either on the assembly line or as storage. If clamps or jigs are needed their number will be inversely proportional to the speed of cure. For very fast curing adhesives such devices can be eliminated.

Safety. Acceptance of adhesives in new application areas demands that they be capable of being used safely. Adhesive systems requiring a minimum of auxiliary equipment to ensure safe operation obviously have a considerable advantage over other systems, such as solvent based adhesives, that require ventilation or solvent equipment.

Low-energy requirement. Today the increased cost of energy is favouring adhesive systems that require little energy to form a bond. Oven curing, which heats not only the adhesive, but the whole bonded assembly is clearly at a disadvantage.

One of the key advantages of acrylic chemistry is that its range and versatility make possible a controlled balance of properties that can be adjusted to meet the requirements of different application areas. Recent developments in acrylic adhesives are excellent samples of such molecular design engineering.

ACRYLIC CHEMISTRY

Practical acrylic adhesive systems can be divided by the method by which the base monomers are polymerised. Virtually all proceed by free radical mechanisms, with one particularly important exception, the cyanoacrylates.

Cyanoacrylates

Cyanoacrylate monomers have the structure:

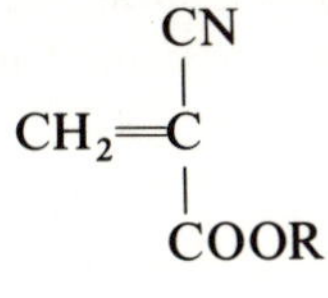

where R is an alkyl group.

Although cyanoacrylate monomer can be polymerised by a free radical mechanism it is the anionic mechanism which gives these materials such attractive properties as adhesives. The mechanism of polymerisation of α-cyanoacrylates, described in detail by Coover,[3] is shown below.

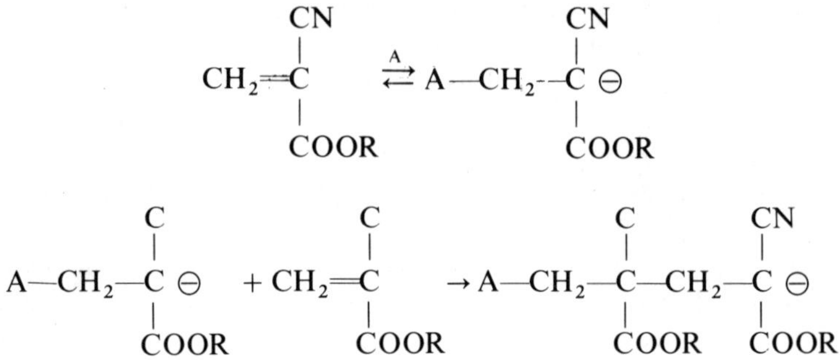

and so on.

The strongly electronegative character of the nitrile and carboxyl groups attached to the α-carbon make the monomer very susceptible to anionic polymerisation. This can be initiated by even weak bases (A^-) such as water. Properly formulated cyanoacrylate adhesives cure very rapidly in contact with the small amounts of moisture normally present on most surfaces.

The cured adhesives are uncrosslinked homopolymers with glass transition temperature (T_g) and surface bonding characteristics determined by the alkyl ester group. The cyanoacrylates are thus inherently (almost ideally) suited for use as adhesives as they are, but their properties are not easy to modify or control over a wide range. This is in direct contrast to the methacrylate and acrylate polymers.

Methacrylates and Acrylates

Methacrylic and acrylic polymers have been used for about 50 years in the coatings industry. As a result the versatile chemistry of these materials has been developed into a sophisticated technology, and the monomers on which the various polymers are based are readily available at low cost. (Table 1 shows some of the acrylic monomers available.)

These monomers are primarily esters of methacrylic and acrylic acid:

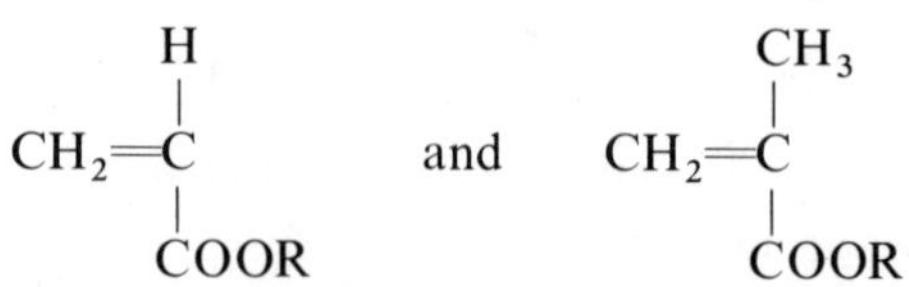

where R is an alkyl or other functional group.

Brindley, Calder, and Wetzel[5] summarise a number of commercial means of synthesising these materials.

The polymers are made from the monomers by free radical polymerisation using either bulk, solvent, or emulsion techniques. Riddle[6] and Leonard[7] give detailed descriptions of the various processes involved. The resulting polymers have the following structure:

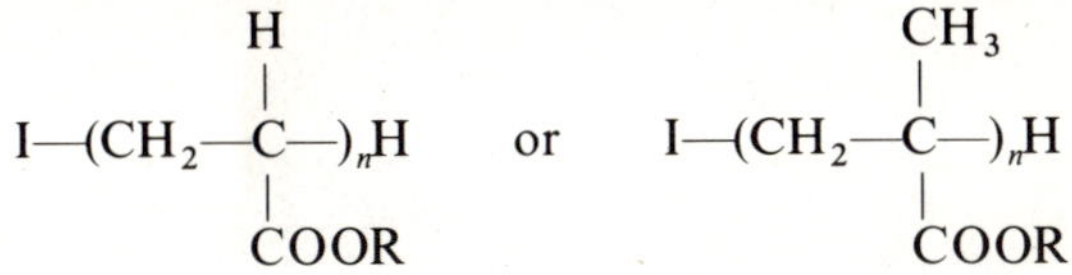

where I is the free radical initiator, usually a peroxide. The mechanical properties of the polymer are strongly affected by whether or not it is an acrylate or a methacrylate and by the size of the alkyl side chain R. Table 2 shows the glass transition temperature, T_g, for a number of acrylate and methacrylate polymers.[6] Methacrylates, for a given ester group have a higher T_g than acrylates. The value of T_g in Table 1, assume polymerisation results in very high molecular weights. If low molecular weights are obtained, for example by the use of chain transfer agents, lower T_g values can result. The T_g determines if the polymer is hard and brittle with a high useful temperature limit or soft, extensible, and rubberlike at normal temperatures. The methacrylate and acrylate monomers can be copolymerised with each other or even other vinyl monomers to give predictable T_g values, which greatly extends the already large number which can be obtained from the many monomers available.

A second set of polymer properties of importance to adhesives such as tack, adhesion, and surface compatibility, are difficult to define briefly yet unambiguously, but can be varied by the addition of monomers other than alkyl esters into the interpolymers. Acrylic and methacrylic acids, hydroxy alkyl acrylates and methacrylates, and acrylimides and methacrylamides can be polymerised with acrylate and methacrylate esters to vary adhesion to different substrates.

Solubility of acrylate and methacrylate interpolymers may be varied by the addition of carboxylic acid groups or alkylene oxide esters to make possible water-based emulsions.

A further basic method of controlling physical properties is by introducing cross-links between the polymer molecules. A small degree of cross-linking will retard or prevent sagging or creep in rubbery polymers, while extensive cross-linking will give hard, temperature and solvent resistant materials of typical thermoset structures. While this cross-linking

TABLE 1
PHYSICAL PROPERTIES OF SOME ACRYLIC MONOMERS
(Data after Luskin[4])

Monomer	*Formula weight*	*Boiling point (°C) at 760 mm H_2O (unless stated otherwise)*	*Flashpoint (°C)*	*Solubility in water (parts/100 parts of water)*
Acrylates				
Methyl	86·1	80·5	10^a	5
Ethyl	100·1	99·6	10^a	1·5
n-butyl	128·2	147	49^a	0·2
2-Ethylhexyl	184·3	216	90^b	0·01
2-Ethoxyethyl	144·2	174	66^a	5·5
Hydroxyethyl	116·1	82	104^b	Soluble
Hydroxypropyl	130·1	77	99^b	Soluble
Glycidyl	128·1	57	60^a	<0·01
Dimethylaminoethyl	171	116	90^b	Hydrolyses

Methacrylates				
Methyl	100·1	101	13[a]	1·5
Ethyl	114	117	35[b]	Insoluble
n-butyl	142·2	163	66[b]	Insoluble
2-Ethylhexyl	198	110 at 14 mmHg	96[b]	Insoluble
Lauryl	262	160 at 7 mmHg	133[b]	Insoluble
Stearyl	332	205 at 5 mmHg	>150[b]	Insoluble
Hydroxyethyl	130·1	95 at 10 mmHg	109[b]	Soluble
Hydroxypropyl	144·2	96 at 10 mmHg	121[b]	Slightly soluble
Glycidyl	142·5	75 at 10 mmHg	84[a]	Insoluble
Dimethylaminoethyl	157·2	69 at 10 mmHg	74[a]	Insoluble
Ethylthioethyl	174·3	102 at 15 mmHg	80[a]	Insoluble
Alkyl	126·2	43 at 15 mmHg	—	Insoluble
Acrylamides				
n-t-butyl	127	128–130 (m.p.)	—	0·7
n-Methylol	101	79 (m.p.)	—	Soluble
Methacrylamide	85	110 (m.p.)	—	Soluble
Acrylic acid	72·1	142	68[b]	Soluble
Methacrylic acid	86·1	163	77[b]	Soluble

[a] Tagliabue open cap.
[b] Cleveland closed cap.

TABLE 2
GLASS TRANSITION TEMPERATURES FOR VARIOUS POLY(ACRYLATES) AND POLY(METHACRYLATES)

Polyacrylate	T_g (°C)	*Polymethacrylate*	T_g (°C)
Methyl	3	Methyl (atactic)	105
		Methyl (isotactic)	45
		Methyl (syndiotactic)	120
Ethyl	−23	Ethyl	47
n-Propyl	−51·5	*n*-Propyl	33
n-Butyl	−70	*n*-Butyl	17
		n-Octyl	−70
		n-Decyl	−70
n-Tetradecyl	+20	*n*-Tetradecyl	−9

may be carried out during polymerisation by introducing polyfunctional materials such as ethylene glycol dimethacrylate or trimethylolpropane triacrylate the resulting polymer is intractable and the method is only suited to liquid adhesives that are cured in place. Other cross-linking systems may be formed by introducing functional groups along the polymer chain during polymerisation which will react with different polyfunctional compounds at a later stage. Examples of these are discussed below.

ACRYLIC ADHESIVES

Acrylic Polymers as Adhesives

Adhesives are made either from liquid monomers that cure in the bondline or from high molecular weight polymers formed outside the bondline which inherently have the required adhesive and cohesive strength. By far the largest proportion of acrylic adhesives sold today belong to the latter category and a great deal of research has and is being done to optimise the properties of the polymers used.

A major application area is in pressure-sensitive adhesives for the manufacture of tapes, labels and decals. Polyacrylate homopolymers normally lack the proper balance of tack, cohesive strength, and resistance to creep or flow. By changing the molecular structure of the polymer these properties can be changed.

Control of Adhesion

Tackiness can be improved by incorporating monomers with polar elements into the interpolymer. Examples of these are shown below.

A novel way of introducing adhesive promoting groups is demonstrated by Knapp.[8] An interpolymer of vinyl acetate, ethylacrylate, and 2-ethylhexyl acrylate is made with maleic anhydride in a solvent. The anhydride groups are then hydrolysed to form the carboxylic acid groups.

A non-acidic adhesion promoting system is shown by Pietsch and Curts.[9] This is based on acrylate interpolymers containing N-vinyl lactams such as N-vinyl-2-pyrrolidone and N-vinylcaprolactam. Adhesive tapes made from these polymers show good tack and are suitable for use in electrical applications because of the absence of acid moieties which can cause corrosion under the influence of an electric field.

An illustration of the scope of acrylic chemistry is given by Kosaka *et al.*[10] Maleic anhydride and an alkyl acrylate (e.g. 2-ethylhexyl acrylate) were copolymerised in the presence of an ethylenevinylacetate copolymer. The growing copolymer attaches itself to the ethylenevinylacetate copolymer to form a graft copolymer. This combination gives a pressure sensitive adhesive composition with improved tack.

Another example of graft polymerisation is given by Bartz *et al.*[11] In this case, the basic polymer is a thermoplastic hot melt such as polypropylene, polyethylene, or ethylene–propylene copolymer. A continuous extrusion reactor system is described by which acrylic acid is graft polymerised to form a material with improved adhesion.

Cohesive Strength

One of the problems in making pressure sensitive tapes, labels, or decals is that a low viscosity, easily spreadable liquid is needed to form a film over the carrier tape, but once formed a flow resistant, coherent structure is desired. These properties are conflicting since a high molecular weight polymer needed to give a strongly coherent film will have a high viscosity in solution. To solve this problem methods of cross-linking the adhesive after it is on the carrier have been devised. There are many ways of achieving this; some examples are described below.

Several methods of cross-linking acrylic interpolymers containing carboxyl groups are possible. Addition of polyfunctional alcohols with an acid catalyst and heat will form a cross-linked structure.[12] Similarly, Peterson,[13] describes water soluble acrylic acid, 2-ethoxyacrylate copolymers that can be cross-linked by adding polyfunctional amine such as hexamethylene diamine. Incorporating both acrylic acid and glycidyl methacrylate to give cross-linking in an acrylic interpolymer is described by Knapp.[14] Acrylate interpolymers containing carboxyl, epoxy, or hydroxyl functions can be cross-linked with polyfunctional isocyanates of low[15] or high[16] molecular weight that are mixed in with the interpolymer, spread over the carrier tape, and then heat cured.

Interpolymers containing acrylic acid and hydroxylated amides such as N-methylolacrylamide are described by Coffman.[17] Heating to 140 °C cross-links the system, apparently by the reaction of the N-methylolacrylamide groups with each other to form methylol ether or methylene bisacrylamide linkages.

Other, novel, systems are possible. Sanderson and Zdanowski[18] describe a polymer system having improved resistance to shear. One particularly interesting example is an emulsion copolymer formed of 2 % methacrylic acid with 98 % *n*-butylacrylate to which is added zinc ammonium glycinate and ammonium hydroxide. After the formulation is spread on the carrier tape, ammonia is driven off by heating and the polymer is ionically cross-linked. The use of titanium acid esters is described by Andersen *et al.*[19] in which interpolymers containing hydroxy- or carboxyl functions are reacted with metal alkoxides. Chelated titanium esters are preferred and improved shrink resistance is claimed. In a related patent, McKenna[20] describes interpolymers containing hydroxyl, carboxyl, or enolisable keto groups along the chain. Cross-linking is accomplished by adding a chelate ester of orthotitanic acid and heating. Other methods may be found in the literature. It is clear that many methods have been devised to make acrylic polymers that can be cross-linked in a separate step after polymerisation is complete.

Solubility

With solvent based systems becoming increasingly unattractive, the use of acrylic polymers that are water soluble or that are in emulsions is being explored. Bohme and Illagen[21] describe interpolymers of water insoluble alkyl acrylate esters, methacrylic or acrylic acids, and water-soluble polyalkylene oxide esters of acrylic acids. The polymer can be made water soluble by adding potassium hydroxide to the reaction mixture. Another selectively soluble system is described by Gander.[22] A copolymer of 2-ethylhexylacrylate with dimethylaminoethylacrylate was neutralised with ethanolic hydrochloric acid forming an amine salt. The resulting polymer made into an adhesive tape shows good adhesion to skin with very little irritation, but can be removed with water. A similar requirement resulted in the formation of an interpolymer of alkylacrylate, and hydroxyalkylacrylate by McKenna and Gardner.[23] The adhesive makes labels that give strong and water-resistant bonds, but which can be removed easily in a hot (65 °C) solution of 10 % sodium hydroxide.

Emulsion polymerisation requires the use of emulsifying agents. Samour[24] described the emulsion polymerisation of N-alkyl amic acids such as those derived from maleamic acid, with alkylacrylate or alkylmethacrylate monomers. At least a portion of the N-alkyl maleamic acid is present as an ammonium salt. The alkyl group is a tertiary alkyl of 12 carbon atoms. The result is that it acts as a polymerisable emulsifying agent. The resulting emulsion can then be made directly into a pressure sensitive tape or the polymer can be coagulated by adding acid. A similar, if less direct result is obtained by Gander.[25] A copolymer of an alkylacrylate with dimethylaminomethacrylate is formed by an emulsion process. The presence of the dimethylaminomethacrylate apparently counteracts the normally deleterious effect of the emulsifying agents on long term adhesion to substrate. Samour and Satas[26] describe emulsion polymers that are an extension of those described above.[24] In addition to the amic acid in salt form, the interpolymer contains acrylamide, methacrylamide, acrylonitrile, methacrylonitrile, cyanoethylacrylate or cyanoethylmethylacrylate. When the polymer is coagulated by adding acid to convert the salt carboxylates to the free acid state it can then be cross-linked by heating to about 150 °C for 3 min.

Radiation Curing of Monomers

While emulsion polymerisation techniques circumvent some of the problems associated with solvent based polymers, radiation curing of monomers on the carrier base attempts to eliminate completely the operation of applying the polymer. Fukukawa *et al.*[27] describe copolymers of butylacrylate with acrylic acid formed by irradiation with 300 kV electrons. Pressure-sensitive tapes with properties comparable to conventional tapes were obtained. A similar system described by Brookman *et al.*[28] uses a protective release layer to prevent oxygen inhibition of the polymerisation. Line speeds of about 15 m min^{-1} were obtained. Christenson and Anderson[29] used 450 kV electrons irradiation on pressure-sensitive adhesives applied as hot melts. The radiation treatment cross-links the polymer sufficiently to improve creep resistance, strength and solvent resistance.

It will be appreciated from the brief review given above that the acrylic polymer systems provide a wide range of control of important practical variables such as tack, cohesive strength, and solubility. In many cases these variables can be controlled independently of one another.

Although many molecular designs are possible there are other practical and economic constraints which have not been considered here. An overall perspective of pressure sensitive adhesives and their formulation can be found in Chapter 5 of this book.

Acrylic Monomers as Adhesives

In contrast to adhesives based on acrylic or other polymers are those used as liquids and which cure or polymerise in the bondline. These include epoxies, polyurethanes, and monomeric acrylates. Undoubtedly the most spectacular of these '100% solids reactive fluids,' as they are called, are the cyanoacrylate adhesives.

Cyanoacrylates

The basic chemistry of cyanoacrylates has been described above. Only the methyl, ethyl, butyl and isopropyl esters have been developed for extensive use commercially. Some properties of these monomers and the softening temperatures of their polymers are shown in Table 3. The adhesives are

TABLE 3
PHYSICAL PROPERTIES OF CYANOACRYLATE ESTERS
Data after Coover and Wicker[30]

Property	*Methyl*	*Ethyl*	n-*Butyl*
Boiling point (°C) (2–3 mmHg)	48–49	54–56	92–94
Viscosity at 25 °C	$2{\cdot}2\,mN\,m^{-2}\,sec^{-1}$	—	—
Vicat softening temperature of polymer (°C)	168	128	109

formulated with the addition only of thickeners (high molecular weight polymers), free radical and acidic stabilisers, and optionally, plasticisers. Stabilisation of the adhesives requires balancing shelf life (in a water-impermeable package) against speed of cure. The molecular structure which leads easily to very rapid polymerisation makes the addition of modifying materials without significant loss of either stability or speed of cure, relatively difficult.

In spite of these limitations a range of commercially distinct cyanoacrylate adhesives has been developed. Table 4 describes the properties of methyl, ethyl, and butyl cyanoacrylate esters formulated into adhesives. The methylcyanoacrylate gives higher bond strengths on steel and aluminium than the ethyl and butyl esters. Ethylcyanoacrylate, with somewhat lower bond strength, is faster curing, particularly on plastic and

TABLE 4

TYPICAL PROPERTIES OF CYANOACRYLATE ADHESIVE BONDS[a]

Property and material	*Methyl cyanoacrylate*	*Ethyl cyanoacrylate*	*Butyl cyanoacrylate*
Tensile shear strength (ASTM D-1002 fixture time)			
Grit blasted mild steel	22 MN m^{-2} (20–40 sec)	17 MN m^{-2} (10–30 sec)	15 MN m^{-2} (60–80 sec)
Etched aluminium	22 MN m^{-2} (60–80 sec)	17 MN m^{-2} (20–40 sec)	15 MN m^{-2} (60–80 sec)
Polycarbonate	15 MN m^{-2} (40–50 sec)	15 MN m^{-2} (10–15 sec)	9 MN m^{-2} (40–50 sec)
PVC	100 MN m^{-2} (10–20 sec)	10 MN m^{-2} (3–5 sec)	6 MN m^{-2} (15–25 sec)
ABS	15 MN m^{-2} (35–45 sec)	15 MN m^{-2} (10–15 sec)	10 MN m^{-2} (35–45 sec)
Butyl rubber	7 MN m^{-2} (4–6 sec)	7 MN m^{-2} (3–4 sec)	7 MN m^{-2} (20–30 sec)
Impact (steel/steel; (ASTM D-950)	180–250 kJ m^{-2}	180–250 kJ m^{-2}	180–250 kJ m^{-2}
Maximum operating temperature	>80 °C	>80 °C	>80 °C
Moisture resistance (8 weeks; 40 °C; 95% relative humidity) tensile shear strength			
steel/steel	60% retained	60%	60%
P/Carbonate	95%	95%	95%
PVC	75%	75%	75%
ABS	60%	60%	60%

[a] Data are from commercial literature[31] and are believed to be typical of cyanoacrylates but some variation in bond properties with different grades of material are observed.

rubber substrates. According to Lennox[32] the ethyl monomer requires less surface preparation for bonding plastics than does methylcyanoacrylate, although the methyl is more resistant to solvents such as acetone or methylethylketone. Butyl cyanoacrylate is much slower to cure than either ethyl or methyl cyanoacrylate, which can be a useful property if time is required to position parts precisely before the adhesive cures. The higher molecular weight butylester also has a lower vapour pressure which reduces somewhat the ventilation requirement for a bonding area.

Moisture resistance of cyanoacrylate bonds to metals is generally not good, but bonds to plastics and rubbers can show excellent resistance to moisture and outdoor exposure.[33] Durability of bonds of plastics and rubbers appears to be determined more by the environmental resistance of the polymeric substrate than by that of the adhesive.

Commercially-important modifications of these basic formulations have produced cyanoacrylate adhesives designed specifically for bonding EPDM rubber and for temporary bonding. A relatively-weak adhesive that forms easily disassembled bonds can be very useful for holding parts for grinding and similar operations.

Other, more fundamental, changes are difficult to achieve with cyanoacrylates, but a number of developments have been reported.

Improved temperature resistance. Adhesives believed to be based on allyl cyanoacrylates have been available commercially.[34] Where other commercially available cyanoacrylates are usable only up to 80–100 °C these show substantial retained strength after ageing 7 days at over 200 °C.

One very interesting application for cyanoacrylate adhesives is in underwater bonding. When a part to be bonded, with a drop of cyanoacrylate adhesive on it, is put into water the surface of the adhesive in contact with the water polymerises, forming an encapsulating shell around the remaining unpolymerised adhesive. When the other substrate is pressed firmly against the first the shell breaks, the adhesive fills the bond gap, cures, and forms a strong bond.

Methacrylates and Acrylates

Anaerobic adhesives. The development of liquid methacrylates and acrylates as adhesives can be said to have begun with the introduction of 'anaerobic' threadlockers and sealants. Krieble disclosed, in a series of patents,[41–43] formulations consisting of low molecular dimethacrylates (e.g. triethylene glycoldimethacrylates) with hydroperoxides as free radical initiators and specific accelerators and stabilisers. Such a formulation could be stored in a container through which oxygen diffused, free radical polymerisation being inhibited by the formation of a polyperoxide. For example:[6]

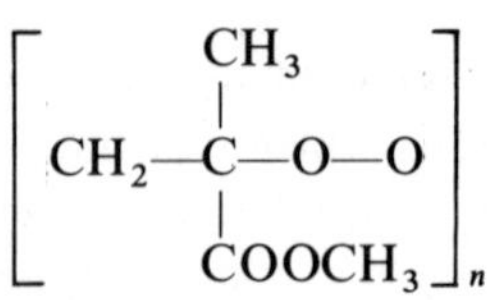

for methylmethacrylate.

This polyperoxide would then decompose to give materials no longer capable of participating in reactions with free radicals. The effect of the oxygen in preventing free radical polymerisation in methacrylates has long been known; the formation of polyacrylates and polymethacrylates requires purging the reaction vessels to remove all oxygen. What Krieble showed was that a careful balance of accelerators and free-radical stabilisers made possible an air stabilised formulation which, when placed between the adherends in an adhesive joint where air was excluded, would cure to form a hard thermoset structure. This phenomenon was exploited in bonding nuts to bolts to prevent vibration from loosening them and also to bond other close-fitting metal parts. The speed of curing in such a system can be greatly affected by the presence of transition metals such as iron or copper which set up a redox reaction with the initiator in the system to form free radicals. Other reducing agents can be applied to the surfaces of the materials to be bonded in order to accelerate the speed of cure with inactive surfaces which do not contain transition metals. Such 'activators' also increase the maximum bond gap that the adhesive can cure through. Since polymerisation initiates at the surface in these systems it is difficult to get a complete cure through very large bond thicknesses. By using a reducing agent which diffuses from the adherend surface into the adhesive this problem can be minimised.

A basic problem with the original 'anaerobics' was that the monomers used were low molecular weight dimethacrylates which, when cured in the bondline, gave very highly cross-linked thermoset-type structures which were extremely brittle (although temperature and solvent resistant) and gave very poor impact and peel strength. These first materials were unsuited to general structural bonding, being used either as threadlockers or retaining adhesives where the principal mode of deformation is shear.

Another approach is described by O'Sullivan and Melody[35] in which maleic anhydride is added to the adhesive formation. This increased the useful temperature to about 120°C. Although the mechanism of the improvement is not clear it is thought to be a retarding of depolymerisation at the high temperatures.

Cross-linked cyanoacrylate adhesives have been described[36] consisting of a mixture of 5–95% cyanoacrylate ester and 95–5% of a difunctional monomer diester of acrylic or methacrylic acid and a polyfunctional alcohol. The formulation includes free radical catalyst.

Medical applications. Because cyanoacrylate adhesives bond skin very rapidly and very well some effort has gone into finding the optimum adhesives for medical applications.[37] For bonding tissue the higher alkyl esters such as butyl or heptyl are found to be faster curing and to give stronger bonds than methyl, apparently because the butyl and heptyl esters spread over the tissue surface better.

In a related area work on cyanoacrylate dental fillings has concentrated on developing controllable means of curing the normally-rapid curing monomers.[38]

Other new adhesive developments are conductive cyanoacrylates filled with dendritic nickel,[39] and silane surface primers to improve moisture resistance.[40]

Extension to general purpose bonding. Developments to improve the performance of liquid methacrylate adhesives have been in two major directions: in one the polyfunctional character of the monomer has been retained, but the molecular structure has been changed to give improved strength and toughness; in the second monofunctional methacrylates are the basis of the formulation, with rubbery polymers being added to give the essential toughness.

(a) Modified polyfunctional acrylate and methacrylate monomers: Gorman and Toback[44] describe urethane–methacrylate monomers formed by the reaction of polyisocyanates with hydroxyalkylmethacrylates or acrylates. The monomers have the general structure:

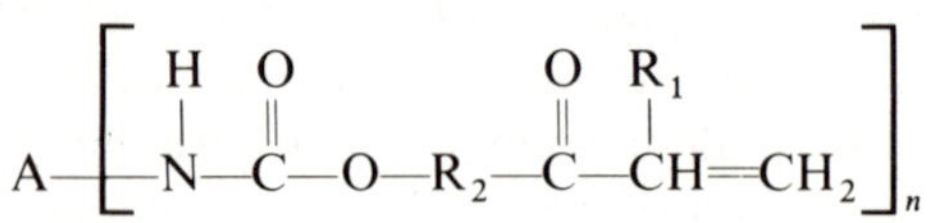

where R_1 can be H or CH_3, and R_2 can be, for example, an ethyl or isopropyl group. The base of the molecule, A, can be any polyisocyanate such as diphenylmethane 4,4′-diisocyanate or toluene diisocyanate or, to obtain a more flexible monomer, an isocyanate-capped polymer such as a polyether or polyester. With such systems significant improvements in bond strength and impact resistance are obtained over the original dimethacrylate structures.

Owston and Howard[45] describe formulations in which the poly(methacrylate) monomers are hydroxyalkylmethacrylate, allylalcohol or vinyl alcohol capped polyisocyanates. These polyisocyanates can be formed by reacting diisocyanates with polymeric diols or triols to give monoisocyanate capped polymers which are then reacted with the alcohols described above. The result is a polymer that is cross-linked during curing with the monofunctional monomers such as methacrylates and styrene that act as reactive diluents in the adhesive. Monomers such as methacrylic acid are included to promote adhesion along with thickeners to control viscosity, and reducing agents. A peroxide initiator was mixed into the formulation just before bonds were made. The adhesive was claimed to be particularly effective in bonding PVC.

A different way to achieve a novel dimethacrylate structure is described by Kusayama *et al.*[46] Instead of using a diisocyanate to act as a connecting link between the polymer base and the methacrylate function a diapoxide is used. In the example given, a commercially-available glycol diglycidyl ether is reacted with an acrylic or methacrylic acid. This is formulated with monofunctional methacrylates, with either acrylonitrile–butadiene or carboxylated acrylonitrile–butadiene rubbers added, to give a strong adhesive that can be cured by conventional free radical polymerisation systems.

(b) Rubber-toughened monofunctional methacrylates: The second approach taken to develop improved liquid methacrylate adhesives has been to incorporate high molecular weight rubbery polymers into formulations in which monofunctional methacrylates are a major portion.

Owston[47] describes adhesive compositions in which the essential ingredients are acrylic monomers such as acrylates and methacrylates to which are added poly(butadiene) or styrene–butadiene or butadiene–acrylonitrile copolymers, and methacrylic acid along with reducing agents to provide free radical initiation when an initiator such as benzoyl peroxide is mixed in. Examples of formulations based on methylmethacrylate and *n*-hexylmethacrylate and butadiene–acrylonitrile rubber show good adhesion to metals and plastics, good shear strength and high peel strength.

Briggs and Muschiatti[48] describe compositions of vinyl monomers (preferably acrylates and methacrylates) with chlorosulphonated polyethylene as the essential toughening ingredient. Examples are given of methylmethacrylate–*n*-butylmethacrylate compositions containing about 50% of a chlorosulphonated polyethylene. Both high tensile shear bond strengths and high peel strength were obtained with adhesives cured either by mixing a peroxide and a reducing agent into the mixture or by applying the reducing agent to the adherend surface with the free radical initiator in the adhesive.

(c) Commercial liquid acrylic adhesives: In recent years a number of liquid methacrylate-based adhesives have become commercially available. Table 5 summarises some key properties of these adhesives and of the bonds that can be made from them. The formulations of the adhesives are believed

TABLE 5

PROPERTIES OF COMMERCIAL LIQUID ACRYLIC ADHESIVES

Adhesive	®*Tame*[49]	®*Versilok*[50]	®*Cavalon*[51]	[a]300 *Series*[52]
Time to fixture	20 min to 24 h	10 min (2-part) 15 min (activator)	2–5 min (2-part, *with* activator) 3–15 min (activator)	10–20 sec or 10 min
Pot life (2-component)	10 min to 12 h	5–25 min	2 weeks	Single component, with activator
On-part life (surface activator)	No activator	30 days	8 h	15 min to 30 days
Bond strength Tensile shear (ASTM) grit blasted steel D-1002	36 $MN\,m^{-2}$	15–18 $MN\,m^{-2}$	15 $MN\,m^{-2}$	15 $MN\,m^{-2}$
T-Peel (ASTM D1876-61T) (etched alum)	490 N/7·6-cm width	No data available	90–220 N/2·54 cm	31 N/2·54 cm

Environmental resistance Maximum operating temperature (°C)	120	120	120	120
Humidity resistance (grit blasted steel) (percentage loss after 30 days at 49 °C/95 % relative humidity)	10–15	0–10	10	30–40
Volatile monomer		Yes	Yes	No
Plastic bonding	Very good	Very good	Very good	Fair

®Versilok is a Registered Trade Name owned by Hughson Chemicals, Lord Corp.
®Cavalon is a Registered Trade Name owned by E. I. du Pont de Nemours & Co.
[a] This numbered series refers to the products of the Loctite Corp.

to be of either of the two general types described above. They all have in common the advantage of low energy room temperature cure, with fixturing in less than a minute with the faster formulations. The ®Tame adhesives (of which there are several) require mixing and although bond properties are good they suffer from the same problem as two-part epoxies in that the useful pot-life of the mixed adhesive is less than the time to fixture.

Unlike most epoxy-based adhesives however, very good peel strength can be obtained. The Versilok series of adhesives can be cured either with a surface activator or by mixing a curing agent into the adhesive. Fixing with the aid of the surface activator with the one-component adhesive takes about 15 min. The Cavalon adhesive system fixtures in 3–15 min using a surface activator. The '300 Series' adhesives can give very rapid curing (10–20 sec to fixture) or moderately fast curing (10 min to fixture) depending on the activator used. In this case the very fast activator is susceptible to oxidation and the adhesive joint must be made within 15–20 min of the activator being applied. The other, more slowly curing activators, can be left on the parts from 8 h to 30 days, an advantage that can be more important than cure speed.

Tensile shear strength and temperature resistance for all those adhesives are good. The Tame and Cavalon adhesives show excellent peel strength, and with the Versilok system, good moisture resistance and adhesion to plastic substrates. The Versilok and Cavalon adhesives contain volatile monomers. The 300 Series adhesives show less peel strength, moisture resistance, and adhesion to plastics, but contain less volatile monomers. Recently modifications of the 300-type adhesives have been described that are much tougher,[53] with peel strengths from 15–25 pli.

While the acrylic adhesives described above are representative of those currently available as '100% solids reactive fluids' it is almost certain that new products will become available which show improvement over these. At present, however, this type of adhesive compares favourably with older adhesive technologies such as the epoxies.

Comparison of Acrylic Adhesives and Epoxies

Table 6 compares both cyanoacrylate and methacrylate adhesives to epoxies on the basis of the five criteria discussed earlier for industrial adhesives. From this it can be seen that no 'ideal' adhesive exists although the cyanoacrylates come very close, lacking only the moisture resistance on metal surfaces, and the peel strength necessary to be used as general

®Tame is a Registered Trade Name owned by the B. E. Goodrich Co.

TABLE 6

COMPARATIVE ADVANTAGES OF EPOXIES, CYANOACRYLATES, AND LIQUID ACRYLIC ADHESIVES AS INDUSTRIAL ADHESIVES

Parameter	*Epoxies*[a]		*Cyanoacrylates*	*Liquid acrylic adhesives*
	2-part mixed	*1-part heat cured*		
Bond strength (strength, dissimilar substrates, peel durability)	+ +	+ +	0 to +	+ +
Ease of use (one component, activator, mixing)	–	0	+ +	+
Speed of cure	– –	– to 0	+ +	+
Safety (volatility, flammability, toxicity)	–	0	+	0 to +
Energy requirement	+	–	+	+

[a] 0, No advantage or disadvantage; –, slight disadvantage; – –, great disadvantage; +, slight advantage; + +, great advantage.

purpose adhesives. Their relatively-high cost is generally more than offset by their speed of cure and ease of use. The epoxy adhesives show excellent bond strength to a wide variety of substrates, but lack speed of cure without rapid (high energy) heating systems. In addition, the toxicity of some epoxy curing systems requires the use of special equipment for safe use. The liquid acrylic adhesives show a balance of bond properties, ease of use, and speed of cure that make them very well suited for use in industry. The safe use of some of these adhesives may require ventilation equipment, depending on the vapour pressure and flammability of the methacrylate monomer used in the formulation. As the monomer is cured in the bond line, however, little evaporates in practice.

CONCLUSIONS

Recent developments in acrylic adhesives have exploited the versatility of acrylic chemistry to make polymers suitable for pressure-sensitive adhesive formulations in which tack, cohesive strength, and ease of use can be varied independently. They have also allowed manufacture of liquid (100% solids reactive fluids) adhesives based on flexible high molecular weight polyfunctional methacrylates or on rubber-toughened monomethacrylates which show an excellent balance of industrially important adhesive properties.

The future of these adhesive systems will be determined by how nearly they can meet the requirements of present and new application areas, in comparison with competitive systems. At present they appear to be very attractive, and it can be anticipated that development activity in this area will remain high.

REFERENCES

1. Brief, A. and Skeist, I., in: *Applied Polymer Science*, eds Craver, J. K. and Tess, R. W., 1975, The American Chemical Society, Washington, DC.
2. Estes, G. M. and Kuespert, D., *Chemical and Engineering News*, 23 August 1976, pp. 40–7.
3. Coover, H. W. Jr., in: *Handbook of Adhesives*, ed. Skeist, I., 1962, Reinhold, New York.
4. Luskin, L. S., in: *Vinyl and Diene Monomers*, Part 1, ed. Leonard, E. C., 1970, Wiley-Interscience, New York.
5. Brindley, W. H., Calder, G. V. and Wetzel, L. A., in: *Applied Polymer Science*, eds Craver, J. K. and Tess, R. W., 1975, The American Chemical Society, Washington, D.C.
6. Riddle, E. H., *Monomeric Acrylic Esters*, 1954, Reinhold, New York.
7. Leonard, E. C., ed., *Vinyl and Diene Monomers*, Part 1, Wiley-Interscience, New York, 1970.
8. Monsanto Co., US 3 284 423 (publ. 8.11.66).
9. Beiersdorf, A. G., US 3 728 148 (publ. April 1973, priority DT).
10. Toyo Soda Manufacturing Co., US 3 896 067 (publ. 22.7.75).
11. Exxon Res. & Engineering Co., US 3 868 433 (publ. 25.2.75).
12. Celanese Corporation of America, US 3 258 454 (publ. 28.6.66).
13. Minnesota Mining and Manufacturing Co., US 3 575 911 (publ. 28.4.71).
14. Monsanto Co., US 3 284 423 (publ. 8.11.66).
15. Beiersdorf, A. G., US 3 729 338 (publ. 24.4.73, priority DT).
16. PPG Industries, US 3 532 652 (publ. 6.10.70).
17. B. F. Goodrich Company, US 3 900 674 (publ. 19.8.75).
18. Rohm and Haas, US 3 740 366 (publ. 19.6.73).
19. National Starch and Chemical Corp., US 3 769 254 (publ. 30.10.73).
20. Monsanto Company, US 3 900 610 (publ. 19.8.75).
21. Daubert Chemical Corp., US 3 890 292 (publ. 17.6.75).
22. Johnson & Johnson, US 3 321 451 (publ. 23.5.67).
23. Monsanto Co., US 3 763 117 (publ. 2.10.73).
24. The Kendall Co., US 3 483 155 (publ. 9.12.69).
25. Johnson & Johnson, US 3 475 365 (publ. 28.10.69).

26. The Kendall Co., US 3 492 260 (publ. 27.1.70).
27. Nitto Electrical Indust. Co. Ltd, US 3 772 063 (publ. 13.11.73).
28. Firestone Tire and Rubber Company, US 3 661 618 (publ. 9.5.72).
29. PPG Indust. Inc., US 3 725 115 (publ. 3.4.73).
30. COOVER, H. W., Jr and WICKER, T. H., Jr, *Encyclopedia of Polymer Science and Technology*, Wiley, New York, 1963.
31. *Technical Data Sheet* 60, Loctite Corporation, Conn., USA.
32. LENNOX, A. F., paper presented at the *Conference on Finishing, Decoration and Assembly*, 18–20 June, 1974, Coventry, UK.
33. Technical Data Sheet, Eastman Kodak Co.
34. Technical Bulletin 3-2-910-A, Eastman Kodak, USA.
35. Loctite (Ireland) Ltd, US 3 832 334 (publ. 1.8.75).
36. Johnson & Johnson, US 3 940 362.
37. LEONARD, F., COLLINS, J. A. and PORTER, H. J., *J. Appl. Polymer Science*, **10,** 1966, pp. 1617–23.
38. Johnson & Johnson Ltd, GB 1 442 498 (publ. 14.7.76).
39. Inoue-Japax Research, Inc., JA 76/73036 (publ. 24.6.76).
40. Intercontinental Chemical Co. Ltd, GB 1 373 559 (publ. 13.11.74).
41. The American Sealants Co., US 2 895 950 (publ. 21.7.59).
42. Loctite Corp., US 3 046 262 (publ. 24.7.62).
43. Loctite Corp., US 3 218 305 (publ. 16.11.65).
44. Loctite Corp., US 3 425 988 (publ. 4.2.69).
45. Lord Corporation, US 3 873 640 (publ. 25.3.75).
46. Sumitomo Chemical Corp. Ltd, US 3 870 675 (publ. 11.3.75).
47. Lord Corporation, US 3 832 274 (publ. 27.8.74).
48. E. I. DuPont de Nemours & Company, US 3 890 407 (see also GB 1 436 888) (publ. 17.6.75).
49. COLEMAN, J. F., Durability Studies on Toughened Acrylic Resin Adhesives, *5th National SAMPE Technical Conf.*, Kiamesha Lake, N.Y., 9–11 October, 1973.
50. *Versilok Structural Adhesives*, Product Information Sheet, Hughson Chemicals, Lord Corporation.
51. *Cavalon Technical Data Sheet*, E. I. DuPont de Nemours Co.
52. *Loctite Structural Adhesives Technical Data Sheet*, Loctite Corporation.
53. BATSON, R. E., Anaerobics Adhesives—Future Uses, *Interfast Conference*, Brussels, 3 December 1975.

Chapter 7

THE PRESENT POSITION OF TYRE CORD ADHESIVES

D. B. WOOTTON
John Bright Group, Rochdale, UK

SUMMARY

Many rubber products rely on reinforcement by textile inserts to achieve the required physical properties. This reinforcement can only be obtained when both components are adequately bonded together. A range of adhesive treatments exists to suit most fibre–polymer combinations, as each combination essentially requires a different system to promote adhesion.

Most treatments are applied to the textile, but systems are also available to modify the rubber compound to give adhesion to untreated fabrics: this also enables short discrete fibres to be used for reinforcement. Other compounding ingredients can also affect the bond strengths obtained.

Although the mechanisms whereby adhesion is obtained are not fully understood, both chemical and physical interactions have been shown to occur at the textile–adhesive and adhesive–rubber interfaces.

The adhesive films applied to textiles are susceptible to degradation on exposure, resulting in loss of adhesion, and care must be exercised in storage of treated fabric to minimise this.

Whenever it is required that an elastomeric system be reinforced by a textile insert, the full reinforcement is only obtained when optimum adhesion is achieved between the two components. There exists a wide range of elastomers and of textile fibres and a corresponding range of adhesive systems has been developed to cover the majority of combinations of these.

In selecting a complete system, i.e. elastomer, compound and textile, it is necessary, of course, to ensure that the overall properties will satisfy the service conditions. This requires a knowledge not only of the properties of the compound and of the textile but also of the adhesive system that has been used to unite the two into a coherent unit.

In making the correct choice of adhesive system it is necessary to be aware not only of those available but also of other factors affecting adhesion, such as the effects of compound variation, of the various processes to be used and of ageing effects. Also, where possible, some insight, into the mechanisms whereby adhesion is obtained, is advantageous.

In considering the various adhesive systems, the choice is primarily governed by the type of textile to be used, as the adhesives are, generally, more specific to the textile than to the elastomer used, except when dealing with certain of the speciality elastomers.

COTTON

Generally, no adhesive is required with this fibre, as the adhesion is primarily achieved by mechanical anchoring of the staple fibre ends in the mass of the polymer. Some adhesive promoters may be used with cotton fabrics, such as the *in situ* bonding additives, but very often the purely mechanical adhesion is considered adequate.

The use of aqueous systems, as applied to other fibres, is also rendered more difficult with cotton, as the natural waxes on the cotton interfere with the wetting of the fibre by the adhesive. This can only be improved by the inclusion of some preliminary preparatory process such as scouring; such additional processes, even so, do not add greatly to the levels of adhesion obtained and are therefore economically not attractive.

RAYON (Regenerated Cellulose)

These continuous-filament yarns generally require an adhesive treatment of the resin/latex type. These resin latex systems comprise an admixture of a resorcinol/formaldehyde resin with a latex, designated RFL. Originally, the latex used was natural rubber and with the earlier, relatively-low tenacity, rayons, this gave perfectly adequate adhesion. With the shortage of natural latex during the 1939–45 War, styrene–butadiene rubber latex (SBR or G-RS as it was then designated) was used in place of the natural latex and gave similar levels of adhesion but with the disadvantage of lack of tack, in the treated cord or fabrics, desirable to assist in the assembly of the uncured composite.

TABLE 1
FORMULATION OF RFL SYSTEM FOR RAYON CORD
(80:20 SBR/VP latex)

Component	*Parts wet*	*Parts dry*
Resorcinol	8·0	8·0
Formalin (37%)	14·5	5·4
Sodium hydroxide (10%)	8·0	0·8
Water	115·0	—
SBR latex (35% solids)	204·3	71·5
VP latex (41% solids)	35·0	14·3
Water	115·2	—
Total	500	100

With the development of the higher tenacity rayons, the straight SBR latex systems were found not to give adhesions matching the improved strength of the fibres and the then recently developed 2-vinyl pyridine–styrene–butadiene terpolymer latices (VP latex, originally marketed under the trade name Gen-tac†) were required, blended with the SBR component. Usually, some 20–30% of VP latex was sufficient to give the required adhesion, resulting in a formulation of the type given in Table 1.

The RFL adhesive is prepared by dissolving the resorcinol in water and adding the formalin, with the caustic alkali as catalyst, to give a water soluble resin precondensate. This resin solution, after condensation for about 6 h, is then added to the blended latices. The completed dip is then allowed to mature for a further period before use. The rayon textile is then dipped through the adhesive, excess dip removed by squeeze, suction etc., and the impregnated fabric dried and baked. Baking temperatures in the range of 120–150 °C are adequate to give satisfactory adhesion levels with this system, with exposure times of 1–2 min (depending on the fabric weight and construction).

POLYAMIDES (Nylons 66 and 6)

These synthetic fibres require similar basic systems to those used for rayon, although higher proportions of VP latex are required to achieve the

† Gen-tac is a Registered Trade Name owned by General Tire & Rubber Co.

necessary levels of adhesion. With rayon, replacement of as little as 20% of SBR latex with VP will give significant improvements in adhesion, but with nylon, the levels of adhesion start to fall very markedly if more than 20% SBR latex is used, and often it is considered preferable to use 100% VP latex. This effect is illustrated by the results given in Table 2.

TABLE 2
EFFECT OF VP/SBR LATEX RATIO ON NYLON ADHESION
Test: 2-ply peel test to natural rubber compound. Basic RFL formulation as in Table 1

Latex solids (%)		*Peel adhesion ($kN\,m^{-1}$)*
SBR	*VP*	
100	0	5·8
80	20	11·9
60	40	16·8
40	60	19·6
20	80	21·5
0	100	22·0

Although the two stage mixing process outlined above can be used, it is generally considered preferable to use the 'one-shot' method wherein the resin portion is added immediately to the latex, without precondensation and the completed formulation matured. When using NR latex, it is necessary to allow the resorcinol and formaldehyde to react before addition to the latex, otherwise the ammonia present in the latex would react preferentially with the formaldehyde: however, with the ammonia-free synthetic latices, it is possible to form the resin by condensation *in situ* with the latex. By allowing the condensation to proceed in the presence of the latex, the resin oligomer formed (up to 6 resorcinol units) replaces the stabilising soaps on the latex particles[1] and thereby gives a much improved dispersion of resin throughout the rubber phase (the two components of the resin/latex system are otherwise of low mutual compatibility).

Furthermore, it has been shown[2] that dips prepared by the two routes require different baking conditions in order to achieve optimum adhesion. With the two stage method, optimum adhesion is obtained with a baking treatment of around 105 sec at 150–160 °C while the one-shot system benefits from a much shorter exposure at temperatures around 200 °C. This short, high-temperature baking corresponds to the conditions necessary for the heat setting of the nylon, to obtain optimum cord properties, and is therefore preferable for a combined dipping/heat setting process. Additionally, the one shot dip has a vastly improved shelf life, being suitable for use after several weeks storage rather than the 3–4 days usable life with the two stage preparation.

A typical formulation for a 100% VP nylon dip is given in Table 3, but considerable variations in resorcinol/formaldehyde and resin/rubber ratios can be made without significantly affecting the levels of adhesion achieved.

TABLE 3
FORMULATION OF RFL SYSTEM FOR NYLON CORD (100% VINYL PYRIDINE TERPOLYMER LATEX)

Component	*Wet*	*Dry*
Water	263·0	—
Resorcinol	9·4	9·4
Formalin (37%)	13·8	5·1
Sodium hydroxide (10%)	7·0	0·7
VP latex (41% solids)	206·8	84·8
Total	500	100

POLYESTER (Polyethylene Terephthalate)

For these fibres, very different adhesive systems are required: these differences are variously ascribed to the lower chemical reactivity of the polyester or to the different surface free energy, which precludes the intimate wetting of the fibre surface, by conventional RFL type systems, as required to achieve satisfactory levels of adhesion.

The original systems, for obtaining adhesion to polyester, were based on solvent applications of highly active isocyanates. These could be used either in a solution of rubber compound, after which only evaporation of the solvent is required, or as a straight solution in an organic solvent, to be followed, after removal of the solvent, by treatment with an aqueous RFL system.

The disadvantages of solvent based systems are widely recognised, so that much effort has been devoted to research to find aqueous systems to give adequate adhesion levels. One of the earliest of these was developed by Canadian Industries Ltd[3] and is based on a compounded PVC latex predip, known as TR5, followed by a standard RFL. Although this system gave satisfactory levels of adhesion, the preparation of the first dip was somewhat involved and the storage life of the dip limited, so that its use was not widely adopted.

Another system was the D417 system developed by Shoaf of E. I. du Pont de Nemours.[4] This again is a two stage system, in which the first stage is based on a phenol blocked diisocyanate (blocked to give stability to water etc.) with a small portion of an epoxy, to improve the film forming properties of the dip, as shown in Table 4.

TABLE 4
FORMULATION OF POLYESTER FIRST DIP, DU PONT D417

Component		*Wet*	*Dry*
Phenol blocked isocyanate: 40% dispersion[a]		180·0	72·0
Epoxy resin[b]		27·2	27·2
Gum tragacanth soln. (2%)		40·0	0·8
Water		1 752·8	—
	Total	2 000·0	100·0

[a] Hylene MP. A phenol-blocked methylene bis(phenylene diisocyanate). Hylene is a Registered Trade Name owned by Messrs E. I. du Pont de Nemours, Inc.
[b] Epon 812. A liquid epoxy derived from epichlorhydrin and glycerol. Epon is a Registered Trade Name owned by the Shell Chemical Co. Ltd.

The second dip used with this is again a standard RFL. This system has been widely adopted and is accepted as one of the standard polyester adhesion systems.

Other development has been aimed at single-stage treatments. These developments have generally been aimed at modifications to standard RFL systems. One logical development, from the Shoaf system, is the incorporation of a blocked isocyanate dispersion in an RFL.[5] Although only moderately effective with standard polyesters, useful improvements in adhesion can be achieved by this in conjunction with pretreated polyesters.

Other systems have been aimed at modifications to the resin component of the RFL, as this is generally considered to be the component which promotes adhesion to the textile. In one system,[6] the resin is modified by the replacement of a part of the formaldehyde with an aromatic aldehyde. This is reasonably effective and can be further improved by the addition of a blocked isocyanate. Although giving good levels of adhesion, such systems do not appear to have found any wide acceptance.

Another, more widely accepted, single stage system is based on the co-condensation, with the RF constituent, of a trimer produced by reaction of *p*-chlorophenol with resorcinol: this is the *Pexul*† system developed by ICI Ltd.,[7] the formulation of which is given in Table 5.

An alternative approach, that has been used by a number of the fibre manufacturers, is to modify the surface of the polyester fibre, during spinning, to produce a 'pre-treated' or 'adhesive activated' polyester yarn.

TABLE 5
FORMULATION OF PEXUL-RFL ADHESIVE SYSTEM (ICI LTD)

Component		*Wet*	*Dry*
Water		165·8	—
Resorcinol		9·8	9·8
Formalin		5·9	2·2
VP latex		121·5	48·6
'Pexul' resin solution (20%)		197·0	39·4
	Total	500·0	100·0

These pre-treatments, during the spinning process, may be likened to the first stage of a two stage system, so that the final cord or fabric only requires a single-stage treatment with a standard RFL system. A wide range of pre-treatments have been patented, but the most widely used are based either on epoxy derivatives or on isocyanate systems.

These pretreated polyesters only require dipping in a standard RFL, followed by a moderate temperature treatment, to achieve adequate adhesion. However, it is usually desirable to use higher temperature treatments, to heat-set the yarns to obtain optimum physical properties, but some fibre manufacturers (especially in Continental Europe) combine a heat-stabilising process during manufacture and so eliminate the need for subsequent high-temperature, high-tension treatment.

Apart from the solvent isocyanate systems, all the treatments for standard polyester require high temperatures, to achieve optimum adhesion, usually in the order of 230–240 °C, so that the application of these can be performed at the same time as the heat-setting of the cords or fabric. One disadvantage associated with this, however, is that the higher the temperature of adhesive treatment, the more susceptible is the level of adhesion to variables in the compounding of the rubber to which it is to be bonded.[6]

† Pexul is a Registered Trade Name owned by Messrs ICI Ltd.

ARAMIDS

The newest arrivals, in the reinforcement fibres field, are the aromatic polyamides or Aramids, as exemplified by Kevlar† from DuPont. Although chemically polyamides, the reactivity of these fibres is much reduced compared with the aliphatic polyamides (Nylon 6 and 66), and on account of this, additional treatments, other than standard RFL, are required.

The systems primarily established for polyester systems, either one- or two-stage, can be used to give acceptable levels of adhesion, but somewhat simpler systems may be used and the standard recommendation calls for a first dip of an epoxy based adhesive followed by a standard RFL second stage. The first stage requires a baking treatment of around 60 sec at 240 °C: this obviously can be combined with the heat setting of the yarn, although the mechanism of heat setting the aramids differs from that of nylon or polyester, in that the modulus of Kevlar is determined by the conditions obtaining at the first high-temperature exposure of the yarn and is not significantly affected by subsequent processing, whereas the properties of nylon and polyester can be modified by further processing. Also, at the first high-temperature exposure, Kevlar undergoes a small irreversible loss of moisture.

As with polyester, the high temperature, required to obtain adhesion with the aramids, brings with it some effects on compound compatibility, although reputedly this can be minimised by the incorporation of a carbon black dispersion in the RFL second stage dip.

In situ BONDING SYSTEMS

All the systems mentioned above are treatments to be applied to the textile before its incorporation into the rubber matrix. An alternative system, now in use, is to incorporate the bonding system into the rubber compound: this is the *in situ* or HRS bonding system (the term HRS being derived from the initials of the additives used: hexamine, resorcinol and silica). In this system, the RF resin is formed in the rubber compound, during cure, by the reaction of resorcinol with formaldehyde, generated by breakdown of a formaldehyde donor, such as hexamethylenetetramine (hexamine). The role of the silica in this system is not fully understood, but it is thought to

† Kevlar is a Registered Trade Name owned by Messrs E. I. du Pont de Nemours Inc.

provide reactive sites whereby the resin can bond to the rubber component, by hydrogen bonding.

This system works well with rayon and nylon but, in order to obtain satisfactory levels of adhesion with polyester, it is necessary to use the pre-treated yarns. Care must be exercised when using the *in situ* process with polyester, however, as the amine residues, from the hexamine, can cause severe degradation of the textile, by aminolysis: this degradation can be reduced by the use of alternative formaldehyde donors, such as the trismethoxymelamine derivatives.

The *in situ* process can also be used in conjunction with conventionally dipped fabrics, to give improved levels of adhesion: often this is done not only to give the higher levels of adhesion but also to give some measure of control to the levels achieved, by adjustment of the proportions used.

SHORT-FIBRE REINFORCEMENT

The improvements achieved with the *in situ* bonding systems have contributed directly to significant advances in the field of short fibre reinforcement for rubbers.

In order to obtain the full reinforcement from the incorporation of discrete short fibres in a polymer matrix, it is necessary that these fibres be adequately bonded. Previously, when only the standard aqueous adhesive systems were available, attempts were made to use the principle of short-fibre reinforcement by taking conventionally dipped cords, chopping them to the required length and then mixing these chopped fibres into the rubber compound. Two major problems were encountered in these attempts. Firstly, due to the method of adhesive treatment, adequate dispersion of the filaments was not obtained, as the adhesive film caused the filaments to be held in bundles, rather than allowing them to separate and disperse evenly and secondly, it was not feasible to cut the cords into the ideal length to achieve optimum reinforcement. For optimum reinforcement from dispersed discrete fibres, the length to diameter (L/D) ratio should lie between 100 and 200 and with the conventional textile fibres with a diameter of around 20 μm, the ideal length should be between 2 and 4 mm.

Using the *in situ* bonding system, untreated fibres can be taken and give improved dispersion and good adhesion, although this still leaves the problem of fibre length. However, a novel system has recently been developed by the Monsanto Industrial Chemical Co.,[8] under the registered trade-name Santoweb, which overcomes the majority of the problems associated with short fibre reinforcement. In this system, use is made of cellulosic fibrils, occurring naturally in wood: these fibrils are of the ideal L/D ratio and yet are short enough to be incorporated readily into the rubber matrix, without significant breakdown.

In the patented process, the unregenerated cellulose fibres are treated with certain bonding and dispersing additives, which ensure easy dispersion of the fibrils in the rubber and require the addition of only a formaldehyde donor to the compound to achieve good adhesion.

After incorporation into the rubber, the necessary orientation of the fibres is readily obtained by the normal processing techniques of calendering or extrusion.

This form of reinforcement gives very significant increases in strength and modulus in the direction of orientation, with the effects reducing as the angle of measurement increases to 90° to the direction of orientation.[9]

EFFECT OF COMPOUNDING ON ADHESION

Mention has already been made of the addition of the *in situ* bonding additives to the rubber compound and the marked affect of these on the levels of adhesion obtained. Many of the normally-used compounding ingredients can also affect the levels of adhesion, although the most noticeable of these are generally adverse effects.

The basic polymer used will of course have a significant effect. For the speciality polymers, such as polychloroprene or nitrile rubbers, it is very often necessary to modify the adhesive dip formulation to obtain the required levels of adhesion: often this may be achieved by incorporating at least a portion of the corresponding polymer latex to give the required dip-to-matrix compatibility. With the general purpose hydrocarbon polymers, there are basically no great differences in levels of adhesion achieved, although the styrene–butadiene polymers usually give slightly higher levels than the natural or synthetic polyisoprenes.

The majority of rubber/textile composites are based on carbon-black filled compounds. On the whole, the type of black used does not have a great effect on the levels of adhesion obtained.[10] The reinforcing siliceous fillers also give good levels of adhesion but the non-reinforcing white fillers usually show significantly lower levels of measured adhesion. This is largely due to the physical properties of the vulcanisates, especially lower levels of tear strength and modulus, affecting the measurement of adhesion: such variations in properties can also affect the relative ratings of adhesion

TABLE 6

EFFECT OF CURING SYSTEM ON ADHESION

Test: 2-ply peel test with dipped nylon fabric to natural rubber compound

Curing system		Cure (*min at* 15 °*C*)	Peel adhesion ($kN m^{-1}$)
Ingredient	*Parts per* 100 *rubber*		
MBTS	0·6	15	21·0
S	2·5		
CBS	0·5	12·5	18·7
S	2·5		
MBTS	0·4		
DPG	0·2	12·5	18·0
S	2·5		
NOBS	0·5	15·0	19·1
S	2·5		
MBTS	0·4		
TMTD	0·1	10·0	13·7
S	2·5		
CBS	4·0	15·0	10·5
S	0·5		
CBS	2·0		
S	1·0	12·0	12·0
BDTPTS	1·0		
TMTD	3·0	12·0	2·1
CBS	2·0	15·0	1·7
BDTPTS	3·0		

MBTS = dibenzthiazyl disulphide.
CBS = N-cyclohexylbenzthiazyl sulphenamide.
DPG = diphenylguanidine.
NOBS = N-oxydiethylene benzthiazylsulphenamide.
TMTD = tetramethylthiuram disulphide.
BDTPTS = Bis(diethylthiophosphoryl)trisulphide (Vulcadone 3SN). Vulcadone is a Registered Trade Name owned by Messrs ICI Ltd.
S = Sulphur.

depending on the method of assessment. High-tensile strength compounds give a relatively better measured adhesion when assessed by a peel test rather than cord pull-through test, whereas improved flow characteristics, in the early stage of cure, favour the pull-through assessment.[11]

Probably the most significant effects on adhesion are due to the curing systems used. The best levels of adhesion are obtained with simple thiazole/sulphur systems, but, for many applications, these are too scorchy or too slow curing to be used alone. When activated thiazole systems are used, the levels of adhesion obtained generally decrease as the speed of cure increases, as illustrated in Table 6.

Furthermore, in moving from the conventional sulphur levels to the low-sulphur EV systems, there occurs a further decrease in adhesion and with completely sulphurless cures, either thiuram sulphurless or accelerator/sulphur donor, only very poor adhesion is achieved.

Other compounding ingredients generally have only minor efforts on adhesion. Some antioxidants and plasticisers can show adverse effects on adhesion; this is usually due to migration of these materials to the surface of the rubber, thus preventing intimate contact between the rubber and dip, or, especially with certain plasticisers, migration into the dip film, with consequent reduction in dip film strength, thereby causing premature breakdown of adhesion, during service.

MECHANISMS OF ADHESION

The mechanisms, whereby adhesion is obtained between a textile and a rubber, can basically be considered in two separate stages, firstly, the adhesion between the dip film and the rubber matrix and secondly between the dip film and the textile fibres.

With cotton fabrics, however, where no adhesive treatment is used, the adhesion is obtained basically by the mechanical keying of free fibre ends penetrating into the rubber matrix. On stripping the bond, it is necessary either to pull out these embedded fibres or, if the force required to free them is greater than their tensile strength, to break them off. This mechanism has been studied[12] and a relationship, between number of protruding fibre ends and the adhesion achieved, shown. This mechanism applies to all staple fibre to polymer systems, where adhesive treatments are not used.

When an adhesive dip is present, the adhesion between the dip film and rubber matrix arises primarily from the direct crosslinking of the latex constituent into the matrix polymer network, brought about by migration of sulphur and a curative (or active molecular species derived from these) into the dip film. There is some contribution arising from the resin portion of the dip, due to reaction between the phenolic groups and double bonds in the polymer chain, resulting in chroman formation: the contribution from this is relatively slight due both to relatively low proportion of resin present, usually only around 20% of the dip layer, and the slower rate of the reaction.

This dependence on migration of curatives to the dip film readily accounts for the effects of differing curing systems on adhesion. With fast curing and scorchy systems, there is little time for this migration to take place, and with EV or sulphurless cures, there is a very much reduced availability of sulphur, or none at all. This effect is exemplified in Table 7 by the addition of curatives to the dip, which greatly increases adhesion with a very fast scorchy system.

TABLE 7

EFFECT OF ADDITION OF CURATIVES TO DIP FILM, WITH HIGH SPEED CURING SYSTEM

(Peel test to RFL dipped nylon fabrics; natural rubber compound)

Curing system and cure at 15°*C*			*Fabric treatment*	*Adhesion rating*
Ingredients[a]	*Parts per* 100 *rubber*	*Min*		
MBTS	0·6	15	Standard RFL	100
S	2·5			
MBTS	0·4			
ZDC	0·1	8	Standard RFL	20
S	2·5			
MBTS	0·4		RFL with 1% sulphur	
ZDC	0·1	8	and	55
S	2·5		1% ZMBT	

[a] MBTS = dibenzthiazyl disulphide.
ZDC = zinc diethyl dithiocarbamate.
S = sulphur.

The still-inferior, although vastly-improved, level of adhesion achieved with the modified RFL to the very fast scorchy system, is largely due to the poorer flow characteristics of this compound and the very short induction period before rapid crosslinking occurs.

When considering the nature of the adhesive bond between the dip film and the textile, the situation is very much more complicated: indeed it almost appears that different mechanisms apply with each different fibre type.

With the continuous filament yarns, there still exists a small purely mechanical contribution to adhesion, but this is generally only between 10 and 20% of the total adhesive strength, depending on the surface configuration of the cord or fabric.

With the RFL systems, it is generally accepted that it is the resin portion which is the effective adhesive to the fibre. Various mechanisms for this have been postulated, including diffusion and the formation of direct chemical crosslinks. It has been shown[13] that RF resins, either alone or in RFL systems, will migrate into regenerated cellulose, resulting in swelling and modification of physical properties (increasing modulus and reducing elongation). This can be considered a mechanical anchoring effect, due to diffusion processes, and other work,[14] designed to separate the contributions of various mechanisms to total adhesion, attributes over 30% of the total adhesion strength to this mechanism, with rayon. This work also shows, however, that this contributes virtually nothing to the adhesion with nylon, although the dip and treatment are virtually identical, but that 50–60% of the adhesion arises from direct chemical covalent bond formation, whereas this only contributes some 25% of bond strength with rayon. The residual 20–25% of the total bond strength arises, in both cases, from secondary valency bonds. These resin/fibre mechanisms will also account for the effectiveness of the *in situ* bonding system with both rayon and nylon.

With polyester, however, the mechanisms are rather less-clearly established. With the two-stage systems (including the fibre pre-treatments) the major effect appears to be one of surface modification.

It has been shown[15] that the Shoaf pre-dip reacts primarily to form a rubbery polyurethane film, with the epoxy constituent reacting to provide additional hydroxyl groups, for the isocyanate/polyurethane reaction. It appears unlikely that there is any significant reaction of the isocyanate with the polyester, due to the chemical inertness of the polyester linkages and the very low level of free hydroxyl chain terminating groups. Other work[16] has indicated however, that the cohesive energy densities, of both polyester and the cured polyurethane film from the Shoaf dip, are very similar and this could be expected to allow the formation of an adhesive bond between the two by a diffusion mechanism, as propounded by Voyutskii,[17] especially as the film-forming reactions occur at high temperatures, which would increase the mobility of the molecules concerned, by thermal agitation.

Such systems, either as pre-treatments in the spinning stage or as the first stage of double-dip systems, modify the polyester surface, or perhaps present a new interface, which is then thermodynamically more compatible with or chemically more reactive to the standard RFL adhesives subsequently applied.

With the single-stage systems, where the basic components of an RFL formulation have been modified to adjust the compatibility and cohesive energy densities and to permit of the intimate wetting of the fibre surface, adhesion can arise from a similar diffusion process or by hydrogen bonding. This latter possibility has been studied by Mather[18] who has shown that the active trimer, of the ICI Pexul system, will associate very strongly, by hydrogen bonding, with dimethylterephthalate, his model for polyethylene terephthalate.

This increased compatibility between dip and fibre is also claimed for the system using an aromatic aldehyde.[6] It is also suggested that this modification, and also the additional inclusion of highly reactive materials such as blocked isocyanate, will increase the chemical reactivity of the dip towards the polyester: however, as commented above, it is debatable whether there is any significant contribution to adhesion from direct chemical reaction. It would appear therefore that adhesion to polyester arises either by a diffusion mechanism or by secondary valency linkages (hydrogen bonding) or a combination of both.

That the various systems achieve their adhesion by different mechanisms is supported by differences in behaviour of bonded laminates to certain environmental factors. For example, when bonded samples are immersed in water, the levels of adhesion are affected to different extents, as shown in Table 8.

TABLE 8
EFFECTS OF WATER IMMERSION ON ADHESION
(Test: 2-ply peel test to natural rubber compound)

Textile fibre	*Adhesive treatment*	*Adhesion remaining (%) after 72-h immersion*
Nylon	RFL	97
Polyester	Pexul	52
Polyester	Shoaf 1st stage RFL 2nd stage	90
Pre-treated polyester	RFL	75

The RFL dipped nylon, where adhesion arises largely from direct chemical linkages, is barely affected by water immersion. Similarly, with the Shoaf double-dip system on polyester, where diffusion appears to be the major mechanism, there is only a minor effect, but with the Pexul system, where it is suggested that hydrogen bonding, which could be susceptible to replacement of water molecules, is a major contributor to adhesion, there is a very much greater effect.

ENVIRONMENTAL FACTORS AFFECTING ADHESION

Except with the most modern processing lines, where fabric passes through the dipping and heat-setting unit directly into the rubberising calender, there exists a period of time when the dipped cord or fabric is stored until it is required for the next stage of the production process. It is during this period that certain environmental factors can have significant effects on the subsequent adhesion obtained, due to various ageing effects on the dipped textile.

When stored under favourable conditions, that is cool, dark and dry, dipped textiles will retain their high original adhesion for considerable periods: storage for up to two years without any significant effect on adhesion is not uncommon. However, certain factors will very rapidly cause serious loss of adhesion. Wenghoefer[19] studied the effects of light, ozone and humidity on dipped cords, and found that all of these caused serious loss of adhesion with the most severe losses occurring in the first few hours of exposure: even the u.v. component, emitted by standard fluorescent lights, was strong enough to reduce adhesion levels by one-third after exposure for 48 h.

Iyengar[20] also examined the effects of ozone and u.v. light and found that waxes, added to the RFL, could give some protection, but that chemical antiozonants or solid u.v. absorbers did not. From this it was concluded that the loss of adhesion arose from attack on the butadiene double bonds, on the surface of the RFL film, thereby preventing, or at least drastically reducing, the co-curing of the dip film with the rubber matrix: the strength of dip films was not affected by exposure and bond failure occurred at the dip-rubber interface rather than by cohesive failure of the dip film. Also, stereoscan electron microscopy showed very significant albeit very fine, surface crazing on exposed dip films.

Humidity does not seriously affect adhesion as such, but it has a marked synergistic effect on ozone degradation, and on its own, it can lead to problems such as porosity or blistering, during cure, thereby adversely affecting the final composite. This latter effect is more significant with rayon and nylon, which have relatively high-moisture regain. With nylon, for example, although the equilibrium moisture regain is only around 4 %, this can be reached within 12–18 h after dipping and drying, and after only about half an hour, it can absorb up to 1–1·5 % moisture, which, especially with heavy fabrics, can introduce significant quantities of moisture into the uncured composite, with the possibility of causing porosity or blisters during cure, due to its expansion as steam.

Even though adhesion may be significantly reduced by light or ozone degradation, it is generally possible to regain the original adhesion levels by overdipping with RFL: the resin component of the degraded film is not significantly affected and can cross-link with the resin in the new dip layer, whose surface then has the necessary complement of available double bonds to allow of subsequent cross-linking in the formation of a satisfactory adhesive bond.

REFERENCES

1. ELEY, D. D. and PAWLOWSKI, L. B. J., *J. Coll. Interface Sci.*, **29,** 1969, p. 222.
2. MITCHELL, B., *J. Inst. Rubber Ind.*, **5,** 1975, pp. 151–4.
3. AITKEN, R. G., Canadian Ind. Ltd, US 3 110 644 (1963).
4. SHOAF, C. J., E. I. DuPont de Nemours & Co., US 3 307 966.
5. MEYRICK, T. J. and WOOTTON, D. B., Imperial Chemical Industries Ltd, GB 1 092 908.
6. WEENING, W. and HUPJE, W. H., *Kaut. Gummi. Kunstst.*, **7,** 1972, pp. 321–4.
7. EDINGTON, R. A. and ALDRED, D. H., Imperial Chemical Industries Ltd, GB 1 140 528 (1969).
8. BOUSTANY, K. and CORAN, A. Y., Monsanto Co., US 3 697 364 (1972) and US 3 709 845 (1973).
9. ANTOINE, G., ARNOLD, R. L., BOUSTANY, K. and CAMPBELL, J. M., *1st European Conf., Plast. Rubber Inst.*, Brussels, April 1975. Preprint.
10. MEYRICK, T. J. and WATTS, J. T., *Proc. Inst. Rubber Ind.*, **13,** 1966, pp. 52–66.
11. ERICKSON, D. E., *ACS Div. of Rubber Chem.*, Denver. October 1973. Preprint.
12. BORROFF, E. M. and WAKE, W. C., *Trans. Inst. Rubber Ind.*, **25,** 1949, pp. 39–50.
13. FORD, J. E., *Trans. Inst. Rubber Ind.*, **39,** 1963, T1–T11.
14. SCHOON, TH. G. F. and ZIERLER, L., *Kaut. Gummi. Kunstst.*, **23,** 1970, p. 615.
15. HARTZ, R. E., *J. Appl. Polym. Sci.*, **3,** 1975, pp. 735–46.
16. IYENGAR, Y. and ERICKSON, D. E., *J. Appl. Polym. Sci.*, **11,** 1967, p. 2311.
17. VOYUTSKII, S. S., *Autohesion and Adhesion of High Polymers*, Moscow, 1960 and (translated by Kaganoff, S.; ed. Vakula, V.), Interscience, New York, 1963.
18. MATHER, J., *Brit. Polym. J.*, **3,** 1971, p. 58.
19. WENGHOEFER, H. N., *Rub. Chem. Tech.*, **12,** 1974, pp. 1066–73.
20. IYENGAR, Y., *J. Appl. Polym. Sci.*, **19,** 1975, pp. 855–63.

Chapter 8

FUSIBLE ADHESIVES FOR TEXTILES

L. M. SMITH
Staflex International Ltd, London

SUMMARY

The introductory section of this chapter describes the major areas of application of fusible adhesives for textiles and discusses the main requirements of the adhesive, the mechanism of adhesion and physical evaluation techniques, with particular reference to fusible interlinings.

The current situation is then reviewed with emphasis on the most recent advances. There are sections describing the copolyamides, developed specifically for fusible interlinings, polythene, plasticised vinyl chloride copolymers, polyvinyl acetate and ethylene copolymers. The final sections describe potential areas of development in polyesters, polyurethanes and reactive adhesives.

Adhesives are used extensively for textile applications; the major areas of consumption are:[1] tufted carpet backing (62 %); nonwoven bonding (15 %); fabric backing (8 %); flock adhesives (6 %); miscellaneous (5 %); laminating fabrics (2 %); and fusible adhesives (2 %).

Although fusible adhesives only represent a small proportion of the consumption, they present a unique problem in the formulation of adhesives. Fusible adhesives are used mainly in the garment manufacturing industry, for fusible interlinings and other garment construction aids, fusible sewing thread, shoe linings, fusible embroidery and patches and also for laminating more specialised fabrics such as lace, fur, suede and leather. Of these, by far the most important are fusible interlinings; in 1976 it is estimated that 490 million metres were sold worldwide, of which 142 million metres were consumed in Europe.

A fusible interlining is a woven, knitted or nonwoven fabric coated with a thermoplastic adhesive. The interlining, cut to the required shape, is fused to the reverse side of a garment part to give control or reinforcement to the top fabric, for example, to control and prevent creasing in the front of mens' jackets, to reinforce a collar or cuff, to strengthen a pocket opening or give body to a cheap top cloth. The lamination of the individual parts is normally carried out in a fusing press under the influence of heat and pressure. The main requirements of the adhesive are that it must give adequate and uniform adhesion to both top and base fabrics, and withstand repeated washing and/or dry-cleaning. Commercial considerations dictate that this must be achieved in a fusing press operating for the shortest time and at the lowest temperature and pressure possible. The adhesive should melt and flow sufficiently to bond the fabrics but not so much that it penetrates to the outer surface as this is detrimental to appearance and handle. The softening range of typical adhesives is 90–150 °C and the melt viscosity is between 8–20 N sec m^{-2} (8000–20 000 poise) at the fusing temperature. The adhesive is normally applied to the base fabric as a discontinuous coating so that after fusing the textile hand of the laminate is retained. The fusible interlining must be storage stable as several months may elapse between manufacture and eventual use in a garment factory.

Although, in principle, any thermoplastic material can form a heat seal adhesive, in practice very few are suitable for textiles.[2] Those in current use are: special polyamides; plasticised vinyl chloride/vinyl acetate copolymers; low-density polyethylene (MFI 7–200); high-density polyethylene; polyvinyl acetate (PVA) and PVA/phenolic blends; ethylene vinyl acetate copolymer (EVA), hydrolysed EVA and other ethylene copolymers; and plasticised cellulose acetate. Thermoplastic polyurethanes and polyesters, ionomer resins and acrylics have become available recently but are not yet in general use.

MECHANISM OF ADHESION

Bonding of textiles with fusible adhesives is normally carried out in a fusing press under the influence of heat, pressure and time. The materials to be laminated, which are initially at room temperture, are placed in a hot press, which may typically operate with a head temperature of 150 °C, at a temperature of 20 800 Nm^{-2} (200 g cm^{-2}) and a dwell time of 12 sec. As the temperature of the adhesive increases, the pressure will cause plastic flow to occur which will give way to viscous flow as the softening point is reached and exceeded. The molten adhesive will flow into the interstices and around the yarns of both top and base substrates, and on removal from the press and cooling it solidifies forming a bond between the fabrics. With a press-head temperature of 150 °C the dwell time will be adjusted to give optimum adhesion, depending on the thickness of the fabrics involved. Normally, the final glue line temperature is found to be 10–15 °C lower than the press temperature, i.e. 135–140 °C in this case. At these temperatures, the melt viscosity of most adhesives is high, a typical copolyamide, ®Platamid H005P, has a viscosity of 16·5 N sec m^{-2} (16 500 poise) at 140 °C. With such high viscosities and short fusing times the adhesive is not likely to wet the surface of the fibres and one would not expect flow due to surface tension to be important. There is little evidence that the adhesive penetrates between the fibres of the yarns. Scanning electron micrographs show that the bulk of the adhesive has flowed into the interstices between the yarns not into the yarns themselves (*see* Frontispiece).

The adhesion appears strongly dependent on a mechanical component arising from the irregular surface of the yarn. Some specific adhesion must exist but this may frequently be lower in magnitude due to weak boundary layers formed by weaving oils, electrostatically-held dust or fibre particles.

Most laminates fail due to breakdown of this mechanical key, due to shearing forces or swelling of the yarns or adhesive during cleaning. Scanning electron micrographs of adhesive dots pulled from the top cloth show the impression of the yarns and loose fibres locked into the adhesive.[3] The main exception is plasticised PVC which has low-mechanical strength and often fails cohesively.

PHYSICAL EVALUATION

The evaluation of potential fusible adhesives normally involves measurement of the softening/melting point and the melt viscosity–temperature relationship. If these properties are satisfactory the adhesion to textiles and the resistance to dry-cleaning solvents and laundering is determined. For normal uses the melting point, as measured with a hot stage microscope or Kofler hot plate, is between 110 °C and 130 °C, but may be as low as 80 °C for special low-temperature applications. This softening point determines the minimum fusing temperature possible. The ball and ring softening point (ASTM E28-58T) is typically in the range 120–160 °C. Visual examination

®Platamid is a registered Trade Name owned by Plate Bonn GmbH.

during the test gives a useful indication of melt viscosity at the softening point, and can be used to predict fusing conditions and the extent of strikeback (flow of adhesive to the outer surfaces).

The melt viscosity is crucially important as it controls the extent of flow during fusing. Materials with low viscosity flow easily, will only require short fusing times and can easily be over-fused giving high strikeback and poor adhesion, whereas materials with high melt viscosity will require long fusing times and higher pressures. Materials with melt viscosities which remain fairly constant with increasing temperature allow the most flexible fusing conditions. A melt viscosity of around 30 N sec m^{-2} at 120 °C is typical. Flow rate results from an Extrusion Plastometer (ASTM D1238-62), must be interpreted with caution as the usual pressure in the extruder is 302 000 N m^{-2} (43·7 p.s.i.) whereas that in a fusing press is only of the order of 20 000 N m^{-2} (about 3 p.s.i.). Melt flow data can be converted to viscosity using the method of Edelmann.[4] Brabender Plastograph results can be misleading as the measurements are taken under conditions of high shear. However, they can be used to compare similar materials. This technique can also be used in the formulation of PVC plastisols to measure the gel strength.

Recently, thermomechanical analysis techniques have proved very useful for the evaluation and comparison of hot melt adhesives[5] and fusible interlinings. There are currently 3 models available from Du Pont (UK) Ltd, Perkin Elmer Ltd and Stanton Redcroft. A thermomechanical analyser provides a record of the dimensional and viscoelastic properties of a solid state sample. The Perkin Elmer instrument is shown schematically in Fig. 1. The sample is placed on a quartz sample holder, and a quartz rod probe is counterbalanced to rest on top of it. The position of the probe is monitored by a movable core linear variable differential transformer (LVDT), which gives D.C. output proportional to the linear displacement of its core. This output is fed to a recorder. The core/probe assembly is suspended from above on a float immersed in a high density liquid. Pressure can be applied to the sample by adding weights to the tray on top of the probe. The lower portion of the sample tube is positioned inside a furnace (which can be lowered to provide access to the sample). The furnace temperature controller has direct read out of the temperature of the furnace and provides rates of heating from 0·3125 °C to 320 °C per minute. (In practice, thermal lag limits the fastest rate to 20 °C min^{-1}.) The temperature of the sample is monitored by a thermocouple placed in the sample tube, whose output is displayed on the recorder.

The most useful measurement for fusible adhesives is the determination

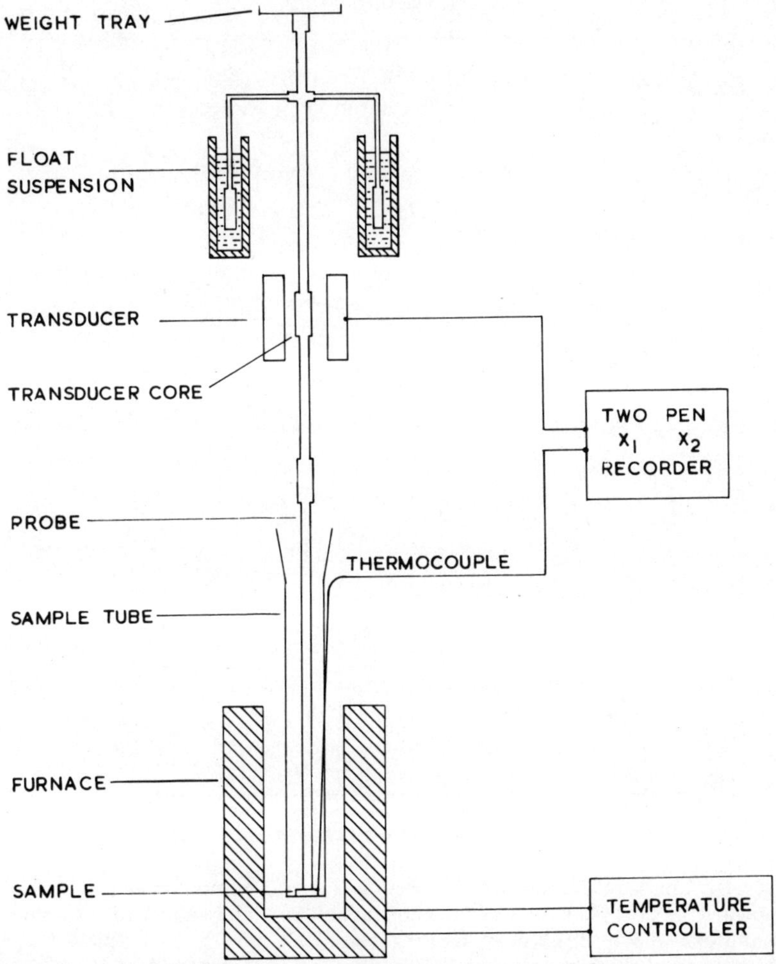

FIG. 1. Schematic diagram of the Perkin Elmer TMS1, thermomechanical analyser.

of the rate of penetration of the probe into the sample as it is heated. This gives an indication of the softening point and the melt viscosity. The probe can be loaded to apply a pressure equivalent to that of a fusing press so that the instrument can be used to indicate the actual behaviour of the adhesive during fusing. Figure 2 shows recordings of three common interlining adhesives, a terpolyamide, low-density polyethylene and plasticised

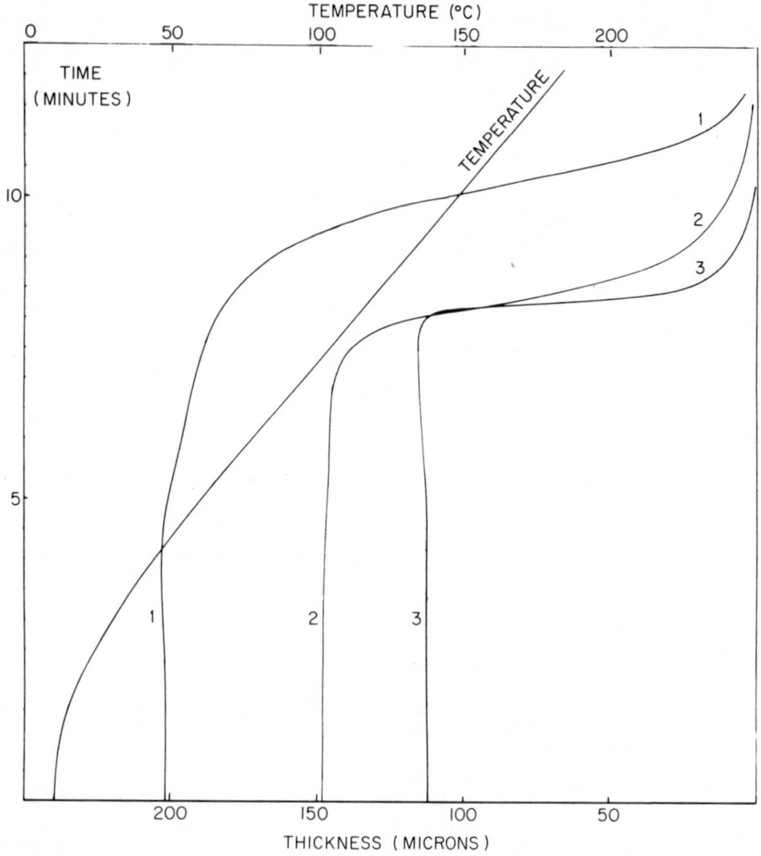

FIG. 2. Chart recordings from a Perkin Elmer thermomechanical analyser, showing the flow characteristics of three adhesives under simulated fusing press conditions. Load $20\,kN\,m^{-2}$, heating rate $20\,°C\,min^{-1}$. 1, |Plasticised| vinyl chloride–vinyl acetate copolymer (Staflex International Ltd, Solidot®). 2, Terpolyamide of nylon 6/6·6/12 (Emser Werke AG 'Griltex 1P'®). 3, Low-density polythene, MFI 70 (ICI Ltd, 'Alkathene 22'®).

polyvinyl chloride. All the samples were hot pressed to form discs 250–500 μm thick and aged for 24 h, before testing. The samples were heated at $20\,°C\,min^{-1}$ with a load equivalent to $20\,000\,N\,m^{-2}$ (approximately $200\,g\,cm^{-2}$). Figure 3 shows the behaviour of four terpolyamides. This technique can also be applied to coated interlinings. The interlining is

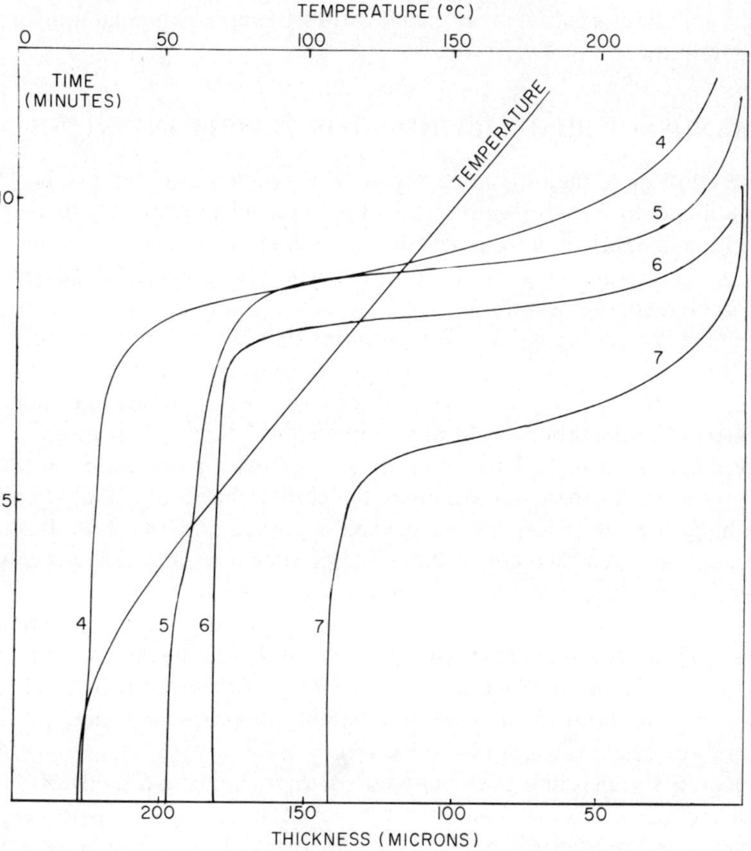

FIG. 3. TMA chart recordings comparing the flow characteristics of four terpolyamides under simulated fusing press conditions. Load 20 kN m^{-2}, heating rate 20 °C min^{-1}. 4, Platamid H105 (Plate Bonn GmbH). 5, Griltex 2P (Emser Werke AG). 6, Platamid H005 (Plate Bonn GmbH). 7, Platamid H103 (Plate Bonn GmbH).

placed coated side up so that the probe covers several dots or particles. Under simulated fusing press conditions the behaviour of the adhesive as it melts and flows into the cloth can be determined from the resultant trace. This technique has been successfully used to differentiate between substrates coated with different polyamides. Also available is an accessory for extension analysis, which permits distortion studies of drawn fibres, polymer films, etc. Here the sample is held between a probe and the sample tube and the elongation or shrinkage caused by various thermal transitions monitored.

MANUFACTURING METHODS FOR FUSIBLE INTERLININGS

There are three major coating methods, random scattering, dry powder printing and rotary screen printing of an aqueous dispersion or plastisol. Random scattering, the original coating method, is the cheapest and easiest method and still widely used for products where the potential disadvantage of uneven coating is unimportant.

In the last few years dry powder printing (particularly from an engraved roller) has been established as the major coating method of interlining for large area fusing. The process depends on a roller engraved with a triangular pattern of indentations each approximately 0·5-mm deep with up to 30 holes per linear inch. Thermoplastic powder (with particles in the range 0–200 μm) is doctored into the holes, hot cloth is then brought into contact with the engraved roller and the powder deposited onto the cloth. Passage through an oven then consolidates the powder into dots. The process is clean, fairly simple in operation and there is little wasted adhesive. Dry powder printing is mainly used for polyamides and high-density polythene, over 50 % of all polyamides used for fusibles are coated by this method. Two major limitations are the difficulty of coating synthetic nonwoven fabrics, due to heat shrinkage and the unreliability of processing some powder blends or externally-modified adhesives.

Rotary screen printing of a plastisol, solvent or aqueous dispersion of fine polymer powder does not have these disadvantages. Nonwovens are easy to process and plasticisers, modifiers, fillers, pigments, etc., can be added to the dispersions to enhance the performance of the adhesive. Recently, significant advances have been made in this area, details of the formulations are given in the sections on polyamide and high-density polythene. The main disadvantages are the alteration of the hand of the substrate due to absorption of dispersing agents and potentially higher wastage of raw materials.

There are several other less important methods of applying the fusible adhesive. Knife coating a continuous layer of emulsion or plastisol is mainly used for polyvinyl acetate and some polyvinylchloride. The adhesive, dissolved in a suitable solvent, can also be applied from a spray. Some adhesives, for example high density polythene and polyamide, can be preformed into a network, which can be used on its own or laminated onto the base fabric to form a fusible interlining.

FUSIBLE TEXTILE ADHESIVES

A discussion of recent developments and the current status of each of the major adhesive types will now be given.

Polyamides

The most important development in fusible adhesives in the last few years has been the dramatic increase in the use of copolyamides. These materials have been specially developed for use in fusible interlinings and their properties have been modified to give optimum results.

Aliphatic homopolyamides have softening points above 200 °C which preclude their use as fusible adhesives. The softening point can be reduced by copolymerisation of amino acids with diamine/dicarbonic acid combinations and introduction of polymethylene chains of different lengths, as this interferes with chain association in the crystalline regions of the polymer. The first developments of this type were with binary and ternary copolyamide systems based on caprolactam, 6·6 and 6·10 salts. Originally the softening points were still too high and were reduced by the addition of external plasticisers. This problem was overcome by the terpolyamides developed in the 1960s by Emser Werke AG[6] and Plate Bonn GmbH.[7] The most successful are terpolymers of caprolactum, hexamethylene diamine adipate and lauroylactam (nylon 6. 6·6 and 12). The use of nylon 12 gives the polymers a much lower softening point and reduces the water absorption potential. The selection of the specific basic monomers and their respective concentrations control formation of hydrogen bonds, thus the degree of crystallinity can be optimised. Similarly, the degree of polymerisation controls the melt viscosity. Figure 3 shows the correlation between the ratio of monomers and the resulting melting temperature.[8] For commercial use polyamides with softening points in the range 110–140 °C are selected. Physical properties of some commercially available polyamides are compared in Table 1, and Fig. 2 shows their behaviour in the thermomechanical analyser. It will be seen that Platamids H005 and H105 have the same softening point but different melt viscosities.

These polyamides have many advantages over polyethylene and plasticised polyvinylchloride. They give very high adhesion to textiles even at low coating weights, which gives the advantage of improved handle of the laminate. Adhesion levels of 1 kg per 25-mm strip are normal in a 180° peel test when fused to wool or cotton outer fabrics. The bonds are so strong that often the fabric tears during the test. A comparison of polyamides with other adhesives is given in Table 2.

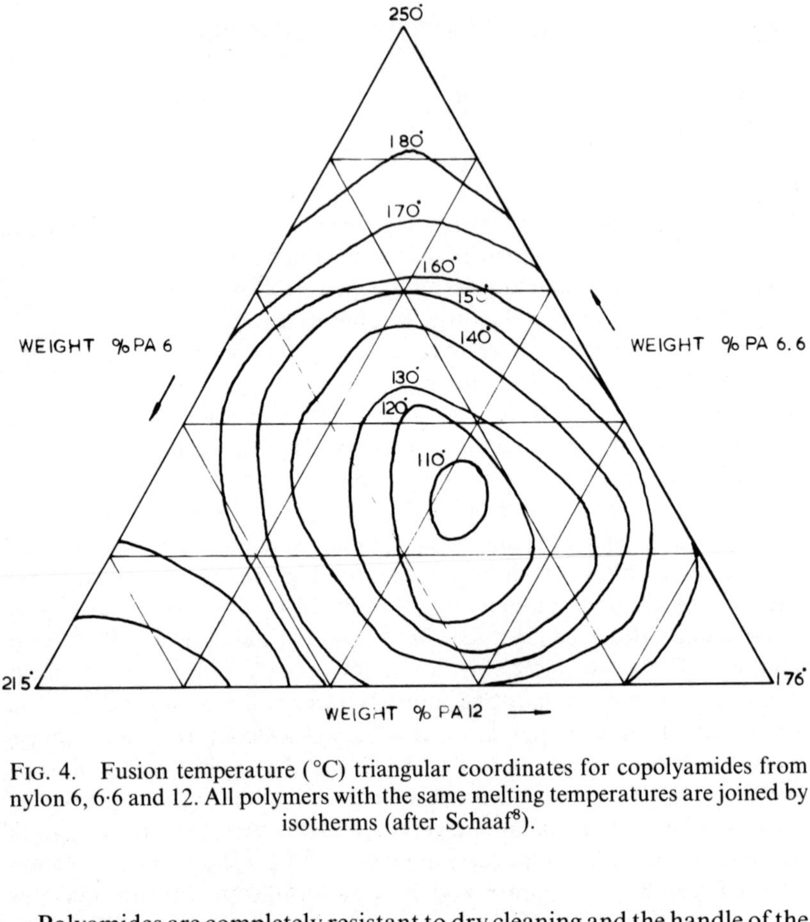

FIG. 4. Fusion temperature (°C) triangular coordinates for copolyamides from nylon 6, 6·6 and 12. All polymers with the same melting temperatures are joined by isotherms (after Schaaf[8]).

Polyamides are completely resistant to dry cleaning and the handle of the laminate is unaffected. This is an advantage over fabrics laminated with polyvinylchloride where the handle becomes firmer after dry cleaning due to plasticiser extraction. The standard polyamides are not, however, fully resistant to washing, and laminates are liable to fail if washed at 60 °C or over. The inherent problem is that these polyamides are plasticised by steam. This can be an advantage as steam assists fusing, allowing lower press temperature to be used, but the steam sensitivity is detrimental in 'pressing off' finished garments and in laundering.

Increased resistance to steam and boiling water can be obtained by increasing the proportion of lauroylactam to 60–65 %. However, this also increases the softening point (Fig. 3) and the commercially-available

TABLE 1
PHYSICAL PROPERTIES OF SOME TYPICAL TERPOLYAMIDES

Manufacturer	Emser Werke AG	Plate Bonn GmbH	Plate Bonn GmbH	Emser Werke AG	Plate Bonn GmbH	Bostik Ltd
Product code	Griltex 1P	Platamid H005	Platamid H105	Griltex 2P	Platamid H103	HM509X
Polyamide type	6/6·6/12	6/6·6/12	6/6·6/12	6/6·6/12	6/12/—	6/terpolymer
Melting range[a] (°C)	110–120	115–125	115–125	120–130	80–90	110–120
Softening point[b] (°C) (ball and ring)	131	133	140	144	108	131
Melt viscosity[c] ($N\ sec\ m^{-2}$) at 130 °C	30·0	33·0	100·0	20·0	6·5	—
140 °C	20·0	16·5	50·0	10·0	4·0	10·0[e]
150 °C	15·0	10·0	33·0	7·5	2·5	—
160 °C	10·0	6·5	25·0	5·5	1·5	5·5[e]
Relative viscosity[d]	1·5	1·55	1·65	1·4	1·4	—
Comments	Standard polyamide	Standard polyamide		Good wash resistance	Very low melting	

[a] Quoted by manufacturer.
[b] Determined in the laboratories of Staflex International Ltd, by ASTM E28-58T.
[c] Quoted by manufacturer, determined by ASTM D1238-62 at 2·16 kg.
[d] Quoted by manufacturer, determined in 0·5 % solution, in *m*-cresol at 25 °C.
[e] Determined in the laboratories of Staflex International Ltd.

TABLE 2

TYPICAL ADHESION LEVELS IN A 180° PEEL TEST WHEN VARIOUS ADHESIVES ARE USED TO FUSE A COTTON BASE CLOTH TO WOOL SUITING

Adhesive	*Coating type*	*Softening point (°C) (ball and ring) ASTM E28-58-T*[a]	*Coating weight* ($g\,m^{-2}$)	*Fusing conditions*			*Adhesion*[b] *(kg per 25-mm strip)*
				Press temperature (°C)	*Pressure* ($kN\,m^{-2}$)	*Time (sec)*	
Terpolyamide	Dot	133	17	150	20[c]	12	0·6–1·2
Low-density polythene	Sprinkle	111	30	140	20	7	0·4–0·6
High-density polythene	Dot	160	20	160	28	20	0·4–0·7
Plasticised PVC/PVA	Dot	154	40	150	20	12	0·5–1·0
PVA/Novolac	Sprinkle	125	35	140	20	7	0·5–0·8
Plasticised cellulose acetate	Sprinkle	163	35	150	20	10	0·5–0·8

[a] Determined in the laboratories of Staflex International Ltd.
[b] Exact value depends on thickness of base fabrics, type of press, etc.
[c] $20\,kN\,m^{-2} = 0{\cdot}2\,kg\,cm^{-2}$.

materials of this type (for example Platamid H104 and ®Griltex 2P) have softening points of 130 °C (Table 1 and Fig. 2).

In 1973, copolyamides with very low softening points, but still retaining resistance to dry cleaning, were made commercially available, for example, Emser Werke AG, Griltex 4P and Plate Bonn GmbH, Platamid H103; typical properties are shown in Table 1 and Fig. 2. These products were aimed specifically at the fur, suede and leather markets where press temperatures are restricted to 100 °C to avoid spoiling the leather. Previously, PVA and EVA were used in these areas because, although they are soluble in dry cleaning solvents, they were the only suitable materials with sufficiently low softening points. Copolyamides with softening points of 80–90 °C were achieved by replacement of some or all of the hexamethylenediamine adipate by other monomers. Patents held by Plate Bonn GmbH[10] and Emser Werke AG (Inventa AG für Forschung und Patentverwertung)[11] describe ter- and tetrapolyamides incorporating the numbered nylon monomers 6, 6·6, 6·9, 6·10, 6·12, 6·13, 11 or 12. Examples are given of copolymers comprising 6/6·6/6·12/12, 6/6·6/11/12 and 6/6·6/6·10/12 with softening points of 95 °, 90 ° and 85 °C, respectively. The softening points can be reduced further, for dispersion printing applications, but the addition of plasticisers such as *o*- and *p*-toluene sulphonamide.

The majority of the polyamides used as fusible textile adhesives are the terpolymers based on lauroylactam, manufactured mainly by Plate Bonn GmbH, Emser Werke AG and Chemische Werke Hüls AG and marketed in various forms by these and other companies. Worldwide consumption of these terpolymers is currently estimated at over 6000 tonnes per annum.[12] Alternative polyamides have been developed which are not dependent on lauroylactam, they are based mainly on nylon monomers, 6, 6·6, 6·9, 6·10, 11 and 6·12. Although these materials have good performance they only command a small share of the market.

The most widely used in Europe are binary and tertiary polyamides, manufactured by Courtaulds Ltd. Their two main products are Celafuse SG, a plasticised copolymer on nylon 6/6·6, and Celafuse 110, a terpolymer of nylon 6/6·6/11 with a softening point of 110 °C.[13] Bostik Ltd have recently made three polyamides available commercially (*see* Table 1). A patent[14] describes polyamides prepared from caprolactam, adipic or sebacic and sebacic or dodecanedioc acids with hexamethylenediamine and/or trimethylhexamethylenediamine. It is claimed that the 6/6·10/6·12 has

®Griltex is a Registered Trade Name owned by Emser Werke AG.

improved resistance to laundering over the 6/6·6/6·12 copolymer and that inclusion of the hindered diamine reduces the chain association in the crystalline regions and so reduces the softening point. Polyamides based on the same materials but in different proportions have also been described[15] and similar materials are available in the USA from the United Shoe Machinery Co. Also in the USA, Du Pont de Nemours Inc. manufacture Elvamids which are thought to be 6/6·10/6·12 type polyamides. Schering AG have recently made available copolyamides based on dimer acids.

In Japan, a large proportion of the polyamides are imported from Europe but some of the 6/6·6/12 type are manufactured under licence. Many other Japanese companies hold patents similar to those described above, for example, a terpolyamide of caprolactam, a HMDA salt with a C_9–C_{20} dimer acid and a trimethyl HMDA salt with adipic or sebacic acid.[16] Other recent modifications include a patent from Emser Werke AG[17] describing a terpolymer of caprolactam (30 %) lauroylactam (30 %) and methyl caprolactam (40 %). Diacel Ltd,[18] disclose that 10–30 % methoxy methylation of 6/6·6/12 terpolyamides gives a lower softening point and better resistance to washing than the untreated polymer. Chemische Werke Hüls AG[19] claim that increased wash resistance results from a 6·6/6·12 copolyamide containing 1 % amides prepared from 3-dimethylaminopropylamine and tall oil fatty acids and 1 % of the reaction product of nonoyl phenol, formaldehyde and tetraethylenepentamine.

Normally, polyamides do not perform well when fused to waterproofed fabrics such as siliconised rainwear; hydrophobic adhesives such as polyethylene and polyvinylchloride give superior performance. However, in 1975 Plate Bonn GmbH introduced a modified copolyamide, Platamid M840, specially designed for this area which is moderately successful.

For scatter coating and dry powder printing, changes in the nature of the adhesive coating, such as lowering the softening point, can only be achieved by modifications to the polymer itself. Dispersion printing offers scope for additives such as plasticisers, fillers, etc., which, although applied externally, effect the properties of the molten adhesive during fusing. The basic ingredients of a dispersion are polyamide powder (maximum particle size 80 μm), an approximately equal proportion of water and dispersion agents and/or protective colloids; for example, polyacrylate thickeners such as ®Primal ASE 60 or ®Viscalex AH15, polyvinyl pyrrolidone solutions such as ®Collacral VL or hydroxyethyl cellulose thickeners. The working

®Primal and Collacral are Registered Trade Names owned by BASF (UK) Ltd,
®Viscalex is a Registered Trade Name owned by Allied Colloids Ltd.

viscosity for a printing paste is typically in the range 0·02–0·05 N sec m^{-2} (20–50 poise). This basic dispersion can then be modified as required. Solid and liquid plasticisers, usually sulphonamides, can be added up to 25% by weight, to reduce the softening point and allow less severe fusing conditions. Waxes, fillers, pigments and u.v. fluorescing agents may also be added to improve handle, reduce surface tack, improve appearance and aid inspection of the finished goods. Obviously, the systems are selected to give the best printing qualities and ease of processing with the least impairment of the performance of the adhesive. The system may be modified depending on the type of substrate to be coated, for example, whether it is a woven or nonwoven fabric. A great deal of work has been done recently in perfecting these aqueous dispersions and most polyamide suppliers distribute details of suitable formulations or supply ready-made dispersing systems.

For coating nonwoven fabrics, Plate Bonn GmbH recommend:[12]

2·2 pts. Dispersing agent Platamid M785
0·7 pts. 1:1 mixture *n*-ethyl and *o*-*p*-toluenesulphonamides
3·4 pts. Water
3·1 pts. Platamid H005PA powder 0–80 μm.

This is pre-mixed, then just before use:

1·5 pts. 0·5% solution polyethylene oxide (®Polyox COAG) is added.

The viscosity should be 3×10^{-2} N sec m^{-2} (30 poise).

The dispersing system M785 is believed to be an aqueous emulsion containing about 15% polymeric thickeners and about 25% C_{22-36} wax acid, such as oxidised montan wax. This type of formulation was described in a recent patent.[20]

Chemische Werke Hüls AG[21] describe an aqueous dispersion based on an aminoamide and its salt, for example, 3-dimethylaminopropylstearoylamide mixed with its acetic acid salt, glycerol and hydroxyethyl cellulose Chemische Fabrik Stockhausen & Cie offer a dispersion system which is itself a heat-sealable compound. Many interlining manufacturers design their own systems to suit their particular processing equipment, base cloths and customers.

Despite all the factors in their favour, polyamides have a severe disadvantage, the price. Currently (1977) at well over £2000 per tonne, polyamides are, weight for weight, more than four times as expensive as low density polyethylene and plasticised PVC. Apart from being cryogenically ground to form fine powders, polyamides can be formed into thread or film.

®Polyox is a Registered Trade Name owned by Union Carbide Ltd.

Fusible thread, both mono- and multifilament, has been used to bond hems and as the underthread for manufacturing fusible embroidery, patches, motifs, etc. A special polyamide thread manufactured by Emser Werke AG breaks up in steam and is used as the draw thread for separating knitted parts. A polyamide film, cut to form a network and manufactured by Xiro AG, is used for hemming and laminating, e.g., fur fabrics to leather. Labels, appliqués and emblems can be fused to garments with polyamide film thus eliminating the cost of sewing. Polyamide film applied to zip tapes can be used to heat-seal them into garments.

Polyethylene

Low-density polyethylene is still widely used in fusible interlinings, despite its relatively poor solvent resistance. This is because it is so cheap; ground powder is still less than £500 per tonne. It is mostly applied in powder form as a random scatter coating for women's and children's wear, or as a continuous film for shirt-collar interlinings. The most widely used grades have a density of about 0·916 with MFI 7, 20, 70 or 200 depending on the fusing conditions required. These materials fuse easily and are resistant to washing up to 50 °C. However, they are not resistant to dry-cleaning if the temperature of the drying cycle is above about 60 °C. The extent to which polyethylene is swollen by hot perchlorethylene and hence its resistance to dry cleaning is dependent on its density and degree of crystallinity. Better results are obtained with polyethylene of density 0·924 and MFI 20; more of this is being seen in the market, particularly coated onto nonwoven fabrics, as a cheap, if not reliable, alternative to polyamides.

Powdered LD polyethylene and polyamide are often blended to give a medium priced product with some of the advantages of polyamide. A recent patent[22] describes a hot-melt blend of LD polyethylene and a polyterpene resin, for example, a β-pinene, which is then ground to a powder. An 80/20 blend of polyethylene (MFI 23) and β-pinene is claimed to give five times the bond strength of polyethylene alone and increase the dry-clean resistance.

High-density polyethylene has good adhesion, wash and dry clean resistance but, because of its high softening point and high melt viscosity, requires relatively severe fusing conditions. Most high-density polyethylene is used in shirt collar interlinings where good wash resistance is essential. Typical fusing conditions would be 160 °C for 25 sec at 170 kN m^{-2}. Because the top fabrics used for shirts are very fine, closely-spaced small dots of adhesive give fused laminates with the best appearance. The coating may be applied by screen printing of an aqueous or solvent dispersion or dry powder printing from an engraved roller. Recent advances in processing technology now permit 25 or 30 mesh (154 dots per cm^2) coatins as standard, and even 40 mesh printing is available. The most commonly employed polymers have density of 0·953–0·964 with MFI 6 or 20, and are applied at a coating weight of 20–30 $g\,m^{-2}$.

Some fusible shirt interlinings are manufactured using a high-density polyethylene network.[42] In this process, the polythene film is embossed and biaxially stretched to form a network of dots connected by thin strands. When applied to hot-base cloth the strands break leaving a regular pattern of dots adhering to the base cloth. This process can also be used to make interlinings for other end uses provided the high fusing conditions can be tolerated, see Table 2.

Recently available on the market are some high-density polythenes with high melt flow indices. These materials have good adhesion properties, are resistant to washing and dry cleaning and because of their lower melt viscosity can be fused at lower temperatures and pressures than other HD polythenes. These materials are likely to be increasingly important, particularly for womens' wear, in the future as the price is just over half that of polyamide.

Plasticised Copolymers of Vinyl Chloride and Vinyl Acetate

The first dot-coated fusible interlinings were produced on a commercial scale by screen printing PVC plastisols, and these products are still widely used. Normally, copolymers of vinyl chloride with 7–10 % vinyl acetate are blended 1:1 with phthalate or polymeric plasticisers and the resulting plastisol printed onto the base cloth. The dots of plastisol are then gelled under carefully-controlled conditions. PVC-coated interlinings have good adhesion and resistance to laundering and dry-cleaning, but the coating weights required are fairly heavy which can adversely affect the handle of the laminate (*see* Table 2). After coating weight differences have been taken into consideration, PVC systems are still only one-third the price of polyamides and, although slightly more expensive, have superior performance to LDPE. PVC coatings are used extensively for rainwear where the performance of polyamides is poor. By selecting a lower level of plasticiser, vinyl chloride copolymer blends can be prepared that are dry powders at room temperature. These are used for random-scatter coating. Base fabrics coated with a continuous film of plasticised PVC are also known, particularly for shirt-collar interlinings.

Ethylene Copolymers

Ethylene vinyl acetate copolymers have low melting points typically in the range 60–95 °C, but are not normally used for fusible interlinings because of their limited solvent resistance. They are widely used for fur, leather and suede where low-fusing conditions are necessary, although the special low melt polyamides are competing in this area. They are used in the upholstery trade and extensively in the automobile and allied industries for backcoating moulded carpets.

Partially-hydrolysed EVA, in which the acetate groups have been converted to alcohol, has better resistance to dry-cleaning and washing and is used in fusible interlinings to a limited extent, particularly in Japan. For example, an EVA with 20–45 % vinyl acetate when saponified to more than 80 %, is claimed to give good performance as a fusible adhesive for textiles.[23] Another patent[24] describes a dispersion for screen printing of a 90 % saponified EVA (with 28 % vinyl acetate) 33 parts, water 60 parts, thickeners 1 part and nitrile rubber latex 10 parts.

Ethylene ethyl acrylate polymers are used as hot melt adhesives[25] and have advantages over EVA and LDPE for packaging applications, but are unsuitable for garment manufacture because of the lack of solvent resistance. A range of EEAs is marketed by Union Carbide. A recent patent[26] describes a blend of hydrolysed EVA with a saponified copolymer of ethylene and methyl acrylate or methyl methacrylate. This is claimed to show improved resistance to dry cleaning, when used in fusible interlinings, than EVA alone. Another patent[27] discloses copolymers of ethylene/glycidyl methacrylate and ethylene/acrylic acid which, when blended, give adhesives suitable for textile applications.

Polyvinyl Acetate

Polyvinyl acetate is extensively used in areas of textile bonding where low-temperature fusing or short dwell times are essential and solvent or water resistance is relatively unimportant, for example, for furniture upholstery and leather.

The use of PVA/Novolac blends has declined with the rise in popularity of polyamides. These blends have limited resistance to washing and dry-cleaning. They are mainly used where steam fusibility is essential and, in many cases, this type of material is used as a garment construction aid rather than a true fusible interlining.

Interlinings coated with fusible adhesives which only give temporary bonds are used as construction aids to save basting fabrics together and having to remove the stitches after assembly of the garment. There are three main types, either the fabrics are only lightly bonded so that they can be pulled apart easily, or the adhesive is so low melting that the bonds disappear during pressing off with steam, or the adhesive is water soluble and complete detachment of the interlining from the top fabric occurs in the first wash. The last type is extensively used in the Far East, particularly for fine fabrics where the dots of adhesive on normal interlining would show through and automated collar production demands that the parts be fused together. Polyvinyl acetate, modified PVA and polyvinyl alcohol are all suitable for this application. Patents are known covering interlinings coated with a saponified and acetalised PVA with 30 mole % acetate groups and 20 mole % acetate groups;[28] a dried polyvinyl alcohol foam;[29] the sodium salt of a 24:16:57:3 copolymer of 2-ethylhexylacrylate, methacrylic acid, vinyl acetate and polyethylene glycol;[30] and a 69:20:11 copolymer of ethylacrylate, methylmethacrylate and sodium acrylate.[31] All are coated at about 5 $g\,m^{-2}$ and fused with a steam press or iron for 5–15 sec.

Polyesters

Copolyesters of terephthalic acid with aliphatic acids and diols are used extensively as hot melt adhesives, particularly in the shoe industry. The melting point, normally in the range of 100–200 °C is largely determined by the mole fraction of terephthalic acid, whilst the glass transition temperature is directly related to the aliphatic content. The proportion and type of comonomers is thus varied to give the required physical characteristics. These types of materials are not normally used in fusible interlinings for garments as the wash and dry-clean performance is inadequate.

In the last few years several copolyesters have been made available in which the laundering and solvent resistance has been improved without increasing the softening point unduly. An example is Dynamit Nobel's S1218 with a ball and ring softening point of 134 °C and a melt viscosity of about 7 $N\,sec\,m^{-2}$ (7000 poise) at 140 °C. However, polyesters are not yet serious competitors for polyamides as they offer little price or performance advantage.

There are many patents describing copolyesters as adhesives for fusible interlinings but it is not clear if any are available commercially. Most appear to be based on terephthalic and isophthalic acids and esters, ethylene or polypropylene glycol and/or 1,4-butanediol, several include adipic acid and caprolactam. Some recent examples of these are patents from Toray Industries Inc.,[32] Eastman Kodak Ltd,[33] National Starch and Chemical Corp.[34] (all copolyesters), Nippon Synthetic Chem. Ind., Co., Ltd[35] (a poly(ester amide)) and Toray Industries Inc.[36] (a block poly[ether ester amide]).

Polyurethanes

Thermosetting polyurethanes are used extensively to produce laminated fabrics, particularly using low-cost knitted materials. The laminates are formed in one operation, the urethane prepolymer solution is printed in a discontinuous layer onto one fabric and immediately brought into contact with the second fabric. The adhesive is then dried and thermoset. The 'pot-life' of this type of urethane is so short that they are not suitable for fusible applications.

Thermoplastic polyurethanes are available but tend to have softening points and melt viscosities which are too high for use as fusible textile adhesives. A few of these materials can be successfully fused under conditions similar to those used for high density polyethylene. For example ®Estane 161-48, with a ball and ring softening point of 165 °C. Tests show that the resistance to washing is not as good as HDPE. An interlining coated with a continuous film of polyester urethane aimed at work wear garments, overalls, etc. is available commercially. In 1973, B. F. Goodrich Chemical Co., produced Sharnet, a random network of polyurethane filaments for laminating, but the cost was high and the fusing conditions too severe. Many patents disclose thermoplastic urethanes; recently Bayer AG[37] described the heat-sealing of textiles with powdered polyurethane and polyurethane–ureas prepared from aliphatic polyols. For example, adipic acid–1,4-butanediol polyester reacted with 5 % of a polyisocyanate and a small proportion of an aliphatic diol. This gave a powder with a softening point claimed to be in the range 100–120 °C and which bonded textiles at 170 °C. Another[38] claims a polyurethane elastomer film, which laminated textiles when fused for 40 sec at 110 °C, prepared from an OH-terminated 1,4-butanediol–sebacic acid prepolymer and xylene diisocyanate.

Miscellaneous

Many other polymers have been tried as fusible adhesives with limited success. Thermosetting acrylate adhesives which are so important in producing laminating fabrics are unsuitable for fusible application because their pot-life is so short. Other polyacrylates are not resistant to dry-cleaning and the melt viscosity tends to be too low leading to overfusing.

An interesting patent[41] describes the preparation and use of terpolymers of styrene (3–30 parts), acrylonitrile (20–50 parts) and a C_{1-4} alkylacrylate (20–77 parts). These materials are said to have softening points in the range

®Estane is a registered Trade Name owned by B. F. Goodrich Chemical Co.

60–150 °C, show high initial adhesion and good resistance to washing and dry-cleaning. To date, none of this material has been seen commercially by the author!

FUSIBLE THERMOSETTING ADHESIVES

A fusible adhesive with the capability of cross-linking on fusing to form a permanent bond is a very attractive idea. High-fusing pressures would not be necessary as the initial polymer could have a low melt viscosity and flow easily. This should also favour high adhesion as the low viscosity should allow the polymer to wet the fibres of the yarns more readily and form more effective bonds. The usual type of thermosetting adhesives used for textile laminating, such as solvent polyurethanes and polyacrylates, are not suitable as once supplied their shelf life is too short. To be suitable for fusible interlinings the catalyst/cross-linking system employed must be stable for several months after application to the base cloth and then be activated by temperature in the range 140–160 °C during fusing. A further and more severe restriction is the dwell time in the fusing press, normally less than 20 sec. As most of this time is taken up with heating the system from ambient to the press temperature, the total time available for reaction is limited to a few seconds. Therefore, it is probable that the extent of cross linking is limited by the dwell time in the press.

Systems of this type are available commercially and have been described by Staflex International Ltd,[39] and Raduna AG.[40] Both systems use a suitable powdered polyamide as the base adhesive. This is incorporated into an aqueous dispersion which also contains thickening agents, a catalyst and a formaldehyde donor as cross-linking agent. Once coated onto cloth to form interlining and fused, the adhesive is claimed to give bonds with improved performance over normal polyamide. The fusing condition quoted in the latter patent is 180 °C for 40 sec which is considerably more severe than for normal shirt interlinings.

FUTURE TRENDS

Of all the fusible adhesives currently available for textiles, polyamides are closest to the ideal system, as they can be modified to suit the requirements of the industry. However, at their current price it is not economic to use polyamides for all applications; cheaper materials will continue to be used despite their inferior performance. What is needed is an adhesive with properties as good as polyamides at a price closer to that of polyethylene! It is claimed that the two systems which are potentially nearest to fulfilling these requirements are polyesters and polyurethanes.[8] However, it is doubtful whether these materials would be significantly cheaper than polyamides due to the huge cost involved in development and capital investment.

One thing that is certain is the adhesives will be of increasing importance in clothing manufacture as garment design is modified to eliminate costly and time-consuming sewing operations.

ACKNOWLEDGEMENT

The author's thanks are due to Staflex International Ltd, for the use of data obtained in their laboratories and for permission to publish, and other members of the Development Department.

REFERENCES

1. Hurwitz, M. D., *Adhesion Science and Technology, Part B*, ed. Lee, L. H., 1975, Plenum Press, New York and London.
2. Stukenbrock, K. H., *Clothing Inst. J.*, **19,** 1971, p. 183. The British Interlining Manufacturers' Association, Choice of Resins Related to Garment Manufacture, 1976, London.
3. Smith, D. K., *Textile Chemist and Colorist*, **1,** 1969, p. 619.
4. Edelmann, K., *Kunststoffe*, **21,** 1968, p. 429.
5. Fyans, R. L., *Adhesives Age*, **18**(3), 1975, p. 19.
6. Inventa AG, für Forschung und Patentverwertung, CH 491 150 (publ. 1967), US 3 410 832 (publ. 1968) and US 3 536 780 (publ. 1970).
7. Plate Bonn GmbH, DT 1 253 449 (publ. 1966).
8. Schaaf, S., *Kunststoff-Berater*, **19,** 1968, p. 900.
9. de Jong, E., *Adhäsion*, **19**(11), 1975, p. 317.
10. Plate Bonn GmbH, DT 2 324 159 (publ. 1973). Plate Bonn GmbH, DT 2 324 160 (publ. 1973).
11. Inventa AG für Forschung und Patentverwertung, DT 2 436 430 (publ. 1973).
12. Data supplied by Plate Bonn GmbH.
13. Courtaulds Ltd, GB 1 329 697 (publ. 1970).
14. Bostik Ltd, GB 1 449 680 (publ. 1973).
15. Badische Anilin und Soda Fabrik AG, GB 1 443 768 (publ. 1973), IDEM and DT 2 445 167 (publ. 1975).
16. Toyobo Ltd, JA 75/18,535 and JA 75/18,534.
17. Inventa AG für Forschung und Patentverwertung, DT 2 446 054 (publ. 1973).

18. Diacel Ltd, DT 2 542 632 (publ. 1975).
19. Chemische Werke Hüls AG, DT 2 448 344 (publ. 1975).
20. Plate Bonn GmbH, DT 2 407 505 (publ. 1976).
21. Chemische Werke Hüls AG, DT 2 507 504 (publ. 1976).
22. Carl Freudenberg, GB 1 421 072 (publ. 1976).
23. Sumitomo Chemical Co. Ltd, JA 75/101,643.
24. Tokyo Printing Ink Mfg. Co. Ltd, JA 75/11,966.
25. MAYER, W. P. and JOHNSON, N. W., *Adhesives Age*, **18**(8), 1975, p. 37.
26. Sumitomo Chemical Co. Ltd, JA 76/32,872.
27. Sumitomo Chemical Co. Ltd, JA 74/92,372.
28. Nippon Synthetic Chem. Ind. Co. Ltd, JA 74/93, 368.
29. Kurashiki Spinning Co. Ltd, DT 1 610 645 (publ. 1976).
30. Shikishima Spinning Co. Ltd, Showda Polymer Co. Ltd, JA 76/6,260.
31. Toyobo Co. Ltd, JA 75/16,456.
32. Toray Ind. Inc., JA 75/160,344 and 160,562.
33. Eastman Kodak Co. Inc., US 3 948 859 (publ. 1976).
34. National Starch Chemical Corp., US 3 926 920 (publ. 1976).
35. Nippon Synthetic Chem. Ind. Co. Ltd, JA 76/22,755 (publ. 1976).
36. Toray Ind. Inc., JA 76/19,034.
37. Bayer AG, DT 2 842 149 (publ. 1975).
38. Honny Chemicals Co. Ltd, JA 75/34,577.
39. CHAPMAN, F., *Brit. Clothing Manuf.*, **9**, 1972, p. 26. Staflex International Ltd, JA 74/99,344.
40. Raduna AG, US 3 922 418 (publ. 1976).
41. Carl Freudenberg, GB 1 372 676 (publ. 1973).
42. Staflex International Ltd, US 3 464 876 (publ. 1969).

Chapter 9

THERMOPLASTIC SBS RUBBERS AS ADHESIVES AND ADHERENDS

A. R. CARTER and D. PETTIT
Shoe and Allied Trades Research Association, Kettering, UK

SUMMARY

The basic structure of the SBS copolymer is dealt with briefly. Special reference is made to the two-phase structure and the importance of this in compounding. The compounding of contact, pressure-sensitive and hot melt adhesives is discussed and the effect of the various compounding additives on the properties of these adhesives is described.

The SBS compounds are very versatile and are used as adherends as well as adhesives. A major use of such compounds is as a footwear soling material and the methods of bonding to these substrates are described in detail.

INTRODUCTION

The thermoplastic block copolymers based on SBS namely polystyrene–polybutadiene (low *cis* type, e.g. 45 % *cis*, 46 % *trans*, 9 % 1,2-vinyl addition)–polystyrene were introduced in the mid 1960s. These block copolymers exist in the linear and radial forms shown in Fig. 1.

An unusual feature of these polymers associated with their block configuration is, unlike the random styrene–butadiene copolymer SBR, the occurrence of two distinct phases. This is demonstrated by the observation of two different glass transition temperatures (T_g).

The T_g of SBR is *ca.* −55 °C but the T_g of the block copolymer gives a value at −65 °C due to the polybutadiene block and another at +90 °C due to the polystyrene. In this structure the polystyrene and polybutadiene blocks exist as separate phases instead of the single phase system associated with the random-copolymerised SBR.

This two-phase system is of paramount importance and leads to the unusual properties of these SBS copolymers. If the polystyrene content of the copolymer is less than about 40% the polystyrene phases will form discrete aggregates of less than 0·03 μm diameter which are known as domains. These polystyrene domains with a T_g much higher than room

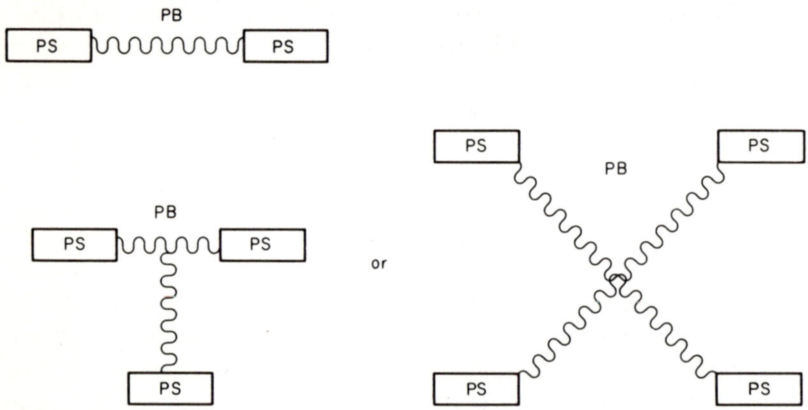

FIG. 1. Linear and radial block configurations of SBS copolymers. (Courtesy: Shell Chemicals Ltd.)

temperature are glassy in nature and are linked by polybutadiene chains which having a T_g below room temperature are mobile and form a continuous rubbery phase. Hence a situation exists where the polystyrene domains reinforce the rubbery structure and act as a filler. The morphology of these block copolymers is quite complex and has been dealt with by Mieras and Wilson[1] and more recently by Aggarwal.[2]

SBS copolymers require no cross-linking system as the polystyrene domains themselves fulfil this function and hold the polybutadiene phase in a reasonably rigid state. In fact the polystyrene domains function as labile cross-links in that on raising the temperature of the copolymer above 90 °C (the T_g of polystyrene) the domains will soften and flow at 140–220 °C. On cooling to less than 90 °C the domains will tend to reform and at room temperature dimensional stability is again attained.

The major useful physical properties of the copolymer may be explained by reference to the structural features outlined above. For example, the compounded copolymer can be processed by normal plastic processing techniques such as injection moulding; extrusion; and blow moulding. Because the T_g of the polybutadiene continuous phase is about $-65\,°C$ the copolymer and compounds prepared from it have excellent performance properties at low temperatures and normal compounds show virtually no tendency for crack-growth to occur at $-20\,°C$. The ratio of polystyrene to polybutadiene in the copolymer may be varied with the resultant change in physical properties from a plastic to a rubber. In compounding it is important to recognise that a two-phase system is present as it is necessary to select compounding ingredients suitable to modify a particular phase. The polymer can be filled using normal inorganic or organic fillers; extended with oils (aromatic oils will soften the polystyrene domains and lead to a weakening of the reinforcing structure); modified by blending with other hydrocarbon elastomers or ethylene–vinyl acetate copolymer, which modify the property of either the copolymer or the additive depending on the relative proportions, or made tacky by using resins commonly employed in adhesives. The science of compounding will be discussed in some detail in the sections dealing with adhesives made from SBS copolymers.

USES OF SBS COPOLYMERS

The applications for this relatively new copolymer are still being extended but an indication of its versatility is shown by the following list of current usage.

1. *Elastic bands.* Polymer crumb is mixed with further stabiliser and pigment and extruded as a circular section which is simply cut into bands. No vulcanisation is necessary.
2. *Stair treads.* The compounds are highly filled and oil extended and similar compounds are used for sound insulation in buildings.
3. *Footwear industry.*[3] Solings—heels and uppers as in wellington boots and some ski boots.
4. *Blends with bitumen*[4]—to extend its high and low temperature service range, in normal bitumen applications.
5. *Modifiers for other thermoplastics*,[4] e.g. improves impact resistance of polystyrene and polypropylene and improves the flexibility and stress crack resistance of both low- and high-density polyethylene.
6. *Adhesives*—contact; pressure sensitive and hot-melt types.

Although not complete the above list of applications demonstrates that compounds based on the SBS copolymer have made a significant impact on technology during the past decade.

This paper will consider the use of the SBS copolymers as the basic polymer in adhesive formulations and as an adherend in the footwear industry. In footwear, the SBS copolymers can be the materials from which soles, heels or even the shoe or boot uppers are made. Adhesives are applied to these components and in these cases the SBS copolymer is functioning as an adherend instead of an adhesive.

CONTACT ADHESIVES

Contact adhesives function by placing two adhesive coated adherends in contact under pressure and these two adhesives then coalesce with elimination of their interface. The bond formation relies therefore on a diffusion process (i.e. autoadhesion). For this mechanism to act the two adhesives must be compatible (and obviously they are being the same) and must have a T_g below the bonding temperature to allow the molecules to be mobile.

Adhesives formulated from SBS copolymers produce useful contact adhesives and have been suggested for the installation of decorative laminates and wall coverings; footwear, e.g. moulding thermoplastic rubber to leather uppers; furniture construction, e.g. fixing dowels; manufacturing plastic laminates of wood and metal; bonding floor panels;attaching insulation in buildings and automobiles; 'do-it-yourself' applications.

The uses are in fact numerous and new markets are being developed. Formulations can be prepared to accommodate all normal application methods, i.e. brush; roller-coating; spraying.

The base SBS polymer is more expensive than general purpose rubbers but the cost of the formulated adhesive is competitive as it may be processed at high speeds. The polymer itself is supplied in a crumb form which does not require cutting or chopping and of course no mastication step is necessary to effect solution. In fact the solubility is very high in common solvents and the polymer is compatible with a wide range of normal compounding ingredients so that a high solids low viscosity solution can be readily obtained.

Choice of Solvent and Viscosity of Solution

The solubility parameter δ for polystyrene is 9·1 and that of polybutadiene 8·4. In practice this allows the use of a range of common solvents to be used having solubility parameters between about 8–9·7. Such solvents include those shown in Table 1.

TABLE 1
SELECTION OF SOLVENTS FOR SBS COPOLYMERS

Solvent	*Solubility parameter* (δ)
Cyclohexane	8·2
Methyl isobutyl ketone	8·4
Xylene	8·9
Toluene	8·9
Ethylacetate	9·1
Methyl ethyl ketone	9·3
Chloroform	9·3
1,1,2,2-tetrachloroethane	9·7

The choice of a solvent in the middle of the range, e.g. toluene, will provide the best solvent and hence give the lowest solution viscosity. However, it is possible to use a solvent blend rather than a single solvent and that blend may contain solvents outside the recommended δ range. This is demonstrated in Fig. 2 where a blend of toluene–hexane is shown.

Hexane has a solubility parameter of 7·3 and toluene 8·9. If the volume proportions fall within the range of 8·0–9·7 then solubility exists and a reasonable solution viscosity may be obtained. In practice, this means that the adhesive manufacturer has a very wide choice of solvents available and that economically attractive solvent blends may be employed.

The viscosity of solutions is increased by increasing the polymer

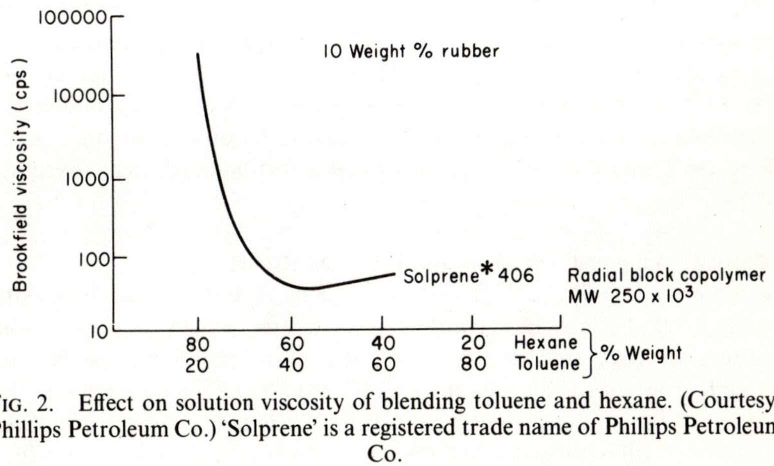

FIG. 2. Effect on solution viscosity of blending toluene and hexane. (Courtesy: Phillips Petroleum Co.) 'Solprene' is a registered trade name of Phillips Petroleum Co.

concentration and reduced by adding resin at a constant solids loading as the resins have a much lower molecular weight than the base polymers. This feature may enable lower viscosity solutions to be made with solvents of low solubility parameter. The molecular weight of the SBS block copolymer can be varied and the effect on viscosity in a toluene solution is shown in Fig. 3.

Additionally, the ratio of styrene to butadiene influences the solubility in the way that would be expected. The higher the polystyrene content then the more difficult it is to dissolve the copolymer in solvents of low solubility parameter or in blends containing non-solvents of δ less than 8·0.

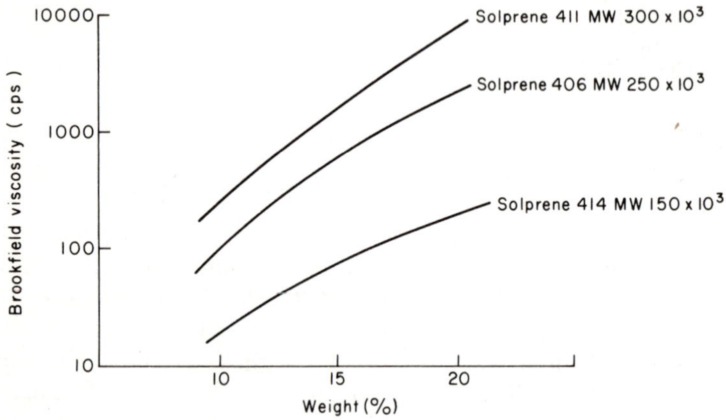

FIG. 3. Effect of molecular weight of SBS radial block co-polymer on solution viscosity. (Courtesy: Phillips Petroleum Co.)

Thickening agents such as ethyl cellulose, clays and calcium carbonate may be used to increase the viscosity of the solutions. One interesting thickening agent that is claimed to increase the solution viscosity by a factor of ten, is a 3 % addition of a 12 % dispersion of very fine silica in methyl ethyl ketone.

Adhesive Compound Variables and Effect on Properties

In compounding the adhesive the two-phase system of the base SBS copolymer is of utmost importance as it is essential to know which phase the additives are associated with. If they associate with the polystyrene then the strength properties of the adhesive are affected whereas association with the polybutadiene affects the rubbery nature of the adhesive film. The SBS copolymer itself is not tacky and must therefore be compounded to produce optimum properties. The relative amounts of polystyrene and butadiene are important as increasing the polystyrene content results in less tack.

The molecular weight of the base polymer does not influence the peel strength measured at room temperature[5] in that a change through molecular weight within the range 150×10^3 to 300×10^3 only realises a 12% change in peel strength. However, the results of creep test

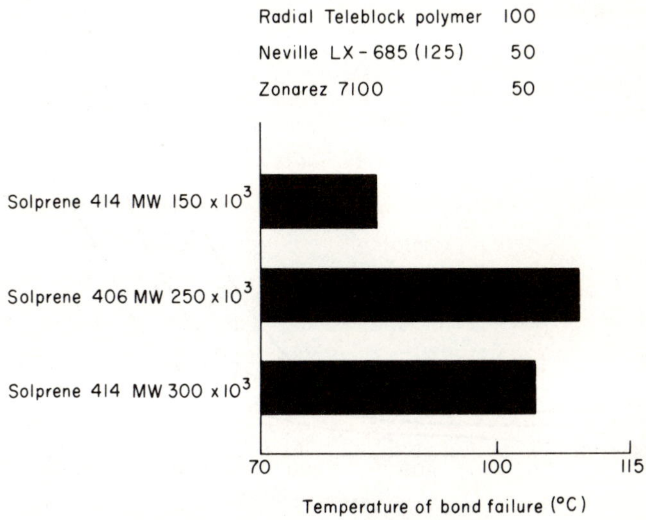

FIG. 4. Effect of molecular weight of SBS copolymer on heat resistance. (Courtesy: Phillips Petroleum Co.)

measurements shown in Fig. 4 under a load of $39\,N\,m^{-1}$ show that the higher molecular weight compounds have the highest heat resistance.

When adding resin it is important to consider with which phase of the SBS copolymer the resin associates. A recommended method of assessing this is to form a film of a 1–1 blend of polystyrene and resin or polybutadiene and resin and examine the film for clarity. Good clarity indicates good compatibility and with this knowledge the appropriate resin can be added. Apparently four situations can exist in that the resin may associate with the polystyrene or the polybutadiene or both or may form independently a third phase.

Usually, aromatic resins such as coumarone–indene associate with the polystyrene phase and can expand these polystyrene domains until they make contact to form a weak continuous network of resin–polystyrene.

This results in an increase in the stiffness producing higher impact and creep resistance. Modified wood rosins associate with the polybutadiene phase and so reduce the modulus, increase the elongation and can also tackify the compound as it is present free in the matrix. It is not unusual to use two resins in practical formulation, one reacting with the polystyrene and the other with the polybutadiene phase. The incompatible resins such as the phenolics form a separate phase and can actually form a separate layer during storage.

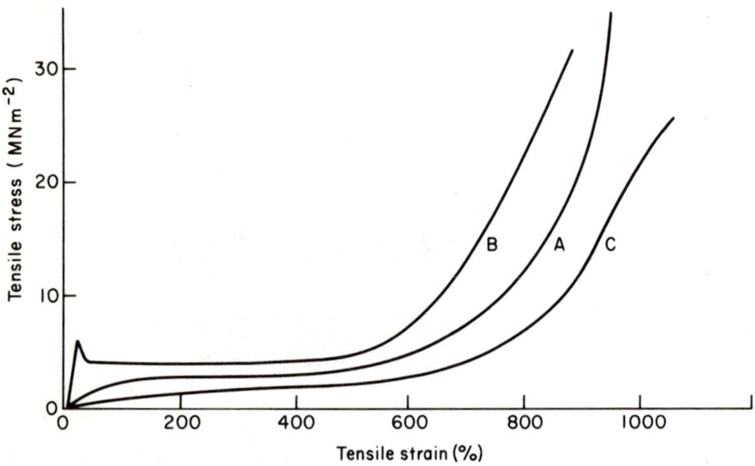

FIG. 5. Effect of resin type on stress–strain response. A: Cariflex TR-1101 block copolymer; B: Cariflex TR-1101 plus coumarone–indene resin (50 phr); C: Cariflex TR-1101 plus modified wood rosin (50 phr). (Courtesy: Shell Chemicals Ltd.) 'Cariflex' is a registered trade name of Shell Chemicals Ltd.

The melting point of the resin is important as it can alter the T_g. Using polystyrene compatible resins the change in T_g is dependent on the resin loading and the melting point and by using 75 parts per hundred parts resin (phr) of methylindene resin with a softening point of 25 °C the T_g of the polystyrene is reduced about 60 °C. This of course has a very significant effect on the temperature at which the two adhesive films coalesce together, and presumably also on the heat resistance properties.

The type of resin also influences the stress–strain relationship of the adhesive as shown in Fig. 5.

The coumarone–indene resin which is associated with the polystyrene phase causes an high initial stress at a very low extension before the cold draw region. This initial high modulus is useful in preventing creep at low stress. The wood rosin additive associates with the polybutadiene and this tends to reduce the initial modulus and lowers the creep resistance.

Again, plasticisers associate with either phase depending on the type. Aromatic oils and phthalates will associate with the polystyrene domains and soften them resulting in a loss of cohesive strength. Non-aromatic and mineral oils associate with the polybutadiene to soften the compound leading to a reduction in modulus and higher elongation. It has been suggested[6] that asphalts may be used as plasticisers and blends containing them have been used for bonding paper. Also ethylenevinylacetate copolymer has been mentioned as an additive to improve adhesion to polar surfaces.

The adhesive manufacturer has many variables to consider from the base copolymer composition and molecular weight through the differing effects of resins and plasticisers to the additional polymers which may be added for specific applications. There is therefore a very large number of formulations possible and hence the range of adhesives can be very versatile.

The canvas-to-canvas peel adhesion results are in excess of 5 kN m^{-1} and even after ageing for a month at 70 °C there is no significant reduction in strength. The resistance to the effect of water would be expected to equal that of normal rubber-based adhesives. There is however, the question of the resistance to oxidation ozone and u.v. radiation. However, in most applications the bulk of the adhesive is protected from direct attack and the effects of these agencies is therefore likely to be minimal leaving perhaps the effect of solvents and oils as the major disadvantage. In fact, the latter have been shown to have a slightly deleterious effect on the bond when an SBS-based adhesive is used in contact with an SBS adherend. In such a case, the oil from the adherend may migrate into the adhesive and so increase the plasticisation which reduces the strength.

PRESSURE-SENSITIVE ADHESIVES

Pressure-sensitive adhesives differ from contact adhesives in many respects; the main one probably being the permanent tacky nature of the adhesive layer. As the SBS block copolymer is inherently non tacky it requires compounding to ensure development of this basic requirement of tack. In use, bond formation is a slow process and therefore another requirement is that the adhesive should have a low viscosity at low deformation rates to enable the adhesive adequately to wet the substrate under application of light pressure. The destruction of the bond during peeling is a fast process and ideally the adhesive should have a high modulus at a high rate of deformation. Coupled with these properties are the additional requirements of good holding power, i.e. low creep; sufficient cohesive strength to withstand stress, and good resistance to ageing of the adhesive before and after bond formation.

As a pressure-sensitive adhesive possible applications are quite wide and include the traditionally accepted uses of tapes, labels, carpet tiles, and wall coverings. It is necessary to mention in this respect the use of an alternative block copolymer based on styrene–isoprene–styrene. This has some inherent tack as a base copolymer and is therefore favoured for pressure-sensitive applications.

In the preparation of pressure-sensitive tapes using the general purpose elastomers the normal procedure is to prepare a solution of the elastomer and apply this to the primed backer. The solvent is evaporated and recovered before the tape is rolled. This accepted procedure has the usual disadvantages associated with the flammability; toxicity; and general atmospheric pollution problems associated with solvents. Compounded adhesives based on the SBS copolymer have the distinct advantage of being able to be processed as melts. Rather than applying the adhesive as a solution, a 100% solids melt can be extruded to the tape and therefore the problems associated with solvents are eliminated. The general-purpose elastomers currently used as pressure sensitive adhesive do not have this processing advantage as they cannot be processed as melts at temperatures low enough to avoid degradation. The processing of melts will be dealt with further in the section dealing with hot melt adhesives.

The compounding of the adhesive follows a similar trend to that described in the section on contact adhesives. Hence, if the formulator recognises the fact that a two-phase system exists in the base copolymer additives may be added which associate with the required phase and the adhesive can thus be tailored for specific applications.

The more general question of the importance of two-phase systems with rubber–resin mixtures to the development of tack in pressure-sensitive adhesion is discussed in Chapter 4.

EFFECT OF COMPOSITION ON PROPERTIES

The basic copolymer properties influence the tack of the adhesive and this is shown by Kraus *et al.*[7] by reference to Fig. 6.

With a high polybutadiene radial copolymer the tack is high and is not

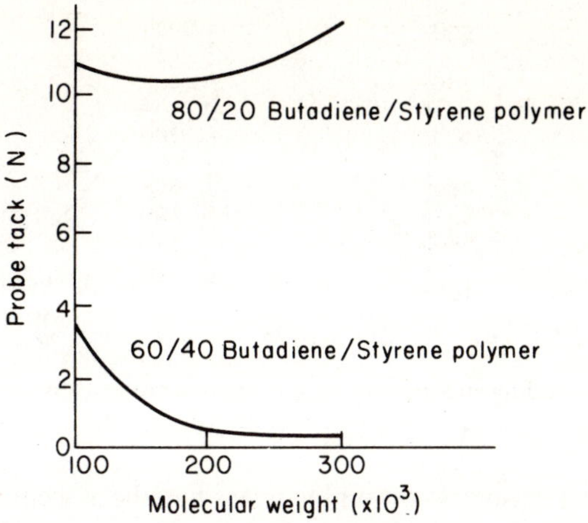

FIG. 6. Effect of molecular weight and polybutadiene content on tack properties of compounded adhesives. (Courtesy: Phillips Petroleum Co. and *J. Adhesion*.)

influenced markedly by changes in molecular weight. At a lower polybutadiene content the increase in molecular weight causes a considerable reduction in tack. The effect of the polybutadiene to styrene ratio is of major importance and this is more clearly shown in Fig. 7 where there appears to be a definite peak in the tack at about a 70 % butadiene level under the formulation conditions used.

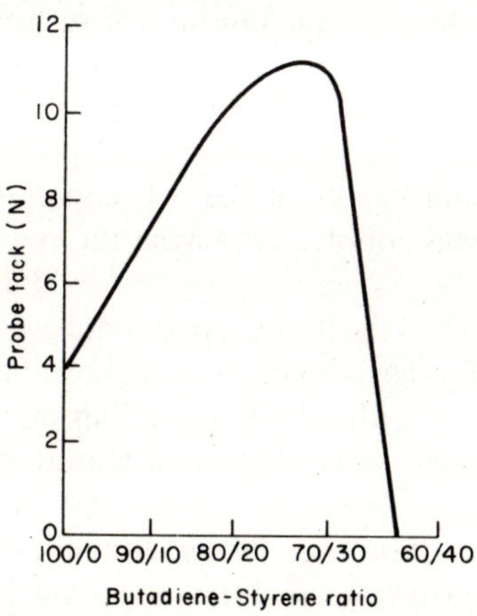

FIG. 7. Effect of polybutadiene–polystyrene content on tack properties of compounded adhesives. (Courtesy: Phillips Petroleum Co. and *J. Adhesion*.)

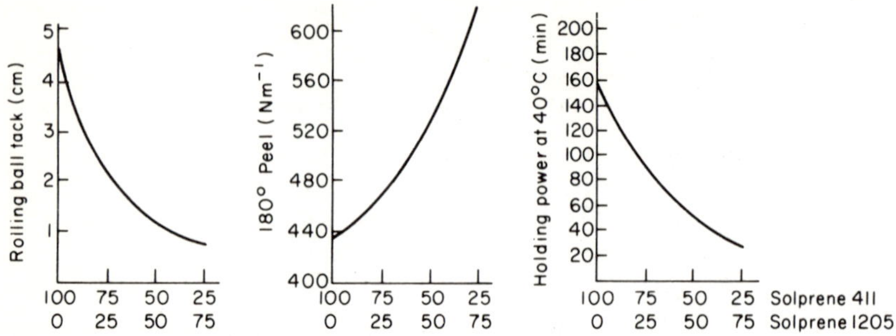

FIG. 8. Effect of adding an S-B block copolymer to a radial copolymer. (Courtesy: Phillips Petroleum Co.)

Some theoretical aspects concerning resins have been mentioned briefly but in practice it is essential to ensure the resin is compatible with the polybutadiene phase. This is demonstrated by Kraus *et al.*[7] with a 80/20 polybutadiene–styrene radial copolymer of molecular weight 150×10^3. Rosin ester gives a probe tack test result of 8·3 N whereas a coumarone–indene resin, presumably at the same loading, gives no measurable tack under the same conditions. As mentioned in the section on contact adhesives, the rosin types associate with the polybutadiene and the coumarone–indene types associate with the polystyrene phase. Non-aromatic oils are used in formulations but as they have a tendency to migrate it is recommended that part of the oil is replaced by a second resin in some applications without reduction in tack or holding properties.

S-B block copolymer may be used in conjunction with a radial block copolymer to improve the tack properties and typical results are shown in Fig. 8.

The polystyrene end blocks of the S-B copolymer associate with the polystyrene in the SBS copolymer leaving the polybutadiene chain ends free.

The addition of the S-B block copolymer improves the tack and adhesion strength properties but has an adverse effect on the creep resistance. It is therefore necessary, as with all compounding ingredients, to design the adhesive for the required use by careful consideration of the relative levels of addition.

It is, of course, important that the performance of the pressure-sensitive adhesive is maintained during a realistic service life. Evidence is presented in the trade literature showing that adhesives left open to the atmosphere for four weeks show no significant change in tack or adhesion performance.

HOT-MELT ADHESIVES

As indicated in the previous section, melt adhesives are becoming more attractive as they are not affected by the environmental limitations associated with the solvent based adhesives. Melt adhesives are therefore a growth area and much development effort is being applied in this field. The SBS-based hot-melt adhesives are currently used for pressure sensitive applications for book binding and packaging.

Melt adhesives initially are in a solid form which on application of heat should become a low viscosity liquid which will wet an adherend and then re-solidify on cooling. The solidified form should be tough and often flexible so that the joint can withstand a degree of flexing or impact. Furthermore, they must ideally be stable in the melt form without change in viscosity if maintained at a constant temperature for prolonged periods. The SBS block copolymers fulfil these requirements in that between about 150–200 °C they become low viscosity liquids. As previously mentioned they regain their physical properties on cooling in that the polystyrene physical cross-links give them the desired cohesive strength which is coupled with a high degree of flexibility even at very low temperatures.

The SBS copolymers do not show a normal Newtonian behaviour in the melt form. As is shown in Fig. 9 the viscosity decreases as the shear rate is increased.

The presence of resin decreases the overall viscosity and this is demonstrated in a more quantitative manner in Fig. 10.

In this case the added resin is a coumarone–indene type which associates with the polystyrene phase and therefore makes no contribution to tack but increases the cohesive strength. If high melting point resins are used the permissible upper service temperature is also increased. It is obvious that addition of a compatible resin will reduce melt viscosity due to the lower molecular weight of the resin and the same argument applies to non-aromatic oil and wax additions which are more effective in this respect. In general, larger additions of resins and oils are added to melt adhesives than to solution adhesives because the major requirement is for the low melt viscosity.

Unsaturation in the base copolymer renders it susceptible to oxidation and it is essential to consider methods of stabilising the adhesive formulation. This is especially important with melt adhesive where high temperatures increase susceptibility to attack. Additionally, any unsaturated resin or oil added to the formulation can create stability problems and therefore this should be recognised by the formulator. These stability

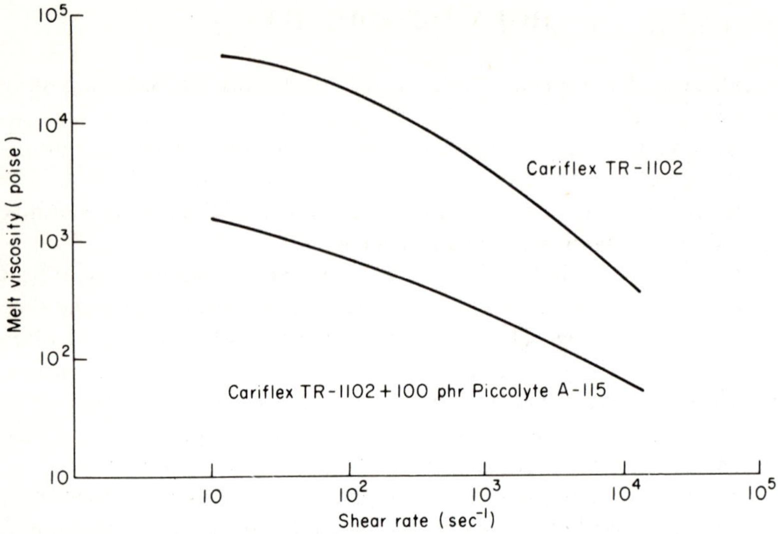

FIG. 9. Effect of shear rate on melt viscosity. (Courtesy: Shell Chemicals Ltd.)

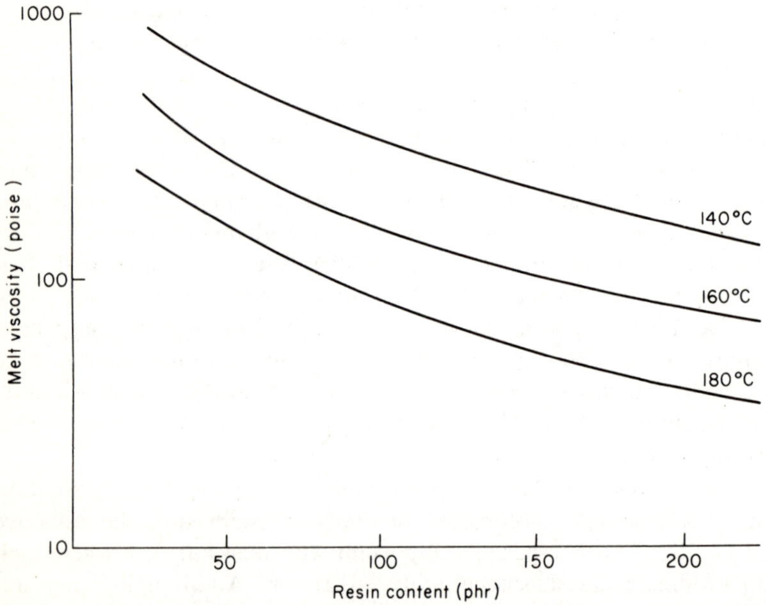

FIG. 10. Effect of coumarone–indene resin on melt viscosity. (Courtesy: Shell Chemicals Ltd.)

aspects are important at the compounding; the processing and performance stages.

During mixing oxidation is the main concern and the mixing time and temperature should obviously be kept to a minimum. In no case should the melt temperature be allowed to rise above 220 °C as the polymer degrades very quickly. Typical melt formulation and mixing conditions are:

SBS copolymer	100 phr
Resin	250 phr
Oil	0–150 phr
Antioxidant	5

The antioxidant and 50 parts of oil are heated to about 120 °C and the copolymer is added continuously in crumb form so the temperature does not fall below 120 °C. After about 15 min a homogeneous blend is obtained and the remaining oil and resin are added slowly maintaining the temperature between 120 and 150 °C. The complete cycle is claimed to take 45–50 min and the melt can then be cast into the appropriate physical form. Z-blade mixing is claimed[8] to cause less polymer breakdown than planetary types and if a nitrogen blanket is used degradation is negligible. Extruders or other types of continuous mixers which function under anaerobic conditions may be used to avoid oxidation.

In practice the melt may suffer from being melted and cooled for many cycles during application if bulk heaters rather than limited zone melting

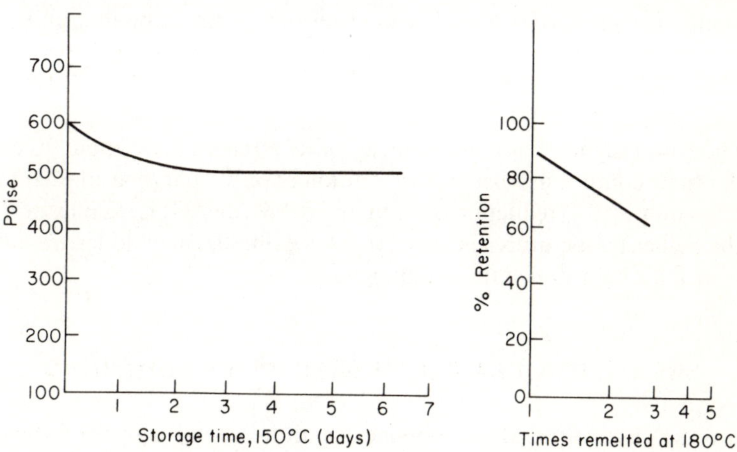

FIG. 11. Changes in melt viscosity on storage and reheating in the presence of zinc dibutyl dithiocarbamate (ZDBC). (Courtesy: Shell Chemicals Ltd.)

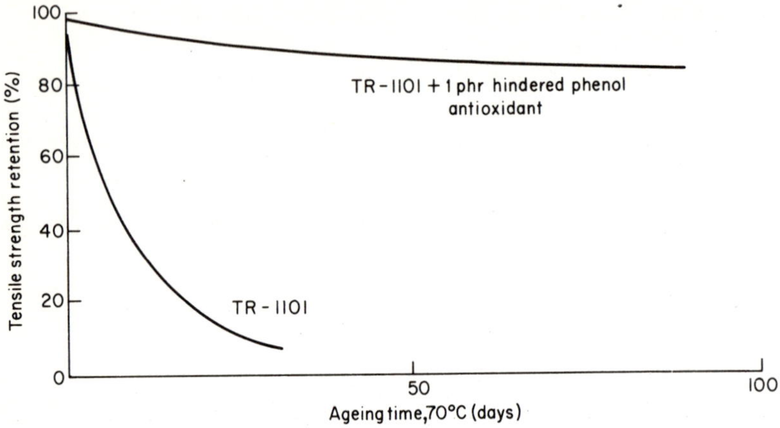

FIG. 12. Effectiveness of antioxidant on adhesive film stability. (Courtesy: Shell Chemicals Ltd.)

heaters are employed. Such heat history may cause degradation and result in viscosity changes complicating the application process. The effectiveness of zinc dibutyl dithiocarbamate as an antioxidant is shown in Fig. 11.

Apart from an initial reduction, the ZDBC is effective in stabilising the viscosity of the compound held at 150 °C but less so when temperature cycling occurs. Of course, with less stabilisation the application could be complicated by the tendency of the melt to run out of extrusion heads, and the adhesive properties would also deteriorate. Antiozonants and u.v. stabilisers may be used if required and these are conventional types.

The stability of the SBS-based adhesive films to oxidation is shown in Fig. 12.

The base polymer loses the majority of its strength after about 20 days under these conditions whereas the addition of only 1 phr of an antioxidant permits only a 10 % reduction in strength after 90 days. It is significant that at the higher temperatures involved in processing the melt adhesives a five fold increase in antioxidant is required.

SBS THERMOPLASTIC RUBBER AS AN ADHEREND

The widest use of the SBS copolymer as an adherend is in the footwear industry as a soling compound where it has been used for almost 10 years. For compounding it is necessary to choose additives that are compatible with either the polystyrene or the polybutadiene component parts if extension is required or conversely it is possible to form a third phase by adding components incompatible with both existing phases.

(a) Polystyrene or poly-α-methyl styrene is added to change the stiffness or hardness. Replacement of the polystyrene by high styrene resin (a styrene–butadiene random copolymer with rather more styrene than SBS rubber) can be beneficial[2] as it appears to reduce the stress concentration at a point on flexing and disrupt the orientation caused by flow during moulding.

Added polystyrene would of course be expected to associate with the polystyrene domains in the block copolymer. However, surprisingly this does not occur and a separate polystyrene phase is formed.[9] Recent microscopy studies at Loughborough University[10] have demonstrated that under the shear conditions encountered in injection moulding this separate polystyrene is stratified, i.e. it is smeared across the structure. This can have an adverse effect on the laminar tear strength as it can form a weak continuous layer. Also the effect may be important to adhesion because as will be shown later the presence of polybutadiene at the bonding surface is necessary and a continuous smear of polystyrene may effectively cover the polybutadiene.

Post annealing of an injection-moulded sample exhibiting the stratified polystyrene effect results in a retraction of this layer.[10] This has the effect of causing a shrinkage in the moulding, although this may have no practical significance as the annealing temperatures are unlikely to be met in normal service conditions.

(b) Extending oils based on paraffinics or naphthenics are incorporated to improve processing characteristics and also result in a reduction in hardness and stiffness. Aromatic oils should not be used as these virtually dissolve the polystyrene domains and cause a severe reduction in strength.

(c) Fillers based on carbonates or silica may be added and form a third phase. The loading is usually below the 10% level and they are used as extenders or to improve the aesthetic properties, although a significant amount will increase the hardness. The presence of whiting is claimed[4] also to reduce tensile strength abrasion resistance and melt flow.

(d) An additional continuous phase may be formed by the addition of ethylene vinyl acetate or polypropylene. The ethylene–vinyl acetate improves the ozone resistance of the compound and the polypropylene addition results in increased stiffness and improves the upper service temperature.

(e) The compounds are stabilised to the effects of oxygen, ozone, and u.v. radiation.

(f) For special applications azo-type blowing agents can be incorporated to produce cellular structures.

Some typical compound properties for solid footwear soling materials are shown in Table 2.

TABLE 2
PROPERTIES OF SBS COMPOUNDS

Specific gravity	0·95–1·10
Shore 'A' hardness	40–95
Tensile strength MN m^{-2}	3·9–12·8
Elongation at break	300–900
Ross crack growth -5 °C	0

The SBS based compounds may exhibit stress softening and this has in fact been observed when peel testing joints in which this material forms one of the adherends. This may be due to a change in the orientation within the compound but it has been suggested that applied stress could result in some disruption of the polystyrene domains.

This feature of stress softening is also apparent in some flexed samples when a noticeable reduction in hardness is observed. Such a phenomenon can of course markedly alter the physical characteristics during the service life.

Depending on the compound the injection moulding temperature varies between 160–200 °C. A cycle time of about 3 min is used and a very low shrinkage is obtained which is advantageous in unit sole moulding where a range of sizes is being produced. The compounding variables enable a wide range of different types of solings to be manufactured from the high-fashion types which reproduce the appearance of natural crepe rubber to the hard stiff units used for sports footwear. An additional advantage is its ability to allow reprocessing of waste compound.

These units have to be attached to the upper of the shoe using adhesives. The footwear industry uses solvent based urethane or polychloroprene adhesives to make this bond and at the present time the urethane adhesives predominate.

The method of use for normal sole attaching is as follows:

Upper Material —prepare surface—apply adhesive—leave for $\frac{1}{2}$ h—1 h open time

Soling Material —prepare surface—apply adhesive—leave for 1 h —2 weeks open time

Heat reactivate adhesive on soling material to *ca.* 80 °C

Bond under pressure of *ca.* 0·5 MN m^{-2} for 10 sec.

When compounds based on the SBS copolymer were introduced, as with all new materials, it was necessary to evaluate various methods of surface preparation and to develop a system which was both effective and practical. This exercise was done and the results obtained are given in Table 3.

TABLE 3

INITIAL ATTEMPTS TO ADHERE TO SBS COMPOUND WITH URETHANE AND POLYCHLOROPRENE CONTACT ADHESIVES

Typical surface preparation	*Polychloroprene adhesive*	*Urethane adhesive*
None	Poor adhesion	Poor adhesion
Solvent wiped	Poor adhesion	Poor adhesion
Wire brush roughed	Poor adhesion	Poor adhesion
Isocyanate primer	Poor adhesion	Poor adhesion
Acid primer	Poor adhesion	Poor adhesion

Examination of the original and split surfaces of compounds using ATR infrared methods failed to reveal the presence of surface contaminants. A weak surface layer effect therefore did not appear to account for the poor adhesion, but rather a definite lack of compatibility between the compound surface and the polychloroprene or urethane adhesive.

A great deal of work was done on the subject of adhesion to these thermoplastic rubbers in the late 1960s concentrating on the adhesion of polychloroprene and urethane adhesives. If the SBS compounds were to be used in footwear it was imperative that they should be capable of being bonded with at least one of these adhesives. Alternative adhesives were not seriously considered as the footwear industry is not usually willing to accept special adhesives for different adherends. Although the early SBS-based contact adhesives were examined but at that time they did not have the necessary properties of a shoe adhesive for sole attaching. In 1968 it was observed that if the surface of the SBS compound was treated with bromine or chlorine the subsequent adhesion of the treated surface was excellent with a urethane adhesive and also showed improvement with a polychloroprene adhesive. From this work chlorination techniques were developed[11] for use in conjunction with a urethane adhesive, as they carried economic advantages as well as a somewhat less toxic risk than bromine. The use of chlorine gas was investigated but without expensive plant control difficulties in practice may result in toxic hazards and an aqueous base chlorination process was developed commercially. In this the unit soles made in the SBS compound were dipped for 1 min in an acidified sodium hypochlorite solution containing 0·1 % chlorine. The conditions employed were such that free chlorine was available in the solution. On removal the sole units were allowed to dry and adhesive could then be applied or the treated units could be left in some cases up to 12 months before adhesive application.

This procedure developed in 1968 is still widely used especially in North America. However, in the UK the aqueous dip chlorination treatment was not readily accepted, because it meant that the surface treatment could not be controlled by the footwear manufacturer as it was not a procedure suited for use in a shoe factory. Further development work resulted in 1970 in the introduction of a solvent-based halogenation treatment[12] which was safe to use in shoe factories provided the usual adequate air extraction units were employed. This solvent halogenation system is applied by a soft brush to the surface of the unit and allowed to dry for $\frac{1}{4}$ h before applying adhesive.

TABLE 4

METHODS OF CHLORINATION OF SBS COMPOUND SURFACE—EFFECT ON ADHESION WITH URETHANE ADHESIVE

Treatment	*Bond strength* ($kN m^{-1}$)	*Type of failure*
None	0·7	Adhesion to rubber
0·1 % aqueous chlorine dip	8·2	Rubber failure
0·1 % gaseous chlorine	9·3	Rubber failure
Solvent based halogenation		
Satreat wipe	12·9	Rubber failure

Again, the time between surface treatment and adhesive application may be considerably extended and in some cases a lapse of even 6 months is possible. Typical adhesion results are given in Table 4.

During the past few years several alternative halogenation wipe treatments have been developed which in the main perform similarly to the 'Satreat wipe'. However, there are subtle differences in performance which will be discussed later.

The mechanism of this chlorination process is quite interesting and worthy of limited discussion. It was observed by examination of the infrared spectrum of the SBS compound surface before and after treatment that a significant change occurred in the peak intensity at 10·3 μm. This peak is attributed to the 1–4 *trans* isomer of polybutadiene and after treatment had been significantly reduced. It was therefore inferred that reaction had occurred at the double bond in the butadiene molecule.

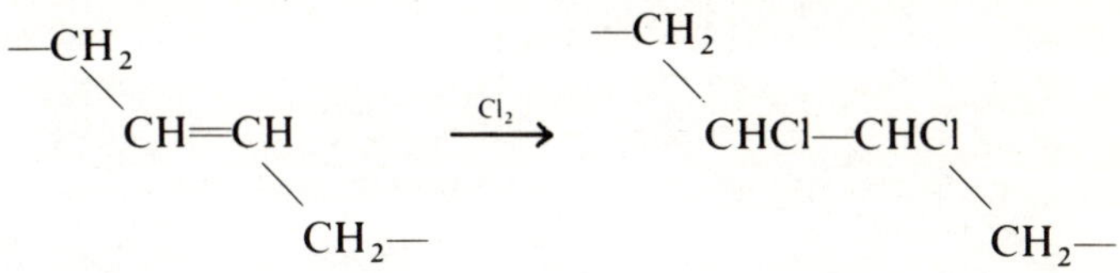

That some chlorination is occurring has been established by elemental chlorine analysis of treated surfaces and it is reasonable to assume that addition to the double bond in the polybutadiene would occur. Substitution of chlorine may also be taking place but essentially the relatively non-polar surface is acquiring some polarity. It has been shown that by using contact angle measurements with liquids of varying hydrogen bonding capabilities that hydrogen bonding may be possible to the chlorinated surface. Greenwood and Wake[13] have speculated that a mechanism for the improvement in adhesion with a urethane adhesive to the treated surface may be due to hydrogen bonding.

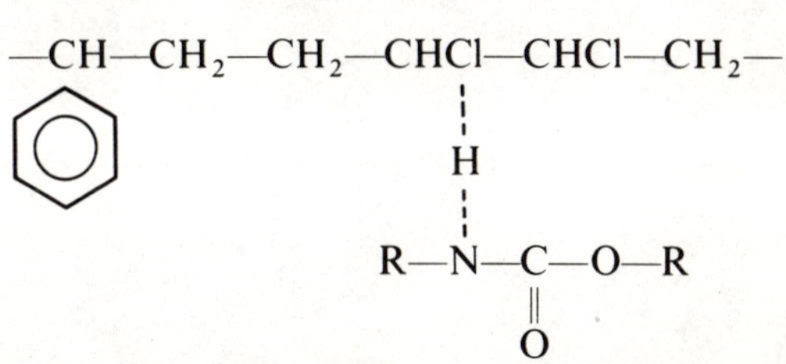

In addition to the chemical modification to the SBS compound surface it is also possible to detect an insoluble layer on the treated surface which extends to a depth of about 5 μm. This could suggest that some crosslinking or cyclisation is occurring within the surface layers.

The stereoscan photographs of treated and untreated surfaces in Fig. 13 also show that some surface crazing is occurring. Whether this is influencing the bonding performance is in doubt as the results given in Table 3 demonstrate that deliberate roughing of the surface does not improve the bond to an untreated surface. It is therefore safe to assume that the surface

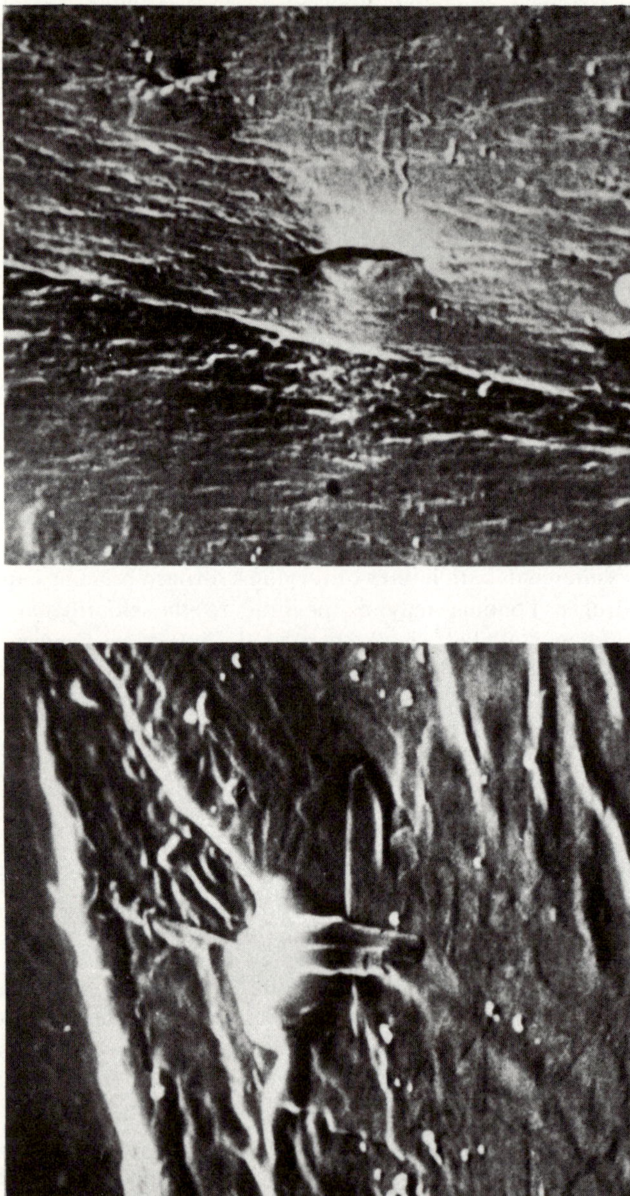

FIG. 13. Stereo electronmicrographs of thermoplastic rubber surfaces before and after halogenation.

cracks caused by the halogenation process are unlikely to lead to a mechanical bond in these circumstances.

Halogenation methods have been employed in the footwear industry for about a decade to prepare the surface of the SBS unit compounds. During that time several practical problems arose which required in depth investigation and which tended to substantiate the proposed reaction between the halogen and the unsaturation in the polybutadiene molecule as being of prime importance.

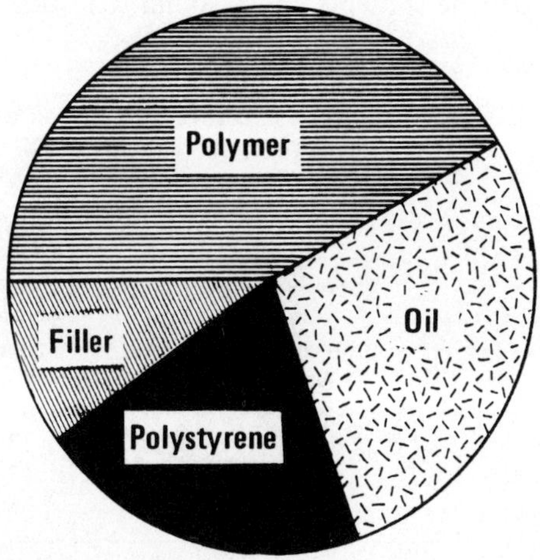

FIG. 14. Diagrammatic representation of thermoplastic rubber compound.

In the first case a compounding restriction was observed. Sole units were being moulded for fashion footwear but problems occurred with them in shoe manufacture because they were too soft. On request, the compounder hardened and stiffened the compound by adding further polystyrene but unfortunately the adhesion on units subsequently moulded was inadequate.

By reference to Fig. 14 it can be seen that the SBS polymer forms only a part of the compound. It is however, the only part of the compound which contains polybutadiene or in fact any degree of chemical unsaturation. As other components in the compound are increased so the relative proportion of polybutadiene is decreased. In the case in point the compounder had increased the polystyrene content to enhance the stiffness of the compound, but unfortunately, in so doing, had effectively reduced the polybutadiene content below a level at which effective surface halogenation was possible.

Alternative solvent halogenation systems to Satreat are available and perhaps the best known one is Tenosol. It was found that Tenosol (an aromatic sulphonamide N-chloro compound) was more effective than Satreat (a heterocyclic N-chloro compound) on SBS compounds having low polybutadiene contents. Unfortunately, the Tenosol was less effective than Satreat on compounds with high polybutadiene contents. The Satreat and Tenosol technologies were therefore combined and a more universal primer for SBS compounds covering a wider range of polybutadiene contents was evolved (*see* Table 5). Current SBS unit sole compounds do not have low polybutadiene contents and respond well with Satreat.

TABLE 5

RELATIVE EFFECTIVENESS OF TWO HALOGENATION PRIMERS ON DIFFERENT THERMOPLASTIC RUBBER (TR) COMPOUNDS

Type of TR	*Primer*[a]			
	Satreat		*SATRA-TNO TR primer*	
	Bond strength ($kN\,m^{-1}$)	*Type of failure*	*Bond strength* ($kN\,m^{-1}$)	*Type of failure*
Low polybutadiene	5·3	AR	13·3	R
Medium polybutadiene	11·5	R	12·4	R
High polybutadiene	7·8	R	8·3	R

[a] AR, adhesion to rubber; R, rubber.

The SBS polymers contain sufficient antioxidant to stabilise them against oxidation during manufacture and storage. However, as they contain unsaturated bonds they are susceptible to oxidation; ozone and ultra violet radiation and the compounds have to be stabilised accordingly. It is suggested that the compounds should be protected where possible from direct u.v. radiation and the use of more than 0·5 phr of carbon black or more than 4 phr of TiO_2 may be used to confer some protection. However, Table 6 shows the adhesion[14] obtained on SBS compounds exposed to sunlight before halogenation treatment.

The results in Table 6 demonstrate how poor adhesion may be inadvertently obtained. The spectra in Fig. 15 show the surface of the compound before and after exposure to sunlight and it is apparent that the 10·3 μm absorption (i.e. that due to trans unsaturation in the polybutadiene molecule) is significantly reduced. Incidentally, it is assumed that the

TABLE 6

EFFECT ON BOND STRENGTH OF DIRECT EXPOSURE TO DAYLIGHT

Before exposure		*After exposure for 6 days*	
Bond strength (kNm^{-1})	*Type of failure*	*Bond strength* (kNm^{-1})	*Type of failure*
7·5	Mainly rubber	1·4	Adhesion to rubber

absorption due to the *cis* isomer would be affected similarly but the absorption due to the polystyrene interferes with measurement. Studies on polybutadiene homopolymer have demonstrated a reduction in the absorption due to the *cis* isomer after halogenation.

Apparently, either oxidation cyclisation or cross-linking has occurred reducing the degree of unsaturation at the surface and hence lowering the response to halogenation. Incidentally, if the surface is chlorinated before exposure to sunlight then the adhesion is unaffected by the same duration of exposure to u.v. The obvious practical recommendation from this

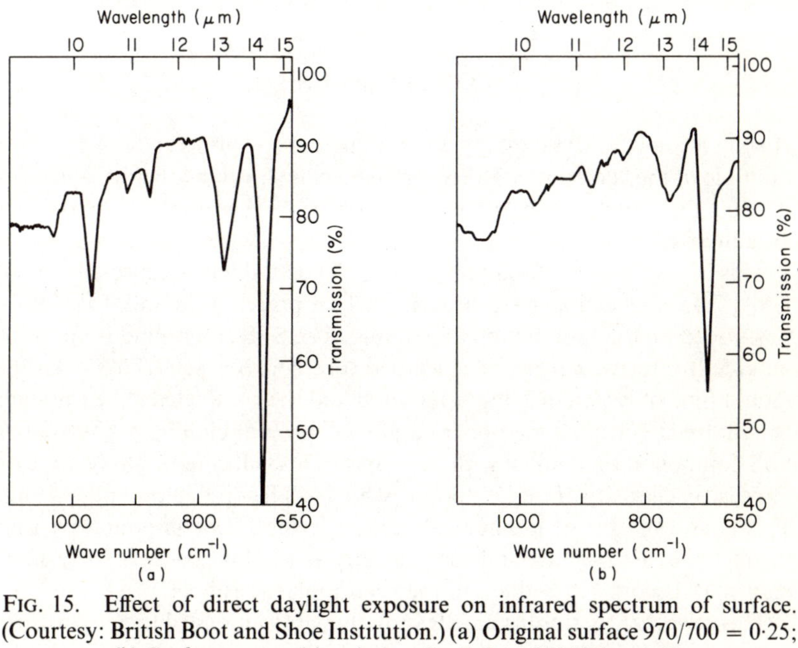

FIG. 15. Effect of direct daylight exposure on infrared spectrum of surface. (Courtesy: British Boot and Shoe Institution.) (a) Original surface 970/700 = 0·25; (b) Surface exposed in window for 7 days 970/700 = 0·08.

investigation is to avoid exposure to u.v. radiation or if this cannot be guaranteed, then surface halogenation should be done as early as possible after moulding.

A further practical problem observed recently with some soft fashion solings is a lack of 'spotting tack'. After reactivating the adhesive on the soling material by heat the sole is attached to the upper by hand pressure before inserting the assembly in a press for final consolidation. It is important that this initial tack is strong enough to bear the weight of the sole unit otherwise positioning the sole and upper becomes rather complicated. The SBS compounds showing this problem contained between 30–40 % of oil and it was shown that some of this oil moved through the adhesive layer during removal of the solvent, and was deposited on the surface of the adhesive. This of course was a classic case of a weak surface layer and adequately explained the lack of 'spotting tack'. To reduce the effect of natural loss of tack which occurs with time, a short open time of the adhesive on the sole unit was advocated. If this was not practical then the resulting lack of tack could be overcome by wiping the urethane adhesive surface with a hydrocarbon or alcoholic solvent about 15 min before heat reactivation.

CONCLUSIONS

Having dealt with SBS copolymers in adhesives and adherends a very good example of the versatility of SBS copolymers is shown by their dual role in the direct moulding of an SBS shoe sole onto a canvas upper without using an adhesive.

This is a widely practised method of shoe manufacture especially in the USA. The canvas upper requires no surface preparation and it is simply positioned on the foot section of a mould. The SBS compound is injection moulded onto the canvas at a temperature between 150–180 °C. An in-mould time of less than 2 min is required and the completed shoe can then be removed. Warming the upper mould will prevent chilling of the molten SBS compound on contact with the upper. This will cause a slower rate of increase in viscosity of the compound which will facilitate flow and wetting. This is an example of mechanical adhesion; the SBS compound passing through the canvas weave and mechanically fixing at the back. Also the fibre ends may become embedded in the thermoplastic rubber.

This review of thermoplastic SBS copolymers is intended to demonstrate the versatility of this relatively new class of thermoplastic compounds. As an adherend they have many useful properties and have penetrated deeply in the footwear industry which is a very critical market. As adhesives they can be formulated for contact; pressure sensitive, or melt adhesives, an uncommon feature for a single elastomer type.

In all the diverse uses it is particularly important to be aware of the compounding possibilities and this necessitates a knowledge of the two phase system which exists. With this information it is possible to use ingredients compatible with the required phase and so design compounds for a very wide variety of applications.

ACKNOWLEDGEMENTS

We wish to thank Phillips Petroleum Company and Shell Chemicals for their considerable help. The major comments on adhesives are taken from both companies' technical literature on the subject and prospective formulators are recommended to obtain copies of these excellent publications.

In addition, the authors wish to thank the Shoe and Allied Trades Research Association for the opportunity to publish the work.

REFERENCES

1. MIERAS, J. M. A. and WILSON, E. A., *IRI*, **7**(2), 1973, p. 72.
2. AGGARWAL, S. L., *Polymer*, **17**(11), 1976, p. 938.
3. VAN HENTEN, K., *Brit. Boot and Shoe Inst.*, **22**, 1975, p. 179.
4. BULL, A. L., *Plastics and Rubber Mat. Appl.*, **2**(1), 1977, p. 27.
5. SIMPSON, B. D. and FOWLER, P. R., *Adhesives Age*, **17**(9), 1974, p. 32.
6. PETERSHAGEN, L. A., *Adhesives Age*, **14**(2), 1971, p. 20.
7. KRAUS, G., JONES, F. B., MARRS, O. L. and ROLLMANN, K. W., *Adhesion*, **8**(3), 1977, p. 235.
8. BORTHWICK, J. B., *International Hot-Melt Pressure-Sensitive Adhesive Course*, November 1976, Amsterdam.
9. BEAMISH, A. and HOURSTON, D. J., *Polymer*, **17**(7), 1976.
10. Private communication.
11. Shell Chemical Company Ltd, GB 1 278 258 (date of application 22.8.68).
12. Shoe and Allied Trades Research Association, GB 1 295 677 (date of application 8.5.70).
13. GREENWOOD, L. and WAKE, W. C., *Adhésion des Elastomères*, *Association Française Ing. du Caout. Plastiques*, Paris 1970, p. 19.
14. CARTER, A. R., *Brit. Boot and Shoe Inst.*, **22**, 1975, p. 184.

Chapter 10

PREPARING LOW-ENERGY SURFACES FOR BONDING

W. A. DUKES and A. J. KINLOCH
Ministry of Defence, Waltham Abbey, UK

SUMMARY

Over the years many empirical surface treatments have been developed, all improving the subsequent bondability of the substrate by organic adhesives. Recently, considerable advances have been made in understanding the mechanisms whereby the various treatments work. In this chapter an introductory summary is given of the main relevant concepts such as surface energetics and wettability, and the associated techniques, such as the measurement of contact angles and Auger and X-ray photoelectron spectroscopy used for surface analysis, are briefly described. Low energy surfaces are dealt with under the headings of fluorocarbon polymers, polyolefins, nylons and other plastics, plastic laminate materials, rubbers and irradiation techniques.

INTRODUCTION

Various reviews[1-5] give details of the empirical surface treatments developed both for metals and non-metals all of which improve the subsequent bondability of the substrate by organic adhesives. It is not our purpose to list again those methods which are particularly applicable to low-energy surfaces; indeed there have been relatively few additional developments recently. However, considerable advances have been made in understanding the mechanisms whereby the various treatments work. A survey of this elucidation of the scientific principles involved is preceded by a brief introductory summary of the main concepts such as surface energetics, wettability, and surface chemistry; together with the techniques used in studying them, such as the measurement of contact angles, and the powerful techniques of Auger and X-ray photoelectron spectroscopy, used for analysis of the surface. Then after a brief description of the main types of surface treatment in ascending order of complexity, particular groups of low-energy materials are considered in turn. Finally, irradiation techniques are reviewed.

Surface Energy

The surfaces of rubbers and plastics have low surface energies (generally less than 50 mJ m^{-2}), in distinction from those of metals, with energies of the order of 1000 mJ m^{-2}. An energetic surface will make the liquid spread on it, while a surface of low energy will let it remain as a discrete drop. So sufficient surface energy is normally needed in order to cause spreading of the adhesive and thus intimate molecular contact between it and the substrate: without that it is clearly impossible to achieve a bond. So poor wetting may be due to contamination, which may be removed by solvent cleaning, or it may be due to the molecular structure of the surface, which may be modified by chemical treatment.

The experimental approach to surface energetics is primarily by observing contact angles between liquids of various surface tensions and the substrate in question. Zisman and co-workers[6] made the first approach to the characterisation of low-energy surfaces: they found an empirical rectilinear relation between the cosines of the contact angles found on a given surface with a series of saturated hydrocarbon liquids, and their surface tensions. They defined a critical surface tension of wetting as the extrapolated value of surface tension corresponding to complete wetting (cosine = unity; zero contact angle). Even if a non-homologous series of liquids were used, the points fell close to a straight line which extrapolated to the critical surface tension, the highest at which a liquid would just spread and give a zero contact angle. This value characterises the wettability of the low-energy surface.

Levine, Ilkka and Weiss[7] found a linear relationship between the critical surface tensions of a series of polymers and the strength of joints between the polymers and an epoxy adhesive, thus emphasising the fundamental importance of wetting in the formation of an adhesive joint. Fowkes[8] introduced the idea of the additivity of intermolecular forces, i.e. that the surface energy of a solid (γ_S) consists of contributions from dispersion (γ_S^D) and polar (γ_S^P) forces, and that the geometrical mean of the dispersion force components $(\gamma_S^D \gamma_L^D)^{1/2}$ across an interface (where γ_L^D refers to the wetting liquid) is a reliable measure of the interaction energy caused by dispersion forces. Owens and Wendt[9] and Kaelble and Uy[10] extended this to include the polar interaction, and established that dispersion and polar forces each only interact across interfaces with their like. Kaelble[11] thus evaluated the dispersion and polar contributions to the surface energy of a number of solids. He defined[12] a wettability envelope for a given substrate, by graphically plotting the contribution of the dispersion forces against that for the polar ones. If the characteristic of any adhesive (γ_L^P, γ_L^D) falls outside the curve, good wetting and hence a good joint will not occur. However, if a surface pretreatment can somehow increase the total surface energy so that the envelope is extended to include the adhesive's characteristic, then good wetting will take place. Many methods of increasing the surface free energy are discussed in later sections: the scientific principles underlying them are becoming well established.

Baszkin and Ter Minassian-Saraga[13] have quantitatively determined the surface densities of the polar sites of polyethylene as the surface is oxidised. They show that as the temperature is raised to about 85 °C there is a decrease in wettability, due to the increase of chain mobility leading to the redistribution of external polar groups into the bulk. Owens[14] has shown that corona-treated polyethylene bonded after heating gives bond strengths which are considerably reduced by raising the heating temperature from 60–80 °C, presumably for the same reason. He attributes the bond to hydrogen bonds (from enolised keto groups formed by the corona treatment) with carbonyl groups. He also has demonstrated[15] in the case of poly(ethylene terephthalate) that molecular motion causes the loss of bondability, which is based on hydrogen bonds between phenol groups created by corona or u.v. irradiation and carboxyl groups in the other surface.

The basic principles underlying the need to prepare low-energy surfaces for bonding also apply to high-energy surfaces, although the emphasis is different. As indicated above, it is helpful to increase the surface energy of low-energy plastics and rubbers. Paradoxically, this is also necessary for intrinsically high-energy materials, because the original unstable high surface energy has normally been masked by the acquisition of contaminants, which must be removed before bonding. Even in the case of noble metals, such cleaning is not necessarily simple. A remnant aqueous monolayer will greatly reduce the spreading of liquid water.[16–18] There is also spectroscopic evidence of persistent contamination by air-borne hydrocarbons.[19]

Even if good wetting is achieved, a strong bond does not necessarily follow. If there is a weak boundary layer on the adherend,[20] the joint will of course be no stronger than the layer.

Surface Analysis

The direct examination of surfaces and contaminants on them has recently become possible by the use of Auger electron and X-ray photoelectron spectroscopy (AES and XPS, respectively). These two techniques are extremely useful in adhesion studies in that both enable the composition of the outer 1 to 5 atomic layers of the surface of a solid to be analysed. The basic technique in AES consists of bombardment of the surface with a beam of electrons in the range of 1–5 keV and analysis of the energies of the ejected electrons, which usually provides only an elemental analysis. In XPS, photoelectrons (and Auger electrons) are generated when the surface is flooded with soft X-rays; the photoelectrons have discrete binding energies whose values depend upon both the element and its state in the atomic matrix in the surface. Therefore, it is possible to determine both the concentration and chemical state in the surface. The two techniques are in many ways complementary. AES gives good spatial resolution since a narrow (1–50 μm) electron beam is used, while with XPS such spatial resolution cannot as yet be obtained, since it depends on X-ray photons to excite photoelectrons. The advantage of XPS is, however, that charging effects are minimised and surfaces of insulators can be more easily analysed.

Surface Treatments

As indicated above, surface preparation for bonding can be of many different types, and these differ depending on the mechanism of amelioration intended. The theories of adhesion and adhesive action have been recently reviewed,[21] and it is clear that adhesion is no simple phenomenon. Thus we will consider the purpose of surface treatments in the light of the theories, starting with the simplest. Probably the first and main purpose is to clean the surface by removing contaminants. The simplest way to remove grease etc. is with an organic solvent, either by wiping with a damped cloth or by liquid or vapour degreasing.

Next is abrasion; either with abrasive paper or wire-brush, or by grit-blasting. This not only removes surface contamination, it removes the surface and perhaps a weak boundary layer as well. It also increases the apparent surface energy. It is often thought that a roughened surface is preferable, as it enables 'keying' of the adhesive into the surface, and there is evidence for this mechanism with fibrous adherends such as textiles, leather and wood. There is evidence also[22] that the mechanism of adhesion to porous anodic films on aluminium involves keying of the polymer into the pores. Surface rugosity also can affect the kinetics of wetting by the adhesive[23,24] and it has been shown[25] that the rate of spreading on a very rough surface is much greater than on a polished surface. However, in practice adhesives are forcibly spread on to substrate surfaces and this driven flow causes the adhesive to outrun any surface capillary spreading phenomenon.[24] It has also been shown[26,27] that random roughness, as distinct from the more uniform surface (of aluminium alloy) achieved by polishing, can increase the strength of a joint, probably by changing the stress distribution at the interface.

However abrasion is usually insufficient as a treatment for low-energy surfaces, and chemical treatment is often needed to increase the surface energy. Specific chemical treatments will be discussed below, for each substrate material in turn.

FLUOROCARBON POLYMERS

The fluorocarbon family includes polytetrafluoroethylene (PTFE) polychlorotrifluoroethylene (PCTFE), fluorinated ethylene–propylene copolymer (FEP), polyvinyl fluoride and polyvinylidene fluoride. They all possess very low surface free energies, about $30\,\mathrm{mJ\,m^{-2}}$ or less, and are extremely inert polymers. The most common surface treatment is etching, i.e. eating away the surface with an aggressive or mordant reagent.

Etching Treatments

The treatment for fluorocarbon polymers is usually based upon metallic sodium dissolved in a solution of naphthalene in tetrahydrofuran. Proprietary solutions are available and it is also possible to purchase PTFE already treated by such chemical processes.

This sodium naphthalene treatment requires immersion of the polymer in the fresh treating solution at room temperature, then rinsing in water, then in acetone, followed by thorough washing in water again and finally drying. The etching process produces a dull brown colour on the fluorocarbon's surface and this coloration intensifies as the etching time is increased. It has been suggested[28] that this colour change may provide a useful quality control check for procurement of uniformly and reproducibly etched material.

Nelson, Kilduff and Benderley[29] have shown that the etching solution attacks only the surface regions of the fluorocarbon polymer. They examined a cut section of etched PTFE and from optical microscopy determined that the depth of the coloured surface region was about one micrometer. However they reported that the bulk electrical properties of the treated and untreated polymer were virtually identical and that their electron diffraction patterns were indistinguishable. They attributed these observations to the thinness of the treated surface layer or to the treated material having the same structure as the bulk material. Benderley[30] has since reported that the treated material does have a substantially lower surface resistivity than the untreated material, so confirming the former explanation. Purvis and Beck[31] have ascribed the brown coloration to a carbonaceous surface being formed as a result of the extraction of fluorine atoms by the etching solution. Their detection of fluoride ions in the expended etching solution provided some evidence for this proposed mechanism. Further support has recently come from the work of Miller *et al.*[32] and Andrews and Kinloch.[33] These latter authors used infrared attenuated total reflectance (ATR) spectroscopy to examine the surface regions of untreated and etched fluorinated ethylene–propylene co-polymer. The spectra were identical and no absorption bonds were observed at the frequencies appropriate to hydrocarbon or fluorocarbon unsaturated groups. Now Polchlopek[34] has reported that a continuous depth of about 1 μm of the material to be identified is required for detection by the type of ATR spectroscopy employed in the above investigation. Therefore, since 1 μm was the etch depth determined by Nelson *et al.*,[29] further characterisation of the FEP surfaces was conducted by placing samples of the fluorocarbon in a 1 mass % solution of osmium tetroxide for six days. Osmium tetroxide reacts with carbon–carbon double bonds to produce a dark appearance. In the case of the untreated material no visible colour change occurred after immersion. However the surface of the etched FEP was considerably darkened; by sectioning perpendicular to the surface and examination in the optical microscope the depth of the darkened surface layer was found to be about 1 μm, in good agreement with the above comments. It is the presence of these conjugated carbon–carbon double bonds in the surface regions that are responsible for the brown coloration. XPS studies[35] have confirmed that unsaturation is introduced by the etching treatment and have shown that

$$\rangle C{=}O \quad \text{and} \quad {-}COOH$$

groups are also introduced.

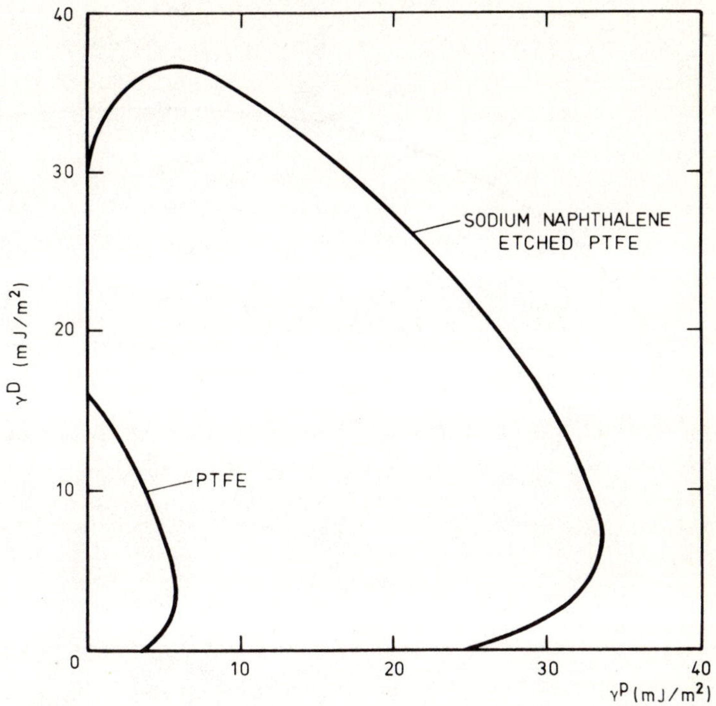

FIG. 1. Wettability envelopes of PTFE and etched PTFE.[12]

Several workers[12,36,37] have reported a large increase in surface free energy obtained by etching fluorocarbon polymers in sodium naphthalene and this is convincingly demonstrated by the wettability envelopes[12] of untreated and etched PTFE shown in Fig. 1. It is undoubtedly the increase in wettability which is usually responsible for the increased adhesion to etched surfaces.

The effect of a proprietary sodium naphthalene etching treatment on the strength of tensile butt joints is given in Table 1. The adhesive was an epoxy-polyamide and the beneficial effects of the etching treatment are obvious.

Similar increases in peel strength are recorded when PTFE is bonded to itself with a neoprene rubber adhesive, as shown in Fig. 2. As may be seen an optimum etching time of about 15–30 sec is indicated.

The effect of etching time is also shown in Fig. 3 for model adhesive joints consisting of a crosslinked styrene–butadiene rubber bonded to a fluorinated ethylene–propylene copolymer.[37,39] These results enable the

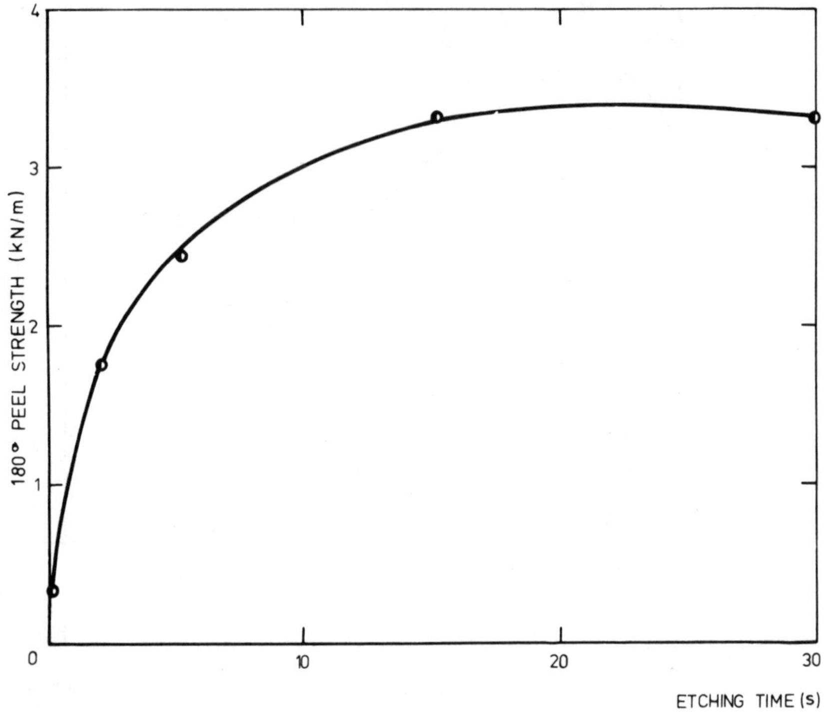

FIG. 2. Effect of a sodium-naphthalene etching treatment on the peel strength of PTFE–neoprene–PTFE joints.[2]

mechanisms of adhesion to be identified since the value of the 'intrinsic adhesive failure energy', P_0, depends solely upon the physical and chemical nature of the adhesive–substrate interface and, if only secondary forces are acting, equals the thermodynamic work of adhesion, W_A. For the rubber–unetched-FEP interface the value of W_A is about 50 mJ m^{-2} and for the rubber–etched-FEP interfaces about 70 mJ m^{-2}. Now in the case of the former interface P_0 is approximately equal to W_A but for the etched substrates $P_0 \gg W_A$. This arises because the etching treatment not only increases the surface free energy of the FEP but also introduces a significant concentration of carbon–carbon double bonds into the etched substrate's surface regions. Therefore, since dicumyl peroxide in the compounded rubber will initiate reactions between the rubber molecules and unsaturated groups in the substrate surface, primary covalent bonding will be established across the rubber–etched-FEP interface during curing of the

TABLE 1
EFFECT OF VARIOUS SURFACE PRETREATMENTS ON THE BONDABILITY OF PTFE AND PCTFE[38]

Fluorocarbon	*Butt joint strength* ($MN m^{-2}$) *for given treatment*			
	No surface treatment	*Abraded*	*Etched*	*Abraded and etched*
PTFE	Fell apart	Fell apart	10·8	13·2
PCTFE	2·6	7·7	19·4	20·7

elastomer. The observed increase of P_0 over and above W_A for these interfaces is thus to be expected since the calculation of W_A takes into account only secondary force interactions (e.g. van der Waals' forces). As etching proceeds, however, the FEP surface layer suffers degradation and this is reflected by the locus of joint failure progressively occurring via fracture of the substrate surface regions. Thus the two effects of etching, namely (i) increasing the concentration of carbon–carbon unsaturation (and hence the likelihood of interfacial covalent rather than secondary

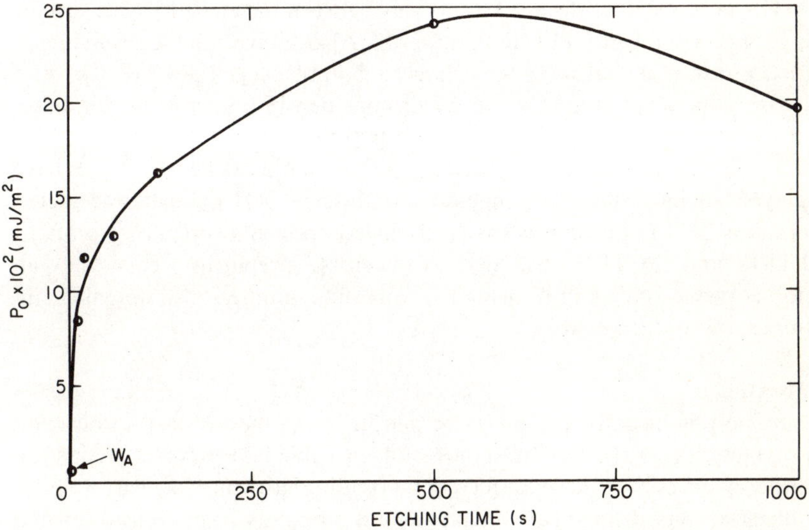

FIG. 3. Intrinsic adhesive failure energy, P_0, versus etching time for rubber–FEP joints.[37,39]

bonds) and (ii) reducing the strength of the etched regions, result in an optimum etching time for maximum joint strength.

Finally, as might be expected, the unsaturated and oxidised etched layer is more reactive, not only to the adhesive but also to ultraviolet radiation, and there are several references to the instability of the adhesive bond between sodium-etched fluorocarbon polymers and other materials in the presence of u.v. light.[28,30,40] The effect of accelerated ageing tests on etched PTFE joints bonded with a polyurethane adhesive and exposed to u.v. light is given in Table 2.

TABLE 2
PEEL STRENGTH OF ETCHED PTFE JOINTS AGED IN A WEATHER-OMETER[28]

Substrate	*Peel strength* ($kN\,m^{-1}$) *after ageing*			
	0 *h*	100 *h*	200 *h*	500 *h*
Etched PTFE	2·00	0·79	0·35	0·00
Etched PTFE: u.v. absorbers	2·00	—	—	1·21

The failure site of the aged joints was between the etched PTFE surface layer and unetched PTFE bulk material. Also shown are the results from incorporating u.v. absorbers, such as carbon black, in the PTFE film and hydroquinone into the adhesive, which considerably increase the durability of the joints upon exposure to u.v. radiation.

Applications in which proprietary sodium naphthalene treatments have played an important role include: bonding of PTFE bushes to metal, bonding PTFE insulation to electrical components, preparing seals to PTFE and PCTFE insulated connectors, preparing seals between polyethylene and PTFE insulated wire and connector terminals, and improving potting seals.

Abrasion

Abrasion is usually a very poor surface treatment for fluorocarbon polymers, as may be seen from the results in Table 1. However, Lerchenthal and co-workers[41–43] have claimed that abrading substrates beneath reactive adhesives may dramatically increase joint strengths compared to normal abrasion methods in air. Results for PTFE bonded to itself and to aluminium with an epoxy-polyamide adhesive are shown in Table 3.

TABLE 3
ABRASION SURFACE TREATMENTS FOR PTFE

Joint	*Treatment of PTFE*			
	Abraded in air		*Abraded beneath epoxy/vinyl acetate mixture*	
	Tensile shear strength (MNm^{-2})	*Locus joint failure*	*Tensile shear strength* (MNm^{-2})	*Locus joint failure*
PTFE–aluminium	0·98	Interfacial	5·71	Interfacial
PTFE–PTFE	0·98	Interfacial	6·67	Interfacial

Whether the advantage of abrasion beneath an epoxy/vinyl acetate mixture arises from better wetting or from reaction of the mixture with free radicals generated in the polymer surface has yet to be firmly established.

Morphological Changes

Schonhorn and Ryan[44] have reported that heterogeneous nucleating and crystallisation of FEP melt against high energy surfaces (e.g. gold, aluminium) produces a transcrystalline surface on the fluorocarbon of high mechanical strength and increased surface free energy. Indeed the critical surface tension may rise from 18·8–40·4 mJ m^{-2}.[45] Other workers[35] have suggested that this is due to the formation of a thin layer of oxygen-containing hydrocarbon material rather than highly crystalline FEP. A surface is produced which is amenable to conventional bonding. The tensile shear strengths of aluminium/epoxy/FEP joints where the FEP has been subjected to this and other surface treatments is shown in Table 4.

TABLE 4
VARIOUS SURFACE TREATMENTS FOR FEP

Surface treatment of FEP	*Tensile shear strength* (MNm^{-2})
None	0·69
Nucleated and crystallised against electrochemically-deposited gold	4·82
Nucleated and crystallised against evaporated aluminium	12·41
Nucleated and crystallised against evaporated gold	19·30
Etched in sodium naphthalene solution for 40 sec	20·00

Schonhorn and Ryan concluded that gold films deposited electrochemically are not effective, perhaps because they are contaminated or complexed with a variety of organic compounds (e.g. cyanide, citrate) normally used in plating baths.

Evaporated films of gold are far more effective in generating a transcrystalline region of high strength and surface free energy. Further, gold is also more effective than aluminium oxide in this respect. The basic difference may be in the density of nucleating sites on the surface. If there are too few, ordered transcrystalline growth may not take place.

Aluminium–Alkali Treatment

Roberts and co-workers[46] have recently described a new method for bonding gold to fluorinated-ethylene-propylene copolymer which yields high joint strength without affecting the electrical properties of the virgin polymer surface. The method consists of depositing a 1000-Å aluminium layer onto the film by evaporation, removing it by washing the sample in dilute sodium hydroxide, followed by deposition of gold by evaporation. X-ray photoelectron spectroscopy (XPS) indicated significant chemical modification of the film's surface, apparently consisting of the generation of a thin and perhaps continuous surface layer of oxygen-containing hydrocarbon material, as a result of the aluminium deposition. In contrast scanning electron microscopy indicated the surface topography was relatively unaffected. It therefore appeared that the increased joint strength could be attributed to the increased surface free energy of the fluorocarbon.

POLYOLEFINS

Like the fluorocarbon polymers discussed above, polyolefins such as low- and high-density polyethylene, polypropylene and poly(4-methyl pentene-1) cannot usually be bonded to give reproducible high strengths unless some form of surface pretreatment is employed. Many treatments have been reported to improve the adhesion to polyolefin surfaces; these include corona[47–49] and glow discharge,[50–53] chemical attack with chromic acid,[54–56] with sulphuric acid[2] or with peroxydisulphate,[57,58] fluorination,[59] solvent etching,[60] exposure to flame and hot-air blasts[61] and melt crystallisation against high-energy surfaces.[62,63] The aims of all these varied surface treatments are to remove any contamination or any weak boundary layer on the polyolefin's surface, increase the surface free energy of the material and make the surface more receptive to bonding.

Solvent Cleaning

The simplest treatment is solvent etching,[60] and the effect of dipping a polypropylene substrate into a trichloroethylene solvent vapour prior to bonding to aluminium using an epoxide adhesive is shown in Fig. 4. The effect of a short exposure time on the joint strength is spectacular but over-treatment may readily occur causing a rapid decline in joint strength. The

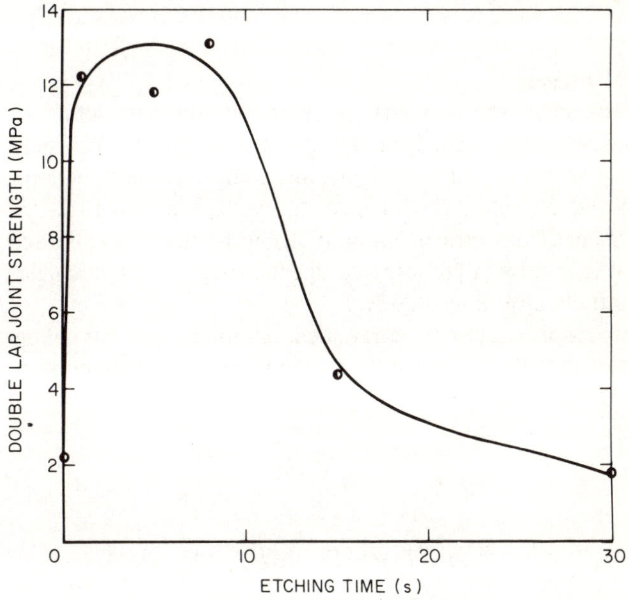

FIG. 4. Effect of a trichloroethylene-vapour surface treatment on the strength of polypropylene joints.[60]

trichloroethylene vapour imparted a dull white finish to the smooth polymer surface, and from scanning electron microscopy studies the authors concluded that the untreated polymer possessed a weak surface layer which was removed by the solvent treatment. This treatment also resulted in a highly porous surface (possibly caused by the solvent preferentially attacking the amorphous regions of the surface). This mechanism for the improvement in joint strength was indirectly supported by the solvent-treated surfaces having a wettability inferior to that of the untreated polymer, as indicated by the contact angle data shown in Table 5.

The decline in joint strength for etching times longer than about 10 sec is probably due to degradation and weakening of the surface layer.

TABLE 5
WETTABILITY OF UNTREATED AND SOLVENT-CLEANED POLYPROPYLENE

Pretreatment	*Contact angle of:*	
	Water	*Epoxide adhesive*
None	90°	26°
Solvent-vapour	98°	34°

Chemical Etching

One of the most common surface treatments for polyolefins is etching in chromic acid and a detailed study of this method has recently been reported by Blais *et al.*[54] They employed chromic acid solutions to etch linear (high-density) and branched (low-density) polyethylene and isotactic polypropylene. Apart from measuring joint strengths they also characterised the etched surfaces by ATR infrared spectroscopy, contact angle measurements and electron microscopy.

Spectroscopic examination revealed extensive chemical changes only in the case of low-density polyethylene, with an increasing concentration of —OH,

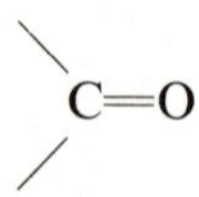

and possibly $-S(=O)_2OH$ groups in the surface layer. However, as pointed out previously, ATR spectroscopy is not a true surface technique and has a relatively large sampling depth compared to the likely etch depth. This is why new surface analytical techniques such as Auger and X-ray photoelectron spectroscopy (XPS) have recently been used to study chemical changes arising from surface pretreatments[64,65] and to accurately identify the locus of failure in adhesive joints.[19,66] Employing XPS analysis Briggs and co-workers[64] detected the presence of —OH,

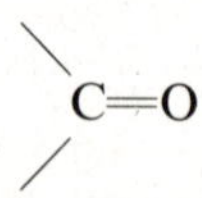

$-COOH$ and $-S(=O)_2OH$ groups in the outermost surface regions of both chromic-acid etched polyethylene and polypropylene. Further, they showed that the essential difference between etching polypropylene for one minute at 20 °C or six hours at 70 °C was in the depth of attack achieved

rather than any increase in the degree of surface oxidation. In the case of polyethylene, however, both the etching depth and degree of surface oxidation increased with etching time. Nevertheless, for both polyolefins considerable improvement in joint strengths were recorded after etching for only one minute at 20 °C but increased etching times did not lead to increased joint strengths, possibly due to the limiting strength of the polyolefin substrate.

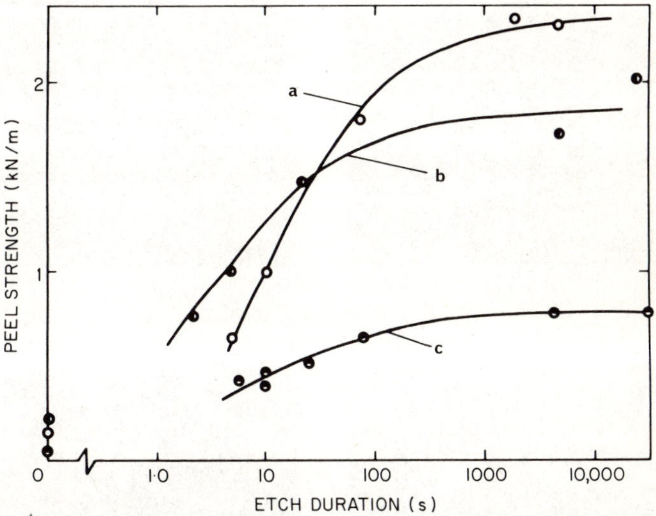

FIG. 5. Peel strengths for etched polyolefin surfaces as a function of etch time. (a): High-density polyethylene; (b) low-density polyethylene; (c) polypropylene.

Blais *et al.* have also studied the effect of etching time. They realised that the enhancement of bonding by a treatment would eventually lead to the situation where joint failure occurs by cohesive fracture of the polyolefin substrate, as observed by Briggs and co-workers. Since measurements under these conditions obviously do not reflect the effectiveness of the surface treatment they therefore used (i) an adhesive which would give an extremely low bond strength to an untreated polyolefin surface, (ii) an adhesive with reasonable cohesive strength and (iii) a joint geometry sensitive to variations in interfacial adhesion. They chose a polyacrylic acid adhesive, plasticised with a little ethylene glycol, and a modified 90° peel test geometry. The peel strength is shown as a function of etching time in Fig. 5. The polypropylene shows an approximately 170-fold increase in peel

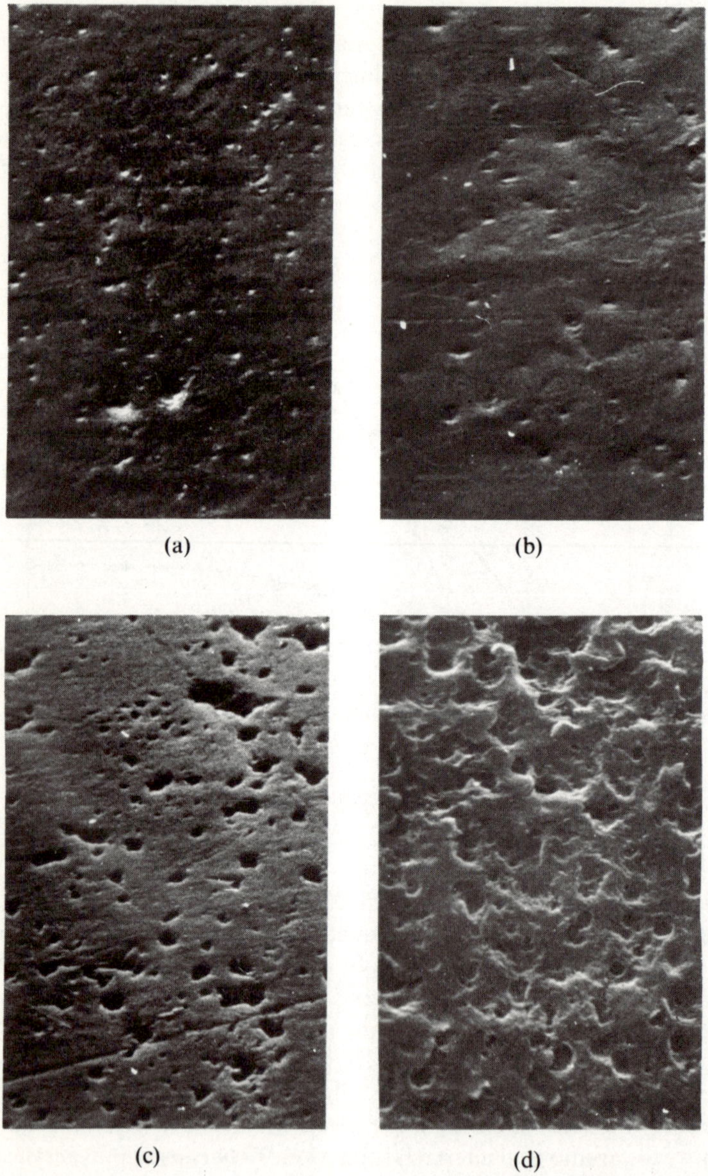

FIG. 6. Surface treatment of polypropylene with chromic acid: Effect of variation in time and temperature.[2] (a): Untreated; (b): etched for 10 min at 20 °C; (c): etched for 20 min at 20 °C; (d): etched for 5 min at 80 °C (×2000).

strength after a 100-sec etch but little improvement for a more prolonged attack, while both polyethylenes exhibit a rapid increase in joint strength with longer etch times.

The work of both Blais *et al.* and Briggs *et al.* indicated that there was little correlation between the contact angles of water on the etched surface and resulting joint strength. However, Shields[2] has shown that chromic acid etching produces a very rough surface and it is well established[67,68] that surface roughness alone may result in contact angle changes. Scanning electron micrographs obtained by Shields are shown in Figs. 6(a–d) and it may be seen that the surface of untreated polypropylene was smooth but marred by the presence of small pits of less than 1 μm in diameter. Changes in the surface topography were insignificant for etching periods of up to 10 min in cold chromic acid. However, after a 20-min cold etch the population of pits had increased and hole diameters enlarged to about 8 μm, though considerable areas of smooth surface remained. Hot chromic acid etching had a more pronounced effect on the surface; the whole surface was of a rugose nature and contained craters with diameters approaching 10 μm. Blais *et al.* found far less increase in the surface roughness of polypropylene upon chromic acid etching but did record interesting observations for a polypropylene sheet which had been initially annealed to produce large spherulitic, crystalline structures. Before etching the sample was quite featureless yet the spherulitic areas became visible after etching, indicating that the less ordered, amorphous material between the spherulites' arms and boundaries were more rapidly attacked than the highly crystalline arms themselves. This observation substantiates some earlier conclusions reached by Garnish and Haskins[60] who found that different grades of the same polymer showed different susceptibilities to pretreatments and suggested that this was linked to the size of the crystalline spherulites in the material. Obviously different polyolefins may exhibit even more marked effects and this adds another degree of complexity to surface pretreatment considerations.

With the chemical and physical changes induced by chromic acid etching elucidated, it might be hoped that a unified mechanism for the improvement in adhesion would be identified. Shields concluded that the effectiveness of chromic acid treatments was mainly due to their ability to produce a roughened surface which increased the surface area and enabled mechanical keying of the adhesive. However, the observation of relatively smooth surfaces by Blais *et al.*, associated with considerable increases in joint strength, suggests that this cannot be the major mechanism operating. Briggs *et al.* ascribed the increases in adhesion to increased surface polarity and some support for this mechanism also comes from the work of Rauhut,[61] shown in Fig. 7, where a correlation is proposed between increased surface polarity, as reflected by the critical surface tension, γ_c, and joint strength. Nevertheless, Blais *et al.* feel that, while increased surface polarity may play a minor role in the case of polyethylene, the major

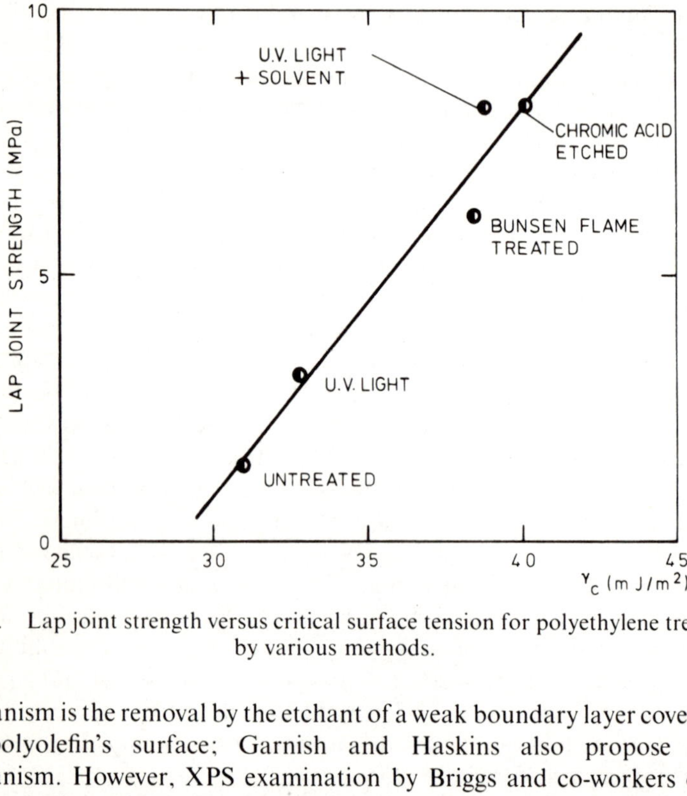

FIG. 7. Lap joint strength versus critical surface tension for polyethylene treated by various methods.

mechanism is the removal by the etchant of a weak boundary layer covering the polyolefin's surface; Garnish and Haskins also propose this mechanism. However, XPS examination by Briggs and co-workers of a fractured joint which had consisted of untreated polyethylene and epoxide adhesive gave no evidence of transfer of the polyethylene to the epoxide; this is evidence against the weak boundary layer concept. Therefore there is no single, readily identifiable mechanism and just as different polyolefins and grades of polyolefins respond differently to etching, due to molecular-weight, chemical and morphological differences, so these various mechanisms may contribute to various extents depending upon the interface involved.

Morris[57,58] has proposed the use of an aqueous ammonium per-

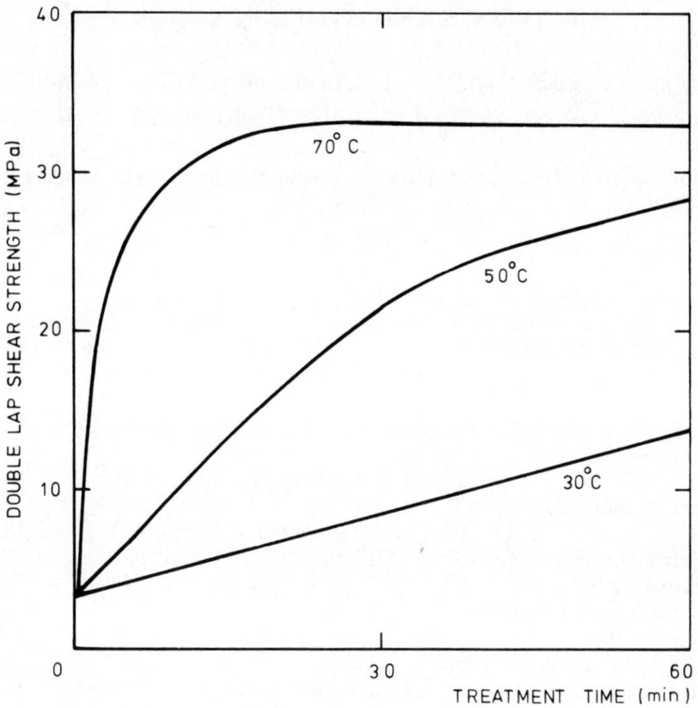

FIG. 8. Effect of treatment time and temperature on adhesive joint strength of polyethylene treated with 0·26-M solution of $(NH_4)_2S_2O_8$.

oxydisulphate solution for treating polyethylene and polypropylene; the effect of immersion time and temperature is shown in Fig. 8. No differences were apparent between the ATR infrared spectra of untreated and treated samples; hence if the polymer was oxidised by the peroxydisulphate the effect was confined to the immediate surface regions. However, the critical surface tension of the treated material was 5 mJ m^{-2} higher. Further, an insoluble gel fraction was present in the treated material, about 1% by weight of the original polymer, but was absent in the untreated samples. Thus, Morris concluded that treatment of polyethylene and polypropylene with a peroxydisulphate solution led to the formation of a surface layer of crosslinked, entangled molecules and that this increased the mechanical strength of the surface region and resulted in higher joint strengths. Again no direct evidence was presented for the presence on the original untreated polymer surfaces of a weak boundary layer, which needed strengthening.

NYLONS AND OTHER PLASTICS

Other plastics such as polyvinyl chloride, polyethylene terephthalate,[69] poly acetals[70,71] and nylons do not present such a severe problem to the adhesive technologist as do the fluorocarbon polymer or polyolefin families. This is because they have somewhat higher surface free energies

TABLE 6

EFFECT OF SURFACE TREATMENT ON TENSILE SHEAR STRENGTH OF NYLON 6·6–ALUMINIUM JOINTS[2]

Surface treatment	*Adhesive*	*Lap shear strength* ($MN m^{-2}$)	*Locus of joint failure*[a]
Solvent cleaning	Epoxy	2·92	C
	Cyanoacrylate	3·95	I/C
Solvent cleaning + abrasion	Epoxy	3·58	I
	Cyanoacrylate	3·37	I/C
15% chromic acid; etched for 7 min at 20°C	Epoxy	4·31	I/C
	Cyanoacrylate	4·76	I/C
50% chromic acid; etched for 30 sec at 50°C	Epoxy	3·21	I
	Cyanoacrylate	6·42	C*
50% chromic acid; etched for 120 sec at 50°C	Epoxy	2·77	I
	Cyanoacrylate	5·72	C*
Acid potassium permanganate; 5 min at 90°C	Epoxy	3·76	I/C
	Cyanoacrylate	6·45	C*
As above + resorcinol–formaldehyde primer	Epoxy	6·05	C*
	Cyanoacrylate	5·56	C*

[a] I = interfacial; C = cohesive in adhesive; C* = cohesive in nylon 6·6.

but nevertheless, in order to obtain high joint strengths approaching the cohesive strength of the substrate, some form of surface pretreatment is often necessary.

This is evident from Table 6 where the effect of various surface treatments on nylon 6·6 is shown for two different adhesive types. For the epoxide adhesive solvent-cleaning and abrading is as effective as the chromic-acid etching treatments and only an acid potassium permanganate etch followed by application of a resorcinol–formaldehyde primer results in the highest joint strengths (i.e. when the cohesive strength of the nylon 6·6 is reached). The interface between the nylon and resorcinol–formaldehyde primer probably has intermolecular hydrogen and covalent bonds established across it and this is reflected in the high joint strengths.

Alternatively, in the case of the cyanoacrylate adhesive a chromic-acid etch, particularly at elevated temperatures, results in high joint strengths. This may possibly arise from (i) the greater penetration of the cyanoacrylate, due to its much lower viscosity, into the etch-pits on the nylon's surface and (ii) the cyanoacrylate's greater reactivity towards the more polar etched surface. These results illustrate that the total interface, i.e. adhesive as well as substrate, must be considered when selecting an effective surface treatment.

Abu-Isa[72] has studied the iodine treatment of nylon materials. This treatment essentially consists of dipping the nylon in an iodine–potassium iodide solution and quickly rinsing the adsorbed iodine out of the nylon surfaces using glycol–water–thiosulphate as a wash solution; this considerably increased the strength of a metal film subsequently plated onto the treated nylon surface. It was found that the strength of the plating was dependent upon the nature of the nylon surface (crystallinity, molecular weight, etc.), the duration of the iodine treatment, and the nature and conditions of the wash solution. Scanning electron micrographs of the nylon surface demonstrated that only a shallow etch pattern was introduced and thus mechanical keying between the polymer and plating was not considered to be of prime importance. However, under certain circumstances the overall degree of crystallinity was decreased and the type of crystalline structure present changed. Further, the polymer was softened, low molecular-weight fractions and monomers extracted and the level of surface polarity raised. These last changes were considered to be responsible for the enhancement of adhesion to the treated nylon material.

PLASTIC LAMINATE MATERIALS

Plastic composites such as glass- or carbon-fibre reinforced composites are being increasingly used in many diverse applications. These materials have low surface energies, will not strongly adsorb water, form oxides or corrode in moist environments. They are usually based upon epoxy or polyester resins and are, therefore, more polar than the fluorocarbons or polyolefins (*see* Table 7) which makes them more receptive to adhesive bonding. This combination of properties results in surface treatments being required which simply remove contaminants such as oils, mould lubricants or general dirt. There are two main techniques used to achieve this,[73] (i) abrasion and solvent cleaning and (ii) the tear-ply method.

For carbon-fibre reinforced plastics it has been reported[74] that rubbing

TABLE 7
EFFECT OF PRETREATMENT ON STRENGTH OF EPOXY–CARBON-FIBRE COMPOSITE JOINTS

Pretreatment employed	*Surface free energy* ($mJ\,m^{-2}$)	*Lap-shear strength* ($MN\,m^{-2}$)
None	33·3	16·5
Light sanding	43·1	20·6
Heavy sanding	58·0	20·6
Nylon tear-ply	46·8	15·3
®Nomex tear-ply	68	16·0
Nitric acid treated	73·4	9·8

®Nomex is a Registered Trade Name owned by E. I. du Pont de Nemours.

with an abrasive cloth of roughness grade 150 and subsequently rinsing with perchloroethylene to remove abrasive dust was sufficient treatment. The abrasive treatment removed the contaminants and the resin film from the laminate surface and exposed the outer carbon fibres. Joints made with this treatment resulted in the carbon-fibre reinforced plastic, rather than the adhesive, failing. Lerchental and co-workers[43] have suggested that for glass-fibre polyester laminates higher joint strengths may be obtained if the abrasion is conducted beneath a 'reactive primer'. The results shown in Table 8 are for lap-shear joints consisting of glass-reinforced polyester

TABLE 8
LAP-SHEAR STRENGTH OF GLASS-FIBRE POLYESTER LAMINATES BONDED WITH A POLYESTER ADHESIVE

Reactive primer	*Abraded in air*		*Abraded beneath reactive primer*	
	Strength ($MN\,m^{-2}$)	*Locus of joint failure*	*Strength* ($MN\,m^{-2}$)	*Locus of joint failure*[a]
1:9 mixture of maleic anhydride + ethylacetate	3·1	Interfacial	5·2	C
1:9 mixture of acrylonitrile + styrene	6·2	Interfacial	9·0	C
1:9 mixture of methyl methacrylate + styrene	6·2	Interfacial	8·8	C

[a]C indicates cohesive failure in surface regions of laminate substrate.

laminates bonded with an unsaturated polyester adhesive. The improvement in joint strengths observed when the abrading was performed beneath a 'reactive primer' was ascribed to thc initiation of chemical bonds between substrate and adhesive via the mechanochemical creation of free radicals on the substrate surface. This is certainly more likely to occur in these examples than when bonding fluorocarbons (*see* earlier) or polyolefins, but improved wetting may still play a role. An interesting feature of these results is that abrading beneath different 'reactive primers' gave various joint strengths but all the joints failed by cohesive fracture in the substrate. This may arise from the adhesive–primer adjacent to the substrate having different mechanical properties (and thus giving rise to different stress concentrations in the surface of the glass-laminate substrate) depending upon the 'reactive primer' employed. This would then result in a range of joint strengths being recorded but all the joints could well fail by cohesive fracture in the surface regions of the substrate.

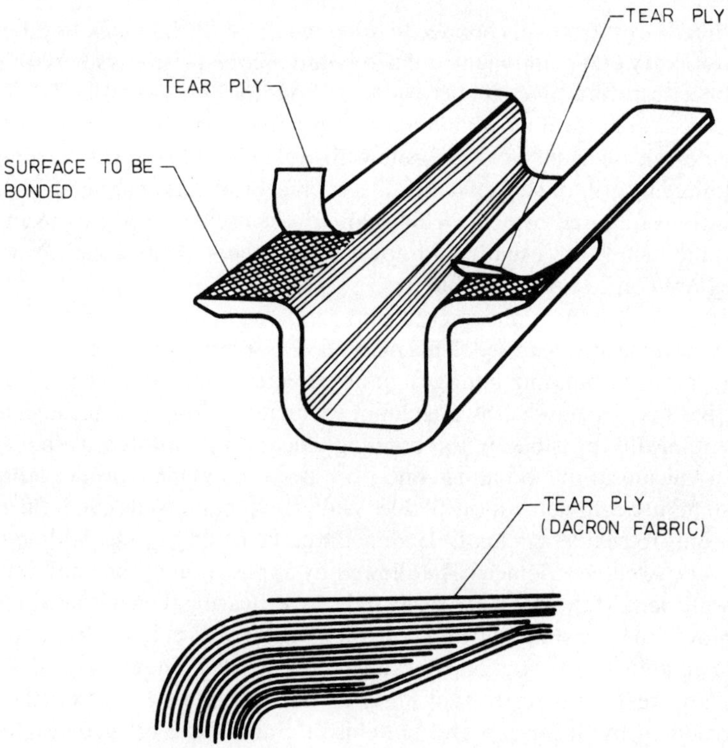

FIG. 9. Structural-reinforced plastic laminate with tear ply.

The second technique is the tear-ply method. In designing the laminate one ply of fabric, such as Dacron or equivalent, should be installed at the bonding surface as illustrated in Fig. 9. Just prior to bonding the tear-ply is removed and a fresh, clean bondable plastic surface is exposed.

Crane *et al.*[75] have compared several pretreatment methods for carbon-fibre composites and their results are given in Table 7.

There is no obvious correlation between surface free energy and joint strength and the authors concluded that a tear-ply serves only to protect the composite surface from contamination between manufacture and bonding (its intended purposes, of course!). The only pretreatment studied that significantly improved bond strength over that of the untreated surface was abrasion.

RUBBERS

Rubbers are most easily bonded to other materials during vulcanisation and proprietary one-component and two-component adhesive systems for such rubber-to-rubber and rubber-to-metal bonding are available.[76–78] The uncured rubber surface should be clean and free from contaminants such as mould-release agents and the metal substrate should be vapour-degreased followed by grit- or shot-blasting. Degreasing by means of aqueous alkaline solutions followed by acid etching can also be used; special treatments for specific metals are usually laid down by suppliers of the adhesives where they are considered advisable.[79]

In some instances, however, it may be necessary to bond rubbers after they have been vulcanised. This method of bonding is referred to as post-vulcanisation bonding and a main advantage is that complicated rubber shapes may be bonded by a technique which would not be technically or economically possible by the bonding/vulcanising one-step method. For post-vulcanisation bonding one- or two-component primer/adhesive systems are commercially available. Wake[80] has suggested that, if the bond has only to bear a light load, cleaning the cured rubber surface with solvent, thereby swelling it somewhat, followed by application of the adhesive may be sufficient. If the bond has really to be load-bearing then it is necessary to remove the existing surface of the vulcanised rubber by abrasion immediately before applying the adhesive to be used. The effects of different surface treatments for the vulcanisate prior to bonding has recently been considered by Spearman and Hutchison[81] and some of their results are given in Table 9.

TABLE 9

BUTT JOINT STRENGTHS ($MN\,m^{-2}$) OF RUBBER–METAL BONDS FOR VARIOUS VULCANISED RUBBERS AND SURFACE TREATMENTS

Vulcanised rubber	*Surface treatment*					
	Solvent wipe	*Ultrasonic degreasing*	*Hot alkaline rinse*	*Abrasion*	*Chlorination*	*Cyclisation*
Natural	4·5	5·4	6·5	9·4	5·6	7·0
Styrene-butadiene	1·8	6·1	6·2	7·9	6·9	4·5
Polychloroprene	5·1	4·9	6·7	6·8	6·9	7·0

Proprietary one-coat adhesive on treated, cured rubber and two-coat primer–adhesive on metal.

Natural rubber is sensitive to the type of surface treatment, with abrasion being the most effective. Neoprene is relatively insensitive. For all three rubbers abrasion is a good pretreatment and results in high joint strengths and cohesive failure of the rubber in the joint. However, other cured rubbers such as butyl and ethylene–propylene–diene rubber are far more difficult to bond than the above, more polar natural, SBR and neoprene materials.

IRRADIATION TECHNIQUES

Irradiation techniques are used on a number of substrates, and are here dealt with together for convenience. These methods include corona discharge in air and oxygen, irradiation with u.v. and gamma radiation and electrodeless glow discharges using a wide variety of gases.

Corona discharge is a long established method for treating polyethylene and is rapid and effective.[47,82] The polyethylene surface is exposed to the corona discharge in air, usually, at atmospheric pressure. Employing infrared spectroscopy Rossman[83] showed that this treatment causes the formation of unsaturated (C═C) bonds and carbonyl

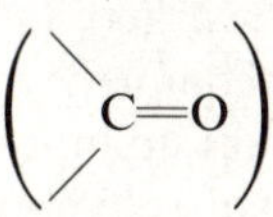

groups and ascribed the improved bonding properties to the increased surface free energy of the polymer's surface (Fig. 10). It has also been demonstrated that corona discharge results in considerable roughening of the polymer's surface[48,84] and at high magnification discrete bumps may be discerned on the surface. These bumps may arise by the migration of low molecular-weight degradation products to charged areas of the surface.

Kim, Evans and Goring[49] have studied the corona-induced autohesion of polyethylene film. They observed that when a low-density polyethylene sheet was treated in a corona discharge and subsequently pressed on to a similarly treated sheet at 45 °C the bond formed between them was much

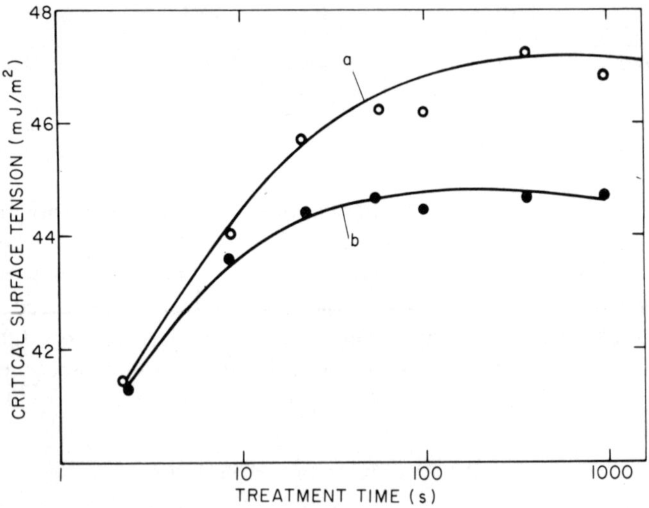

FIG. 10. Critical surface tension versus time of treatment for polyethylene treated in air corona at different temperatures.[49] (a): Treated at 75 °C; (b): treated at 25 °C.

stronger than that between similarly pressed, but untreated, sheets. Their studies suggested that effects such as surface oxidation, cross-linking and rugosity were not responsible but they proposed that the mechanism involved some type of electret formation induced in the polymer sheet by the corona discharge. Electrets are long-lived electrical charges induced when dielectric materials are subjected to strong D.C. fields. Although 60 Hz A.C. was used by these workers they claimed that charge effects produced by one cycle are not completely neutralised by the subsequent cycle and a discrete charge is built up on the surface by application of an A.C. field. They demonstrated that a charge pattern did exist on the corona-treated polyethylene surface by spreading a light dielectric powder on the surface of the polymer and observing a clear 'Lichtenburg figure' pattern. An important feature of this work is that the surface charge due to electret

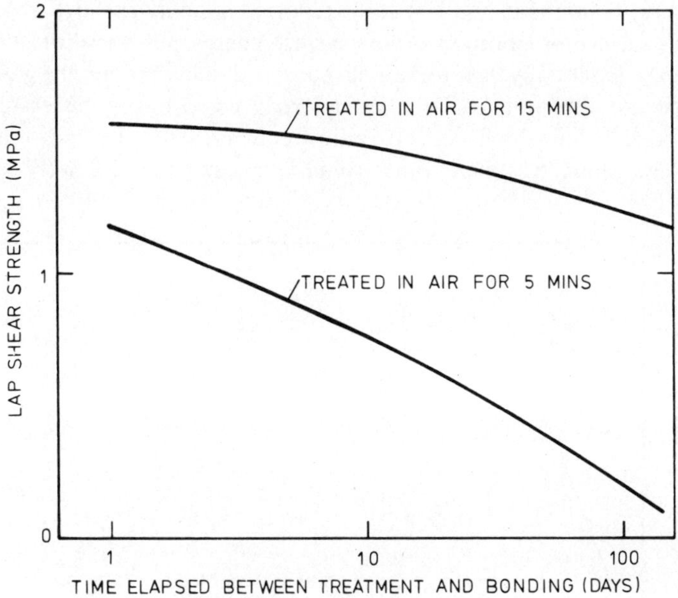

FIG. 11. Effect of time elapsed between treatment and bonding on autohesion of corona-treated polyethylene.[49]

formation is known to decay slowly with time and, indeed, they found the autohesion of the treated polyethylene decayed as the time between treating and bonding was increased, as is shown in Fig. 11.

This mechanism is, however, considered to be vague and physically unacceptable by Owens.[14] His work showed that the adhesion is completely destroyed by the application of hydrogen-bonding liquids to the adhesive joint and the effects of such liquids is completely reversible; joints allowed to dry regain their full, original strength. He proposed that the corona treatment formed keto groups on the polyethylene chain; these groups enolise and the enolic hydrogens bond with carbonyl groups in the adjacent sheet of film when the two sheets are heated together under pressure. Owens[15] extended this work to the autohesion of polyethylene terephthalate films and showed that self-adhesion in these corona-treated films was again due to hydrogen bonding but, in this instance, between carboxyl carbonyl groups in one film with phenolic hydrogens in the other. The phenol groups were created by the corona treatment through a free radical mechanism. Polyethylene terephthalate film could also be made self-adherent by irradiation with ultraviolet light via the same mechanism.

In 1966 Schonhorn and Hansen[85] reported a highly-effective treatment for the surface preparation of low surface energy polymers for adhesive bonding. Essentially the technique consisted of exposing the polymer surface to an inert gas plasma at reduced pressure generated by electrodeless glow discharge (i.e. radio-frequency field). For polyethylene only very short treatment times were necessary (Fig. 12) while longer

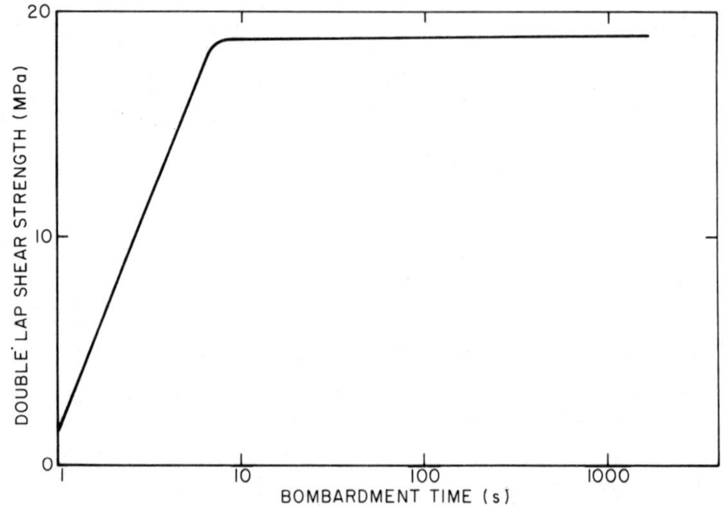

FIG. 12. Lap shear strength of aluminium–epoxide-treated polyethylene joints as a function of bombardment time in helium plasma.[50]

contact times were required for other polymers such as PTFE. In subsequent work, by Hall *et al.*,[86,87] the joint strength of many polymers has been examined as a function of length of exposure to excited helium or oxygen. Polymers examined included low- and high-density polyethylene, poly(4-methyl-1-pentene), polyvinyl fluoride, polyvinylidene fluoride, fluorinated ethylene–propylene copolymer, polyacetals, nylon 6, nylon 6·6, polyethylene terephthalate and polystyrene. Generally, the joint strength increased rapidly initially and then remained nearly constant, perhaps decreasing in some cases at long exposure times. Further work has dealt with the effect of gas employed,[52,88,89] gas pressure,[52] electrode separation[52] and discharge current.[52,89]

Now Schonhorn and co-workers[51,85] reported that this treatment did not change the wettability of the polyethylene, as measured by the critical surface tension, but gel permeation chromatography showed that some highly cross-linked polymer was formed after treatment and the average molecular weights, M_w and M_n, increased from 86 000 to 135 000 and 1800 to 4800, respectively. Also, ATR infrared spectroscopy showed only the formation of transethylenic unsaturation after treatment and a thin intractable skin was readily observed by melting the polymer. Schonhorn concluded that wettability was not the sole criterion for the formation of an adhesive joint but what may be of greater importance is to enhance the cohesive strength of a weak surface region. He suggested that many melt crystallised polymers possess an inherent weak boundary layer, about 0·02–0·10 μm thick, which is composed of amorphous low molecular-weight materials rejected by the crystallisation process which starts in the bulk and proceeds outwards. It was therefore concluded that the increases in joint strength achieved by this technique were primarily due to increasing the cohesive strength of the polymer in the surface regions by the introduction of crosslinks. The surface treatment was therefore called CASING (Crosslinking by Activated Species of Inert Gases).

A major criticism of Schonhorn's proposal that a weak amorphous surface is produced when polymeric films are melt crystallised in air is that his evidence is mainly inferred. Indeed, recent work by Briggs and co-workers[64] using XPS found no evidence of joint failure occurring in the surface regions of the polyethylene substrate. Further, several workers[37,52,53,90,91] have established that the surface free energy of the polymer may be increased by glow discharge treatment; thus surface energetics, surface roughening and the production of chemically active surfaces, to which grafting may occur, may account for the observed increased adhesion.

REFERENCES

1. Shields, J., *Adhesives Handbook*, Butterworth, London, 1970. (Editorial note: 2nd edition 1976 now published, by Newnes–Butterworth, Sevenoaks.)
2. Shields, J., Sira Report, 1972, R500.
3. Snogren, R. C., *Handbook of Surface Preparation*, Palmerton, New York, 1975.
4. Martin, J. J., in *Adhesion and Adhesives* 2, ed. Houwink R. and Salomon, G., Elsevier, New York, 1967.
5. Dukes, W. A. and Kinloch, A. J., (a) Sira Technical Review No. 3, 1976. (b) Propellants Explosives and Rocket Motor Establishment, 1976, TM 169.
6. Zisman, W. A., *Advances in Chemistry*, *Series* 43, ed. Gould, R. F., *A.C.S.*, Washington, 1964,

7. Levine, M., Ilkka, G. and Weiss, P., *J. Polym. Sci.*, **B2,** 1964, p. 915.
8. Fowkes, F. M., *J. Phys. Chem.*, **67,** 1963, p. 2538.
9. Owens, D. K. and Wendt, R. C., *J. Appl. Polym. Sci.*, **13,** 1969, p. 1740.
10. Kaelble, D. H. and Uy, K. C., *J. Adhesion*, **2,** 1970, p. 50.
11. Kaelble, D. H., *J. Adhesion*, **2,** 1970, p. 66.
12. Cirlin, E. H. and Kaelble, D. H., *J. Polym. Sci.*, **11,** 1973, p. 785.
13. Baszkin, A., Nishino, M. and Ter-Minassian-Saraga, L., *J. Coll. & Interf. Sci.*, **54**(3), 1976, p. 317.
14. Owens, D. K., *J. Appl. Polym. Sci.*, **19,** 1975, p. 265.
15. Owens, D. K., *J. Appl. Polym. Sci.*, **19,** 1975, p. 3315.
16. Bowden, F. P. and Throssell, W. R., *Proc. Roy. Soc.*, **A209,** 1951, p. 297.
17. Bennett, M. K. and Zisman, W. A., *J. Colloid. Interf. Sci.*, **28,** 1968, p. 243.
18. Blyholder, G. and Richardson, E. A., *J. Phys. Chem.*, **66,** 1962, p. 2597.
19. Getting, M., Baker, F. S. and Kinloch, A. J., *J. Appl. Polym. Sci.*, 1977, to be published.
20. Bikerman, J. J., *The Science of Adhesive Joints*, 2nd edn, Academic Press, 1968.
21. Wake, W. C., *Adhesion and the Formulation of Adhesives*, Applied Science Publishers, 1976.
22. Malpass, B. W., Packham, D. E. and Bright, K., *J. Appl. Polym. Sci.*, **18,** 1974, p. 3249.
23. De Bruyne, N. A., *Aero Res. Tech. Notes*, 1956 Bull., No. 168.
24. Bascom, W. D. and Patrick, R. L., *Adhesives Age*, **17**(10), 1974, p. 25.
25. Cheever, G. D., in *Interface Conversion for Polymer Coatings*, ed. Weiss, P. and Cheever, G. D., Elsevier, 1968, p. 152.
26. Jennings, C. W., *J. Adhesion*, **4,** 1972, p. 25.
27. Jennings, C. W., *Applied Polymer Symposia*, **19,** 1972, p. 49.
28. Meier, J. F. and Petrie, E. M., *J. Appl. Polym. Sci.*, **17,** 1973, p. 1007.
29. Nelson, E. R., Kilduff, T. J. and Benderley, A. A., *Ind. Eng. Chem.*, **50**(3), 1958, p. 329.
30. Benderley, A. A., in *Symposium on Adhesives for Structural Applications*, at Picatinny Arsenal, N.J., 1962, Interscience, New York, p. 107.
31. Purvis, R. J. and Beck, W. R., US Patent 2 789 063, 1957.
32. Miller, M. L., Postal, R. H., Sawyer, P. N., Martin, J. G. and Kaplit, M. J., *J. Appl. Polym. Sci.*, **14,** 1970, p. 257.
33. Andrews, E. H. and Kinloch, A. J., *Proc. Roy. Soc., Lond.*, **A332,** 1973, p. 401.
34. Polchlopek, S. E., in *Infra-red Spectroscopy*, ed. Kendall, D. N., 1966, Reinhold, New York.
35. Dwight, D. W. and Riggs, W. M., *J. Colloid. Interf. Sci.*, **47,** 1974, p. 650.
36. Allan, A. J. G., *J. Polym. Sci.*, **24,** 1957, p. 461.
37. Andrews, E. H. and Kinloch, A. J., *Proc. R. Soc., Lond.*, **A332,** 1973, p. 385.
38. De Lollis, N. J. and Montoya, O., *Adhesives Age*, **6**(1), 1963, p. 32.
39. Andrews, E. H. and Kinloch, A. J., *J. Polym. Sci., Symposium*, **No. 46,** 1974, p. 1.
40. Cyr, M. C., *Adhesives Age*, **5**(8), 1962, p. 31.
41. Lerchenthal, C. H., Brenman, M. and Yits'haq, N., *J. Polym. Sci., Polym. Chem.*, **13,** 1975, p. 737.
42. Lerchenthal, C. H. and Brenman, M., *Polym. Engng. Sci.*, **16,** 1976, p. 747.

43. LERCHENTHAL, C. H. and BRENMAN, M., *Polym. Engng. Sci.*, **16,** 1976, p. 760.
44. SCHONHORN, H. and RYAN, F. W., *J. Polym. Sci.*, **7**(A2), 1969, p. 105.
45. HARA, K. and SCHONHORN, H., *J. Adhesion*, **2,** 1970, p. 100.
46. ROBERTS, R. F., RYAN, F. W., SCHONHORN, H., SESSLER, G. M. and WEST, J. E., *J. Appl. Polym. Sci.*, **20,** 1976, p. 255.
47. WECHSBERG, H. E. and WEBBER, J. B., *Modern Plastics*, **36**(11), 1959, p. 101.
48. KIM, C. Y. and GORING, D. A. I., *J. Appl. Polym. Sci.*, **15,** 1971, p. 1357.
49. KIM, C. Y., EVANS, J. and GORING, D. A. I., *J. Appl. Polym. Sci.*, **15,** 1971, p. 1365.
50. SCHONHORN, H. and HANSEN, R. H., in *Adhesion, Fundamentals and Practice*. Ministry of Technology, 1969, McLaren London, p. 22.
51. SCHONHORN, H. and HANSEN, R. H., *J. Appl. Polym. Sci.*, **11,** 1967, p. 1461.
52. MALPASS, B. W. and BRIGHT, K., in Aspects of Adhesion–5, ed. Alner, D. J., 1969, Univ. of London Press, p. 214.
53. SOWELL, R. R., DE LOLLIS, N. J., GREGORY, J. H. and MONTOYA, O., *J. Adhesion*, **4,** 1972, p. 15.
54. BLAIS, P., CARLSSON, J., CSULLOG, G. W. and WILES, D. M., *J. Colloid Interf. Sci.*, **47,** 1974, p. 636.
55. RAUHUT, H., *Adhesive Age*, **12**(12), 1969, p. 28.
56. BREWIS, D. M., *J. Mater. Sci.*, **3,** 1968, p. 262.
57. MORRIS, C. E. M., *J. Appl. Polym. Sci.*, **14,** 1970, p. 2171.
58. MORRIS, C. E. M., *J. Appl. Polym. Sci.*, **15,** 1971, p. 501.
59. SCHONHORN, H. and HANSEN, R. H., *J. Appl. Polym. Sci.*, **12,** 1968, p. 1231.
60. GARNISH, E. W. and HASKINS, G. G., in *Aspects of Adhesion*–5, ed. Alner, D. J., 1969, Univ. of London Press, London, p. 259.
61. RAUHUT, H. W., *Adhesives Age*, **13**(1), 1970, p. 34.
62. SCHONHORN, H. and RYAN, F. W., *J. Polym. Sci.*, **6**(A2), 1968, p. 231.
63. BRIGGS, D., BREWIS, D. M. and KONIECZO, M. B., *J. Mater. Sci.*, **12,** 1977, p. 429.
64. BRIGGS, D., BREWIS, D. M. and KONIECZO, M. B., *J. Mater. Sci.*, **11,** 1976, p. 1270.
65. GETTINGS, M. and KINLOCH, A. J., in *Polymer Surfaces Symposium*, 21–24 March, 1977, Durham.
66. WYATT, D. M., GREY, R. C., CARVER, J. C., HERCULES, D. M. and MASTERS, L., *Appl. Spectr.*, **28,** 1974, p. 439.
67. WENZEL, R. N., *Ind. Eng. Chem.*, **28,** 1936, p. 988.
68. JOHNSON, R. E. and DETTRE, R. H., *Adv. Chem. Ser.*, **43,** 1964, p. 112.
69. BREWIS, D. M., EAGLES, A. C. and HURWORTH, N. R., *J. Mater. Sci.*, **2,** 1967, p. 435.
70. BRUNER, W. M., US Patent 3 235 426 (1966).
71. LERNER, R. M., *Adhesives Age*, **12**(12), 1969, p. 35.
72. ABU-ISA, I. A., *J. Appl. Polym. Sci.*, **15,** 1971, p. 2865.
73. SNOGREN, R. C., *A.S.M.E.*, 1966, 66-MD-39.
74. ALTHOF, W., Deutsche Luft- und Raumfahrt, 1971, DLR FB.71-31.
75. CRANE, L. W., HAMERMESH, C. L. and MAUS, L., *SAMPE, J.*, 1976, March/April, p. 6.
76. BUCHAN, S., *Rubber-to-Metal Bonding*, 1959, Crosby-Lockwood, London.
77. ANON., *Rubber J.*, **155**(1), 1973, p. 14.

78. PETERSON, C. H., *Adhesives Age*, **8**(7), 1965, p. 22.
79. COX, D. R., *Rubber J.*, **151**(4), 1969, p. 49.
80. WAKE, W. C., in *Adhesion and Adhesives Vol.* 1, eds Houwink, R. and Salomon, G., 1967, Elsevier, New York, p. 395.
81. SPEARMAN, B. P. and HUTCHISON, J. D., *Adhesives Age*, **17**(4), 1974, p. 30.
82. Visking Corp., British Patent 715 914, Sept. 1954.
83. ROSSMAN, K., *J. Polym. Sci.*, **19,** 1956, p. 141.
84. KIM, Y. C., SURANYI, G. and GORING, D. A. I., *J. Polym. Sci.*, C, **30,** 1970, p. 533.
85. SCHONHORN, H. and HANSEN, R. H., *J. Polym. Sci.*, *B*, **4,** 1966, p. 203.
86. HALL, J. R., WESTERDAHL, C. A. L., DEVINE, A. T. and BODNAR, M. J., *J. Appl. Polym. Sci.*, **13,** 1969, p. 2085.
87. HALL, J. R., WESTERDAHL, C. A. L., BODNAR, M. J. and LEVI, D. W., *J. Appl. Polym. Sci.*, **16,** 1972, p. 1465.
88. SCHONHORN, H., RYAN, F. W. and HANSEN, R. H., *J. Adhesion*, **2,** 1970, p. 93.
89. YASUDA, H., LAMAZE, C. E. and SAKAOKU, K., *J. Appl. Polym. Sci.*, **17,** 1973, p. 137.
90. DE LOLLIS, N. J., *Rubb. Chem. Technol.*, **46,** 1973, p. 549.
91. COLLINS, G. C. S., LOWE, A. C. and NICHOLAS, D., *Eur. Polym. J.*, **9,** 1973, p. 1173.

Chapter 11

SAFETY AND ENVIRONMENTAL LEGISLATION AND ITS EFFECT ON ADHESIVE FORMULATION

P. BOSWORTH
Consulting Chemist, Southampton, UK

SUMMARY

There is growing public concern about safety and the environment. As a result of this concern, a body of relevant legislation is being developed which will affect most manufacturing operations including the manufacture of adhesives.

In this chapter, some of the more important items of legislation for the adhesives manufacturer and user are discussed. The object is to indicate their significance rather than to analyse them in the legal sense. UK Acts, EEC Directives, UN Recommendations and US Food Regulations are considered. All but the US Regulations will have a direct effect, eventually, on the UK position, and the US Regulations will no doubt influence UK developments.

The effect of the legislation on the formulation and use of many types of adhesive is considered.

INTRODUCTION

In recent years, there has been growing concern that individuals and the environment shall not be endangered by manufacturing processes and the resulting products. One outcome of this concern is National and International legislation, which seeks to establish safe and, therefore, acceptable procedures. Another is a redirection of development effort towards devising safer procedures, and this work should be aided by the precise standards created by the laws and the codes of practice.

Risks to the individual can be separated into four categories, viz., chemical, mechanical, electrical and radiation. The major risk of damage to the environment is chemical. The risk that adhesives present to both individual and environment is primarily a chemical one. Hence, it is appropriate to consider adhesives as one type of chemical amongst many that are in everyday use. By comparison with many chemicals, adhesives are low in production volume. It follows from this that their effect on the environment is unlikely to be major. Since their uses and their compositions are diverse and they are handled in environments ranging from well-controlled factories to the home, the same generalisation cannot be made about the effect of adhesives on the individual. In considering these points in rather more detail, it is helpful to bear in mind the characteristics of adhesives. They must wet the substrates, i.e., be liquid at the spreading stage and then subsequently solidify. This is achieved by any one of 4 methods, and sometimes by a combination of these methods, viz:

(a) The adhesive is a solid which is melted by heat for application.
(b) The adhesive is a reactive liquid, and solidifies by chemical action.
(c) The adhesive is a polymer dissolved or dispersed in water or an organic solvent. The water or solvent either evaporates or penetrates into the adherend in the solidification step.
(d) The adhesive is a semi-solid which wets under pressure (pressure-sensitive).

The only major hazard associated with method (a), of liquefaction, is the possibility of a burn. Method (d) has no major hazard but method (c) can involve the presence of toxic and flammable solvent. Method (b) will only involve a risk of irritancy unless a volatile and flammable solvent or monomer is present.

In this chapter some of the important laws and codes relating to safety are reviewed together with the effect of these laws and codes on the use and development of adhesives. No attempt has been made to include every relevant law but the more important ones have been selected so as to indicate the overall picture. Some of these selected items are existing UK laws, others are EEC drafts which will become UK laws. A UN recommendation is reviewed because it is leading to international and national requirements from transport authorities. There is no UK or EEC Food Contact legislation relating to adhesives, so in this case, the US FDA position is indicated. The aim of this discussion is to indicate the nature and implications of the laws and not, of course, to give guidance on exact legal points or to provide any substitute for an examination of the documents cited. The treatment of legislation is inevitably general rather than specific and again, the aim is to indicate broadly the overall position rather than to be comprehensive.

LAWS AND CODES RELATING TO SAFETY

The Health and Safety at Work, etc., Act 1974

The Health & Safety at Work, etc., Act[1] says that every employer shall, 'so far as is reasonably practicable' ensure the health, safety and welfare of his employees whilst they are at work. The aspects to which he must pay special attention are listed and are as follows:

(a) The maintenance of a safe plant and working procedures.
(b) The safe handling storage and transport of article and substances.
(c) The provision of safety information and training to employees.
(d) The safe condition of the place of work including safe means of access and egress.
(d) The provision of a working environment without risks to health and adequate facilities for welfare whilst at work.

The Act also deals with the duties of anyone designing, manufacturing, importing or supplying articles or substances for use at work.

The provisions for articles and substances are very much the same and in the case of substances they are:

(i) Ensuring 'so far as is reasonably practicable' that the substance is safe and gives no risk to health when properly used.
(ii) Carrying out any testing necessary to satisfy the previous provision.
(iii) Ensuring that sufficient information is available about the conditions necessary for safe use.

Moreover, and significantly, the Act says that it is the duty of the supplier of a substance for use at work (there is a similar provision for an article) to carry out, or have carried out, such research as may be necessary to eliminate or minimise the risk to health or safety arising from the use of the substance. As in so many of the provisions of the Act, this requirement is qualified by the phrase, 'so far as is reasonably practicable'.

Besides covering the health and safety of employees in factories, the Act deals with the effect a factory may have on people adjacent to it but not working in it. Specifically, the person controlling the factory must use the 'best practicable means' for preventing emission into the atmosphere of noxious or offensive substances and for rendering any substances that may be emitted harmless and inoffensive.

The provisions described above are the ones which have the major effect on the supply and use of adhesives. There is, in addition, a good deal of information about how the Act will be enforced, usually by the Health & Safety Executive but also by local authorities. In connection with enforcement, it is significant that not only a company may commit an offence and be punished but also any manager, secretary or similar officer of the company, if the offence is shown to have been committed with his consent or connivance or due to his negligence. The way in which Improvement Notices specifying that a change must be made in a particular time and Prohibition Notices stopping a particular operation, are issued, is described. The composition of the Health and Safety Commission and the Health & Safety Executive are defined and their functions indicated.

The Act also deals with matters such as the Employment Medical Advisory Service, Building Regulations, and has special provisions relating to Agriculture.

Two important features of the Act are:

(1) The continual use of phrases such as, 'so far as is reasonably practicable'.
(2) The fact that it is envisaged that the Commission will produce Codes of Practice to provide practical guidance as to how the requirements of the general provisions may be fulfilled. Equally, the Commission may approve Codes of Practice issued by others.

The phrase, 'so far as is reasonably practicable' is an interesting one. It suggests that whoever drafted the Act appreciated that, whilst proper attention should be given to safety, there must be a 'practicable' limit. This concept is a sensible one, perhaps even a bold one, in the present climate of opinion. The thought behind it is presumably that technical development must go on. A total emphasis on safety would somewhat change modern life. For example, automobiles and aeroplanes might well be unacceptable. The full meaning of the phrase can only be established by case law and so will not be apparent for some years.

The Act as written is a general one, but feature (2) indicates that eventually it will be supported by a series of detailed codes and safety regulations which will progressively replace and augment the enactments listed in an Appendix and including, for example, the Alkali & Works Regulation Act of 1906 and the Petroleum Consolidation Act of 1928.

EEC Regulations Relating to the Classification, Packing and Labelling of Dangerous Materials

One of the conditions of EEC membership is that certain types of directives shall become law in all member states. The ones which form the subject of this section fall into this category and the title covers three directives relating respectively to:

(i) Substances[2]
(ii) Solvents[3]
(iii) Paints, varnishes, adhesives and similar products[4]

Of these directives, the first two have been approved by the EEC Council but the third is still in draft form. In the UK, the Health & Safety Executive have, as yet, issued Draft Proposals for Regulations to be made under the Health & Safety At Work Act, relating only to the first directive, i.e., the substances directive.[5] Nevertheless, it is likely that all three will be a part of UK law by 1980. Some of the other European countries, e.g. Germany, already include some or all of the Directives as National Law. These directives relate both to products sold to industry and to the consumer.

Since these Directives are to come under the Health & Safety at Work Act, it will be apparent that they provide some of the detail which will enable that Act to be observed. Unlike the Health & Safety at Work Act, however, they indicate precisely what action a supplier has to take to comply, in regard to the labelling of dangerous materials. The section on packaging, on the other hand, merely makes general points, viz.:

(a) The packages must be so designed and constructed that they do not leak.
(b) The materials from which the packages are constructed must not be attacked by or react with the contents.
(c) The packages must be sufficiently strong to withstand normal handling.

The term, *classification*, in the Directives, relates to labelling. In the case of the Substances Directive, the main risk of the material leads to a classification risk selected from the following list:

1. Toxic	5. Flammable
2. Corrosive	6. Irritant
3. Harmful	7. Explosive
4. Highly flammable	8. Oxidising

and there are appropriate symbols specified for the label.

TABLE 1

Substance	*General nature of risk*	*Risk phrases*	*Safety phrases*
Acetone	Highly flammable	Highly flammable	Keep container in a well ventilated place Keep away from sources of ignition No smoking Take precautionary measures against static discharge
Benzene	Highly flammable and toxic	Highly flammable Toxic by inhalation Toxic in contact with skin Danger of very serious irreversible effects	Keep container in a well ventilated place Keep away from sources of ignition No smoking Do not empty into drains
Toluene	Highly flammable and harmful	Highly flammable Harmful by inhalation	Keep away from sources of ignition—No Smoking Do not empty into drains Take precautionary measures against static discharge

This question of classification is not left to the discretion of the supplier. Dangerous substances are by definition substances quoted in a schedule included in the directive and this schedule includes the term or terms which are to be used for classification. The directive also indicates which risk and safety phrases should be included in the label. By way of illustration, the entries for acetone, benzene and toluene (taken from the Health & Safety Executive UK draft) are listed in Table 1:

The solvents directive refers back to the substances directive. In fact, as will be seen from the examples taken from the substances directive, many of the pure solvents are covered as 'substances'. The solvents directive, however, deals with the case of mixed solvents.

The Paints, Varnishes, Adhesives and Similar Products Directive is the one which is of most concern to suppliers and users of adhesives. The same terms are used for classifying the main risk as in the Substances Directive, i.e. Toxic, Corrosive, Harmful, etc. The classifications depend partly on a schedule in the directive which indicates that if a particular material is present as an ingredient above a fixed level, then a particular classification shall be used. For example, if vinylcyclohexane diepoxide is present at a concentration above 0·1 % by weight, then the T symbol (toxic) must be used. If this chemical is present in the range 0·025–0·1 %, the symbol X_n (harmful) is required.

Deciding the main risk classification, however, also depends on reference to the solvents directive, which in turn indicates the relevant percentage above which the classification is attracted. Reference must also be made to the substances directive and percentages are laid down in the paints, varnishes and similar products directive, depend on the classification in the Substances Directive. These indicate when a particular classification must be adopted under the Paints, Varnishes, Adhesives, etc., Directive.

The labels, besides the main hazard classification and symbol, are required to show:

A. The names of any toxic, harmful, corrosive or irritant ingredients present, using the criteria explained in the previous paragraph.

B. Appropriate risk phrases chosen from those listed in the substances directive (not required if less than 125-ml package).

Not more than 4 risk phrases need be shown, and health risks must be given priority over fire or explosion risks.

C. Appropriate safety phrases taken from the Substances Directive (not required if less than 125-ml package).

Where a preparation is assigned more than one symbol, the following rules are given:

If T required, X and C may be omitted
If C required, X may be omitted
If E required, F and O may be omitted

where T = toxic
X = harmful (X_n) or irritant (X_i)
C = corrosive
F = highly flammable
E = explosive

The three directives, besides setting up the system described above, include details of the size, etc., of the labels.

Once the directives issue as laws, there should be no particular problem in implementing them. There will, of course, be some costs and in the case of adhesives, some loss of commercial secrecy. Perhaps a more serious point, however, is their effect on development. This has been highlighted by the sixth amendment to the substances directive (not yet accepted by the EEC), which emphasises two aspects:

(1) The effect of substances on the environment, and
(2) The procedure required before a new product can be placed on the market.

In this amendment, a new classification, 'Dangerous for the environment' is defined and a three-page annex, indicating parameters for assessing the environmental risks, is included.

One of the articles requires that new substances shall be notified and that the notifier shall have carried out a study on the effects of the substance on Man and on the environment. The minimum information required is indicated in an annex.

As so often in safety legislation, the adoption of a sound principle will lead to the expenditure of a great deal of scientific man-hours and considerable retardation of new developments. The initial reaction of the European Chemical Industry has been adverse. It is to be hoped that further discussions will lead to a better balance.

UN Recommendations Relating to the Transport of Hazardous Products

The two previous items discussed were laws and drafts which will become laws. The UN, on the other hand, merely issues recommendations.[6] These, however, are likely to be accepted wholly or to some extent and enforced by the international transport agencies, viz.:

RID International rail
ADR International road
IMCO Sea transport—both international and coastal
IATA Air transport

It is also possible that they may become the basis of international laws and probable that they will be the basis of national transport regulations.

From the point of view of adhesive manufacturers, the important feature of these recommendations is that by setting a performance standard, including drop height tests, etc., they define the type of containers that are to be used for dangerous adhesives (e.g. flammable, solvent-based adhesives). Some aspects of the recommendations are still under discussion but they will certainly be so framed that more exacting tests are required for more hazardous (e.g. lower flashpoint adhesives). They are also very likely to include viscosity as a factor in deciding the severity of the tests. The argument in this latter case is the simple and sound one that less of a more viscous material will emerge from a specified leak in a given time.

The current position is that IMCO have adopted performance standards including drop tests near the UN level. There is, however, a probability that the RID will be issuing a performance standard with lower drop heights.

One of the problems which the recommendations present, is that they do not differentiate in their requirements between heavy returnable bung-type drums (45-gallon) and lighter, smaller, one-journey containers. This could lead to a cost effect for the adhesive industry as the conventional 5-litre lever-lid cans used in this country will not pass the drop height test. On the other hand, the full-size lid feature is important to adhesive *users* and, in practice, an extensive survey has shown that they do not give rise to accidents.

Food Contact Regulations

The fairly recent ban on cyclamates brought US legislation on Food Additives into universal prominence. Equally, the US Food & Drugs Administration (FDA) seeks to control food contaminants. This is effected by defining what ingredients can be used in materials which will be in contact with food, for example, packaging films, plastic trays, etc.

Adhesives are covered in this way by FDA regulation 121.2520. This covers adhesives in general by one large permitted list of ingredients, i.e., all types of adhesive are covered by one list. The FDA specify, *inter alia*, that the adhesive should either be separated from the food by a functional barrier or, in dry foods, the amount of adhesive that contacts the packaged dry food shall not exceed the limits of good manufacturing practice, or in fatty and aqueous foods, that the contact between adhesive and food shall not exceed the trace amounts at seams or at the edge exposure between packaging laminates that may occur within the limits of good manufacturing practice.

In Europe, not all countries cover food contaminants by permitted lists of the FDA pattern. For example, the UK does not have legislation of this type but Germany and Italy do legislate in this way, although not for adhesives. The EEC Commission is working on directives and the present position (early 1977) is that the Final Draft of the General Directive has been issued, that preliminary drafts have appeared for stainless steel and plastics, and that a Final Draft has appeared for ceramics.

The plastics directive draft is based on a permitted list (although at this stage the list is missing) and maximum levels for overall ('global') migration plus some specific migration levels.

To date, there has been no indication that the EEC are proposing an adhesives directive. Since directives based on permitted lists involve a great deal of scientific work in establishing that constituents are safe, this is probably entirely appropriate. In the vast majority of cases, adhesives do not contact food at all. For example, consider the case of a cardboard box holding biscuits. The only direct contact conceivable would be at any squeeze-out on the seams and good manufacturing practice would ensure this to be minimal. In the few cases where adhesives might contact food, they would probably best be regarded as a part of some other material, e.g. a plastic laminate.

Environmental Legislation

In the UK, the Health & Safety at Work Act covers the question of the emission into the atmosphere of noxious or offensive substances from places of Work. Two of the existing regulations listed in the Act which are relevant to this subject are the Alkali & Works Regulation Act 1906 and the Public Health (Smoke Abatement) Act 1926. The former is implemented by the Alkali Inspectors, now a part of the Health & Safety at Work Executive. They only control workplaces that are registerable and the list covers workplaces producing chemicals such as sulphuric acid, cement, lime, caustic soda, etc.[7] In general, they would not cover adhesive works unless certain types of chemical waste were burnt in these works or diisocyanates were manufactured or polymerised. Adhesive works might be covered, however, by the second Act, and the local authorities who operate this Act sometimes call on the Alkali Inspectors for specialist assistance. In the future, the intention is to cover emissions into the atmosphere by Codes of Practice developed under the Health & Safety at Work Act, and it seems likely that these will be less specific.

The forms of damage to the environment which relate to tips and the possible effects on ground water and rivers are covered by the Control of Pollution Act 1974;[8] this requires Waste Disposal Authorities (often County Councils) to:

(a) Survey the waste that arises in their area
(b) Prepare a plan for the disposal of the waste
(c) Licence sites for waste disposal
(d) Ensure that all waste, whether it be domestic, commercial or industrial, can be disposed of suitably. Both private or public plants are envisaged.

Some types of product have been given special attention. For example, the Department of the Environment recently studied the non-agricultural uses of pesticides in this country.[9] They recommended various improved control systems and, in some cases, suggested the ban of certain pesticides used for certain purposes, e.g., DDT in antifouling paints. In other cases, e.g., mercury compounds, they have pressed that the use should be minimised as far as practicable.

The EEC has prepared a 'black list' of highly-persistant and bio-accummulable substances and is considering supra-national maximum limits on industrial discharges.[10] Indeed, a directive on the limitation of the placing on the market and on the use of polychlorobiphenyls and polychloroterphenyls has already been prepared. The EEC is also proposing to demand information on environmental dangers for new substances via the 6th Amendment to the Substances Directive. The clear implication is control, or even a ban (*see* earlier).

The implication of the legislation and official activity just reviewed is that it will be costly for an adhesive manufacturer, like any other chemical manufacturer, to produce waste, whether it be in the form of a vapour, a liquid or a solid. If the waste is a vapour, the cost is likely to be a capital one, for example, a tall chimney or an electrostatic precipitator. For solid wastes, the manufacturer will have to pay a transport cost. If the waste is particularly harmful to the environment, this cost will be higher, as transport to a 'special' tip may be required. In the extreme case, the manufacturer might be obliged to treat a waste chemically before arranging for tipping. This could well involve both capital and running costs, but possibly savings due to re-use. Liquid waste can sometimes be treated in domestic sewage plants and the cost to the manufacturer will depend on the quantity of liquid and its BOD (Biological Oxygen Demand). In the case of certain wastes, however, the position is complicated. For example, bacterial sewage plants can be used to treat intakes containing a constant and low level of phenol, successfully. An intake with the same level of phenol on a 24-h basis, however, but with most of the phenol content in intermittent peaks, could not be treated, and might kill the bacteria. Thus, manufacturers of phenolic resins are obliged to install tanks for phenolic effluent and regulate flow from them to the treatment plant.

THE EFFECT OF SAFETY CONSIDERATIONS AND THE LEGISLATION STEMMING FROM THEM ON ADHESIVE USE AND DEVELOPMENT

Solvent-based Adhesives

Use

The major hazards in use of solvent-based adhesives normally arise from the presence of the solvent which will be volatile and also usually either flammable or dangerous or often flammable and dangerous. By comparison, the polymer is normally safe as are the various plasticisers, stabilisers, etc., in the formulation. Even in the case of those cross-linking adhesives which contain solvent, the solvent is likely to be the cause of a major part of the hazard although the monomers and prepolymers present in products of this type will usually themselves be more hazardous than the 'finished' polymers present in most solvent-based adhesives (see later).

It is possible to use solvent-based adhesives with complete safety provided that the conditions and equipment are appropriate. The Health & Safety at Work Act specifies, indeed, that they may only be used in the appropriate conditions and demands that the supplier should indicate the safe conditions. In practice, this means that the proper degree of ventilation must be provided so that dangerous concentrations do not arise if the solvent is dangerous. With inflammable solvents, smoking will not be allowed and flame-proof electric motors, etc. will have to be used. Care will have to be taken that noxious or offensive discharges are not produced. This might demand a suitable chimney or, in the extreme case, a solvent recovery plant.

As part of the requirements of the Health & Safety at Work Act, suppliers will soon be required to label solvent-based adhesive containers in a particular way, i.e. according to the EEC Labelling Directives. Such labelling is a part of the information on safe handling which the supplier must provide.

The transport codes which have resulted, or will result, from the UN recommendations will require the strongest type of containers for the lowest flashpoint solvent-based adhesives. The result of this could be that the containers which have to be used are both rather expensive and somewhat less convenient than the user might wish.

If a solvent-based adhesive is used for a food packaging application, the question of contamination of the food must be considered, as well as the question of safe use in the factory. In practice, the dangers of food contact with packaging adhesives is almost always very slight. Nevertheless, the adhesive supplier will clearly choose a formulation where the solids in the adhesive are not toxic. Although not forced to do so by law in this country, the supplier might well formulate from the FDA permitted list. In terms of food contamination, the fact that the adhesive is solvent-based is not normally relevant. If the manufacturing method is sound there will be no solvent remaining in the glue line; in this one respect it is the constitution of the solids which is important.

The method of using solvent-based adhesives is such that the solvent is lost to the atmosphere. Thus, solvent-containing effluents will arise, in relatively small amounts, from two sources, viz. empty drums and mixing vessel washings. The alternatives for disposing of the effluent in these forms are a suitable tip or incineration. Naturally, a cost will be associated with both. As far as tipping (the most widely-used method) is concerned, whoever sends the waste to the tip will have to indicate the nature of the solvents used, both in the case of washings and 'empty' drums. It is likely that not every tip will be prepared to accept solvent-containing wastes, and this may lead to extra transport costs.

Development

With so many precautions to consider and the costs and inconveniences these produce, there is every incentive to replace solvent adhesives by other types. Such approaches are being studied (see later sections) but are not likely to render solvent-based adhesives obsolete, as these products have many technical advantages. Amonst these the following may be listed:

Fast evaporation rate (important for contact adhesives)

Good water resistance (by comparison with aqueous systems; there may be exceptions)
Freeze–thaw stability (wide range of polymers may be used).

The Health & Safety Executive have published a note on Threshold Limiting Values (TLV) and quoted figures for solvents amongst other materials.[11] Sometimes, with the help of this information, an adhesive formulator may develop a somewhat less hazardous product. For example, he might replace some of the toluene or methyl ethyl ketone in a contact adhesive by acetone, which has a lower TLV, and as much aliphatic hydrocarbon as possible. There are limitations in this type of development. For example, aliphatic hydrocarbons are not particularly good solvents, and acetone is very volatile, and this may not only give rise to practical problems relating to speed of drying, but result in the higher TLV being reached more readily in use. Nevertheless, this type of approach is used and can produce advantages in marginal situations.

Again, less harmful substitutes for benzene, which have the right properties, can be found and benzene is seldom if ever used in UK adhesives and, indeed, has not been so used for many years.

The solvent manufacturers are well aware of safety aspects and have carried out work on the development of solvents with lower hazards at the same solvency power. As an example of the results, a particular manufacturer has claimed that a new range of products gives excellent solvency with very low aromatic contents.[12] One practical outcome is that the aromatic contents are low enough to meet the Los Angeles City Authorities Rule 66, which controls the level of photochemically active solvents (aromatics are in this group) to reduce smog. Another outcome is lower TLV figures for a given solvency, and perhaps more important, lower relative ventilation requirements.

Polymers, too, may be modified and a polychloroprene supplier has described work on the production of a polymer which will give stable dispersions in aliphatic solvents.[13] These solvents will be much safer than the aliphatic/aromatic/ketone blends used at present in polychlroprene solutions.

Aqueous Adhesives

Use

Assuming that *no solvent is present*, as is usually the case, aqueous adhesives are inherently safe and, therefore, little affected by the various regulations described earlier. Thus, there will be no particular ventilation requirements (although it is always good practice to have a well-ventilated place of work) and no need to specify flame-proof electric motors. There may be need to specify, e.g. avoid skin contact, if the adhesive contains any toxic additives but usually, this will not be the case. The EEC Labelling Directive will control what precautions need to be highlighted on the label but again, there will often be no need to indicate any precautions.

Materials of this sort are not, in general, dangerous and so not subject to the UN Packaging Recommendations.

If the adhesives are used for food packaging, the adhesive solids are the important factor, just as is the case with solvent adhesives. Toxic products must be avoided and the FDA list of permitted ingredients may be used as a guide.

Effluent problems are less than in the case of solvent adhesives. There will be little danger of the emission of noxious or offensive fumes. 'Empty' drums and washings may be transported to tips with less hazard and there will be less likelihood that the authorities controlling tips will designate 'special tips' as appropriate. The 'washings' indeed, could possibly be handled by a domestic sewage system although there would be a BOD demand which would involve a cost, and the possibility of blockages.

Development

The above comments apply to water based adhesives without solvent additions. The position would not be so favourable if, for example, solvent were present in a polyvinyl acetate emulsion to increase speed of set or give better bonding to rexine (N/L leathercloth). Such additions were very common some years ago. Much development has, however, been carried out on the base emulsions and the best ways to compound these. The effect has been to much reduce the need to make solvent additions and products containing such additions are less common.

The use of very toxic biocides in adhesives would necessitate special precautions. Development work in this area has led to the increasing use of safer materials. Mercurials, for example, are now very seldom used. Nevertheless, this is one aspect of a formulation which is very important in certain applications and which may give rise to hazards.

Because aqueous adhesives avoid the solvent hazard, much effort has been devoted to broadening their field of usefulness. For example, polyvinyl acetate adhesives have been used as wood glues for many years. Until quite recently, however, all contact cements were solvent-based. In the last few years, the properties of emulsion contact cements have been much improved and these products now have a significant share of the market.

This is a field where prediction is difficult, but it seem probable that these products will not totally replace solvent-based contact cements or be equal to them in performance *in all respects.*

Hot Melts

Use

Like aqueous adhesives, hot melts are inherently safe, and so little affected by the regulations reviewed earlier. The same comments apply to the two types of product, except that three hazards may arise because hot melts are applied molten, i.e. at a high temperature. The first of these is the danger of burns to operators, and under the Health & Safety at Work Act appropriate precautions must be taken, e.g. protective clothing and eye protection on occasion. Secondly, there is always the danger that, owing to an electrical defect, the glue bath may overheat and catch fire. Appropriate precautions must be taken to avoid this eventuality and the suppliers must indicate how to treat fires, if they should occur. Clearly, powder extinguishers are appropriate, and water highly unsuitable. The third hazard is toxic fumes at the temperature of application or in particular if there is overheating. Thus, measures will have to be taken to remove these.

The only difference between hot melts and aqueous adhesives in terms of the environment is that one would have to tip or incinerate residues. There could be no question of allowing hot melts to enter a municipal sewage system.

Development

There is just one major aspect in which the formulation of hot melts has been changed in the last ten years, for reasons of safety. Originally, chlorinated diphenyls were used as modifier in many ethylene vinyl acetate polymer-based hot melts. These materials were, in most respects, ideal in their properties. They had excellent thermal stability, gave high tack products, and had good compatibility with the other compounding ingredients. At application temperatures, however, they gave toxic fumes, and although these could be removed by suitable extraction equipment, the risks were considered unacceptable. In current formulations, hydrocarbon resins of various kinds have replaced them. The present position is that hot melts with properties as good as those available from the chlorinated diphenyl compositions, except marginally, as regards pot stability, are available. It is interesting that this particular change took place in advance of the main body of the legislation described earlier.

Hot melts and aqueous adhesives are alternatives in many packaging and bookbinding processes. Since both types of adhesive present few hazards in use, safety considerations have little effect on the choice between the two alternatives. There are some applications where hot melts and solvent-based adhesives are alternatives, and here, safety considerations will favour hot melts. A case in point arises in the pressure-sensitive adhesive field. At a recent British Adhesive Manufacturers' Association Symposium, one of the speakers predicted that the solvent-based rubber adhesives used for pressure-sensitive paper labels and, to some extent, for pressure-sensitive adhesive tapes, would be replaced by hot melts.[14] The reason for the prediction was that the use of hot melts would reduce the hazards involved in manufacturing these items, and the problems of avoiding solvent vapour emissions. The use of hot melts would also eliminate a drying step, i.e. reduce plant costs. To date, the prediction has still to be justified by commercial results but some progress has been made especially in the field of pressure-sensitive labels. The mere fact that hot melts can now be considered for these processes is an indication of the considerable amount of successful development which has taken place in the last few years.

Cross-Linking Adhesives

Introduction

The common characteristic of all these products is that they contain monomers or prepolymers as supplied which react further in the glue line. The reactions are normally produced by hardeners which are added just before use and they may take place in the cold or after the application of heat to the glue line. Monomers or prepolymers are more likely than 'finished' polymers to be irritating to the skin or even toxic. Moreover, the hardeners often have similar effects.

There are various types of product which fall into this broad category. Some are in aqueous solution or dispersion but the majority are 100% reactive. The 100% reactive products both include materials which are essentially non-volatile and materials which are both volatile and inflammable. Thus, the effect of safety considerations on their use and development differs with the type of material. Hence, the remainder of this section will be devoted to a consideration of some of the more important types.

Cross-Linking Materials in Aqueous Solution or Dispersion

The traditional phenol–formaldehyde, urea–formaldehyde or resorcinol–formaldehyde adhesives are the most widely used products in this category. They do not present a fire hazard in use. (Except for the relatively small-scale use of solvent-based versions.) The fumes, however, are objectionable because all the products contain some free formaldehyde. Therefore, proper ventilation must be provided. After mixing with the appropriate catalysts, many of these adhesives will be sufficiently acid or alkaline to be irritant to the skin. Moreover, both phenol and formaldehyde are skin irritants. Thus, skin contact must be avoided.

The EEC labelling Directive says that any adhesive with more than 0·2 % of a substance classified as toxic shall be labelled as toxic. Phenol is classified as toxic and many PF adhesives will contain more than this level of free phenol. All the formaldehyde based adhesives will contain free formaldehyde but the classification of this material in the Substances Directive varies according to its concentration and so it is not very likely that the adhesives will attract a toxic label. Certainly, a harmful/irritant label would seem more appropriate.

As the formaldehyde-based adhesives are not flammable and are usually sold in large returnable drums, the UN Packaging recommendations are unlikely to affect their packaging despite the fact that they cover toxic materials. In general, food legislation is not relevant, as these products are not normally used as food packaging adhesives.

From the environmental point of view, and excluding the phenol-formaldehyde adhesives, tipping should present fewer problems than with solvent adhesives. Indeed, the products could probably be treated successfully but at a cost, in domestic sewage plants. Phenol-formaldehyde resins present some extra problems. For example, even these products can be treated in domestic sewage plants but only providing that their concentration is low. 'Slugs' of phenol kill the bacteria. Again, tipping is entirely possible but appropriate tips must be chosen, as phenol must not be allowed to contaminate rivers, since it is poisonous to fish. Moreover, if water containing even a very little phenol reaches a water-treatment plant, the chlorinated phenols produced, even at extremely low concentration, will taint the water.

Safety considerations have not had much effect on the formulation of phenol–formaldehyde and resorcinol–formaldehyde resins. Basically, the hazards stem from the presence of the phenol and the formaldehyde, and this position cannot be changed. On the other hand, these adhesives are extensively used for joints where weatherproof properties are required, e.g. exterior plywood (WBP) and laminated beams. There are no substitute adhesives readily available.

Urea–formaldehyde adhesives give good water resistance and good resistance to creep. On the other hand, they do not produce WBP plywood and may be replaced in many applications by polyvinyl acetate emulsions. The precautions required with urea-formaldehyde adhesives are perhaps the least for the products in this group. They have been further reduced for many applications by formulating in such a way that free formaldehyde concentration is very low.

Thermosetting polyvinyl acetate emulsion adhesives are a relatively new group of products which fall into this general category. The polymer in this case is high in molecular weight by comparison with the U/F prepolymers present in U/F resins and relatively inert. Care must, however, be taken in handling the hardeners. In general, there should be no inflammability problem or fume problem. The products offer better heat and creep resistance than conventional polyvinyl acetate resins.

Epoxide Resins

These products, in general, are non-flammable and give rise to no irritant vapours, at least at room temperature. If hot cure is involved, localised air extraction systems may be necessary.

The major hazard they present is that they (i.e., the resin–hardener combinations normally used) irritate the skin, that is, in this case, cause dermatitis. They also act as sensitisers in this respect and any individual who has been so sensitised will respond violently (in terms of the degree of dermatitis) to small traces of epoxide resin. Hence, the usual precaution necessary under the Health & Safety at Work Act is to avoid skin contact.

The products are not normally used for food contact applications. In terms of the environment, tipping or incineration would be appropriate methods of disposal. The products would have to be notified but should present rather less hazard than P/F resins. One of the features of epoxide resin technology is that a wide diversity of compounding techniques are available. Thus, for particular purposes, more hazardous mixes than those described above might be developed. Equally, a good deal of attention has been given to developing less irritant resin/hardener systems and this work has been quite successful. In terms of the hazards associated with particular compounding ingredients, there is a good deal of information in the EEC Paints, Varnishes, Adhesives, etc. Directive.

Polyurethane Adhesives

Single-part polyurethanes are used extensively in the shoe industry. The polymers here are quite high in molecular weight and the adhesives may be regarded as normal solvent-based adhesives, with necessary precautions depending on the nature of the solvent.

Two-part polyurethanes with isocyanate curing agents are used in some packaging applications in the UK. Also in Europe, two-part adhesives are used much more widely in the shoe industry. In these cases, apart from the precautions necessary because of any solvents which may be present, the isocyanate curing agents present a hazard.

Few adhesive manufacturers will use volatile isocyanates such as toluene di isocyanate, which attracts a toxic label under the EEC Paints, Varnishes, Adhesives, etc., Directive, if present in a composition at more than 3% by weight. Instead, polymeric isocyanates are used, but even these will give rise to some free volatile low molecular weight isocyanate. Thus, in using these materials, good fume extraction is necessary, whether solvent is present or not, and skin contact should be avoided. A particular hazard presented by isocyanates is that they are respiratory sensitisers. These products present particular hazards to anyone who is liable to asthma. It is also to be noted that if the adhesives are sprayed there is a hazard from inhalation of the spray even if the isocyanates are not themselves volatile. Although considerable precautions are necessary in using two-part isocyanate adhesives, and elaborate cautionary labels, these products are such excellent adhesives for some applications that their use will continue and is safe, provided the appropriate precautions are taken. The required development is more stable polymeric isocyanate hardeners. The use of polymeric hardeners, rather than toluene di isocyanate, was a major step forward in terms of safety, but further developments are likely to be difficult.

Other Thermosets

These will include anaerobic adhesives, two-part acrylics, cyanoacrylates, etc. If one defines this class as excluding non-reactive solvents, then the key to the situation is whether the adhesive contains volatile monomer or not. The anaerobics, in general, do not contain volatile monomers and so give rise to little fire or toxic hazard. Cyanoacrylates have little volatile monomer but there is sufficient to give rise to respiratory irritation in low humidity conditions. Some acrylics contain volatile solvents, and in this case, the danger of fire arises, and also there will be hazards to health, depending on the nature of the monomer. Good ventilation and flameproof equipment will be required. The same comment would be true of the styrene-containing polyesters used in glass laminating, if these were used as adhesives. In fact, few are so used.

In general, this category contains new products for specialised purposes. Safety considerations will have been considered in their formulation but often their performance justifies elaborate handling procedures. Sometimes the very performance of the adhesives gives rise to specific hazards. For example, cyanoacrylates bond most materials, including the fingers, within seconds. Hence, the suppliers go to some trouble to issue clear warnings (labels, etc.), and instructions on how to separate fingers, etc. In fact, this bonding property gives rise to no problem or risk of permanent injury provided that the correct procedures are used (*peel* apart gently after immersion in water).

CONCLUSION

The current preoccupation with safety is shown by the legislation described earlier. The impact of the legislation varies with the type of adhesive and in some cases is considerable. An effect has been the intensification of development work aimed at safer adhesives and some of this work has been successful.

The present outcome appears to be that where their properties are technically necessary, products whose safe handling inevitably requires complex precautions, will still be used, under the appropriate conditions. In other cases, inherently safer products have been developed. The effect has been to slow down—or perhaps, more correctly—to re-direct development work but there is no indication that product improvement has been arrested. There is always a danger that technical progress will suffer excessively in a climate of opinion like the present one, but so far, the situation seems reasonably in balance.

REFERENCES

1. Health & Safety at Work, etc., Act 1974, Her Majesty's Stationery Office, 1975.
2. Directive 67/548/EEC of 27.6.67 (O.J. 16.8.67). Amending Directive 73/146/EEC of 21.5.73. (O.J. 25.6.73). Amending Directive 75/409/EEC of 24.6.75. (O.J. 14.7.75). Preliminary Draft for 6th Amendment, XI/ENV/236/76. Commission Amending Proposal COM(76)433 FINAL (O.J. 5.11.76).
3. Directive 73/173/EEC of 4.6.73. (O.J. 11.7.73).
4. Commission Proposal COM(75)311 FINAL (O.J. 23.7.75).
5. Health & Safety Commission, Draft Packaging & Labelling of Dangerous Substances Regulations.
6. Transport of Dangerous Goods, UN Recommendations, original edn. 1970, new edn.. 1976.

7. List of registrable works and noxious or offensive gases specified in the Acts and Orders—Alkali & Works Regulation Act, 1906; Alkali & Work Orders 1966 and 1971, effective from 1st July 1971, Department of the Environment.
8. Control of Pollution Act 1974, Her Majesty's Stationery Office.
9. The Non-Agricultural Use of Pesticides in Great Britain—a report by the Central Unit on Environmental Pollution, Department of the Environment, Pollution Paper No. 3, HMSO.
10. Chemical Society, Industrial Division, Environment Group—Report on Environmental Group Symposium *Environmental Regulations—an International View*, 25th November 1975.
11. Technical Data Note No. 2 issued by the Health & Safety Executive, July 1976.
12. Carless Solvents Ltd. Technical Bulletin: *Introducing the 'Clairsol' Range*.
13. Lloyd, I. R., *Stabilisation of Polychloroprene Dispersions in Aliphatic Media:* Lecture at Thirteenth Annual Conference on Adhesion & Adhesives, City University, 14–15th April, 1975.
14. Brazier, A. D., *High Performance Pressure Sensitive Adhesives*, paper presented at British Adhesive Manufacturers' Association Symposium, 4th March 1976.

PATENT INDEX

AUTHOR INDEX

SUBJECT INDEX

(N.B. *Trade names are given in italics*)